New National Framework

MATHEMATICS 9+

M. J. Tipler   K. M. Vickers

J. Douglas

OXFORD

UNIVERSITY PRESS

OXFORD
UNIVERSITY PRESS

Great Clarendon Street, Oxford, OX2 6DP, United Kingdom

Oxford University Press is a department of the University of Oxford.
It furthers the University's objective of excellence in research, scholarship,
and education by publishing worldwide. Oxford is a registered trade mark of
Oxford University Press in the UK and in certain other countries

Text © M J Tipler and K M Vickers 2004
Original illustrations © Oxford University Press 2014

The moral rights of the authors have been asserted

First published by Nelson Thornes Ltd in 2004
This edition published by Oxford University Press in 2014

All rights reserved. No part of this publication may be reproduced,
stored in a retrieval system, or transmitted, in any form or by any
means, without the prior permission in writing of Oxford University
Press, or as expressly permitted by law, by licence or under terms
agreed with the appropriate reprographics rights organization.
Enquiries concerning reproduction outside the scope of the above
should be sent to the Rights Department, Oxford University Press, at
the address above.

You must not circulate this work in any other form and you must
impose this same condition on any acquirer

British Library Cataloguing in Publication Data
Data available

978-0-7487-6756-4

10 9 8

Printed in China

Acknowledgements

Illustrations: Harry Venning
Page make-up: Mathematical Composition Setters Ltd

The publishers thank the following for permission to reproduce
copyright material.

Peter Adams/Digital Vision BP (NT): 11; Photodisc 54 (NT): 15; Digital Vision
6 (NT): 18; Digital Vision 15 (NT): 20; Corel 613 (NT): 23; Corel 667 (NT):
27; Casio: 65; Photodisc 19 (NT): 74; Corel 778 (NT): 79; Corel 600 (NT): 90;
Stockbyte 29 (NT): 91; Corel 665 (NT): 154; Photodisc 72 (NT): 188; Corel 89
(NT): 257; M.C. Escher's "Belvedere" © 2004 The M.C. Escher Company – Baarn
– Holland. All rights reserved: 269; Corel 1 (NT): 317; Corel 233 (NT): 368; Corel
62 (NT): 371; Corel 248 (NT): 371; Photodisc 66 (NT): 371; Digital Vision 12 (NT):
378; Corel 725 (NT): 394; Corel 495 (NT): 403

The publishers would like to thank QCA for permission to reproduce extracts
from SATs papers.

Although we have made every effort to trace and contact all copyright holders
before publication this has not been possible in all cases. If notified, the
publisher will rectify any errors or omissions at the earliest opportunity.

Links to third party websites are provided by Oxford in good faith and for
information only. Oxford disclaims any responsibility for the materials
contained in any third party website referenced in this work.

Contents

Introduction   **v**

Number Support   **1**
Place value and powers of ten; Rounding; Integers;
HCF and LCM; Powers; Square roots and cube
roots; Mental calculation; Order of operations;
Estimating; Written calculations; Checking
answers to calculations; Using a calculator;
Fractions, decimals and percentages; Percentage
change; Ratio and proportion

**1   Place Value, Ordering and
Rounding**   **14**
Standard form   15
Calculating with numbers in standard
      form   18
Upper and lower bounds   22
Rounding   24
Summary of key points   29
Test yourself   30

2   Integers, Powers and Roots   **33**
Using prime factor decomposition   34
Common factors of algebraic expressions   36
Powers and roots   38
Index laws   41
Surds   45
Summary of key points   47
Test yourself   49

3   Calculation   **50**
Mental calculation   51
Calculating mentally with fractions,
      decimals and percentages   55
Estimating answers to calculations   57
Written calculation   59
Reciprocals   62
Using a calculator   65
Summary of key points   68
Test yourself   70

**4   Fractions, Decimals and
Percentages**   **73**
Fractions, decimals and percentages   74
Adding and subtracting fractions   78
Multiplying and dividing fractions   81
Adding and subtracting algebraic
      fractions   83
Recurring decimals   85
Summary of key points   87
Test yourself   88

**5   Percentage and Proportional
Changes**   **90**
Percentage change   91
Repeated percentage change   95

Solving ratio and proportion problems   98
Proportional relationships   103
Summary of key points   108
Test yourself   110

Algebra Support   **113**
Equations; Expressions; Formulae; Sequences;
Functions; Graphs; Real-life graphs

6   Algebra and Equations   **124**
Understanding algebra   125
Solving linear and non-linear equations   126
Simultaneous equations   131
Solving simultaneous equations
      graphically   136
Using simultaneous equations to solve
      problems   139
Solving inequalities   142
Linear inequalities in two variables   146
Solving quadratic inequalities   150
Summary of key points   151
Test yourself   153

7   Expressions and Formulae   **156**
Writing and simplifying expressions   157
Expanding two brackets   160
Factorising   163
Working with formulae   166
Substituting into expressions and
      formulae   169
Finding formulae   172
Summary of key points   175
Test yourself   176

8   Sequences and Functions   **179**
Generating sequences   180
Quadratic sequences   182
Finding rules for the nth term of linear
      and quadratic sequences   185
Fraction sequences   189
Spatial patterns   191
Functions   193
Summary of key points   197
Test yourself   200

9   Graphs of Functions   **202**
Linear graphs   203
Graphs of quadratic, cubic and
      reciprocal functions   207
Drawing, sketching and interpreting
      real-life graphs   217
Summary of key points   224
Test yourself   226

**Shape, Space and Measures
Support**   **229**
Lines and angles; Angles in polygons; 2-D shapes;
Circles; Constructions; Locus; 3-D shapes;
Transformations and coordinates; Scale drawings;
Measures; Perimeter, area and volume

10   Lines and Angles, Pythagoras   **246**
Conventions, definitions and derived
   properties   247
Demonstrations and proofs   248
Finding angles   250
Pythagoras' theorem   254
Pythagorean triples   259
Summary of key points   264
Test yourself   266

11   Shape, Construction and Loci   **269**
Constructing triangles   270
Congruent triangles   272
Similar shapes   276
Finding unknown lengths   278
Circles, construction and congruence   283
3-D shapes   286
Locus and construction   287
Summary of key points   294
Test yourself   296

**12   Coordinates and
Transformations**   **298**
Combinations of transformations   299
Enlargement   304
Enlargement and area and volume   310
Coordinates and lines   315
Length of line joining two points   316
Summary of key points   318
Test yourself   319

**13   Measures, Perimeter, Area and
Volume**   **322**
Practical measurement   323
Compound measures   325
Area and perimeter of circles and arcs   332
Surface area and volume   337
Trigonometry   341
Using Pythagoras' theorem and
   trigonometry to solve problems   350
Summary of key points   353
Test yourself   356

Handling Data Support   **358**
Planning and collecting data; Two-way tables;
Displaying data; Mode, range, median and mean;
Stem-and-leaf diagrams; Scatter graphs;
Interpreting and comparing data; Probability

**14   Planning a Survey and
Collecting data**   **371**
Planning a survey   372
Summary of key points   379
Test yourself   380

**15   Analysing and Displaying
Data**   **381**
Mean, median, mode and range   382
Mean, median and range from tables   383
Mean, median, mode for grouped data   385
Finding median and quartiles from
   cumulative frequency   390
Frequency polygons   395
Scatter graphs   399
Interpreting and comparing data   404
Surveys   410
Summary of key points   411
Test yourself   413

16   Probability   **416**
Describing and calculating probability   417
Probability of compound events   421
Independent events   423
Estimating probabilities from relative
   frequency   429
Analysing games   433
Summary of key points   436
Test yourself   437

Test Yourself Answers   **440**

Index   **451**

Introduction

We hope that you enjoy using this book. There are some characters you will see in the chapters that are designed to help you work through the materials.

These are

 This is used when you are working with information.

 This is used where there are hints and tips for particular exercises.

 This is used where there are cross references.

 This is used where it is useful for you to remember something.

 These are blue in the section on number.

 These are green in the section on algebra.

 These are red in the section on shape, space and measures.

  These are yellow in the section on handling data.

Number Support

Place value and powers of ten

Powers of ten

$$10^2 = 100 \qquad 10^1 = 10 \qquad 10^0 = 1$$
$$10^{-1} = \frac{1}{10^1} = \frac{1}{10} = 0\cdot1 \qquad 10^{-2} = \frac{1}{10^2} = \frac{1}{100} = 0\cdot01$$

Multiplying by 0·1 is the same as dividing by 10.
Multiplying by 0·01 is the same as dividing by 100.
Dividing by 0·1 is the same as multiplying by 10.
Dividing by 0·01 is the same as multiplying by 100.

Examples
$$5\cdot9 \times 0\cdot1 = 5\cdot9 \div 10 \qquad\qquad 6\cdot32 \div 0\cdot01 = 6\cdot32 \times 100$$
$$= 0\cdot59 \qquad\qquad\qquad\qquad\qquad = 632$$

Practice Questions 5, 8, 13, 33

Rounding

Rounding to decimal places

1 Keep the number of digits asked for after the decimal point.
2 Delete the following digits.
3 If the first digit to be deleted is 5 or more, increase the last digit kept by 1.

Examples

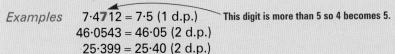

$$7\cdot4712 = 7\cdot5 \text{ (1 d.p.)} \qquad \text{This digit is more than 5 so 4 becomes 5.}$$
$$46\cdot0543 = 46\cdot05 \text{ (2 d.p.)}$$
$$25\cdot399 = 25\cdot40 \text{ (2 d.p.)}$$

> Rounding to the nearest whole number is the same as rounding to 0 d.p.

Sometimes the answer to a calculation is written as a **recurring decimal** instead of being rounded.

Examples
$$47 \div 6 = 7\cdot8333 \dots$$
$$= 7\cdot8\dot{3}$$
$$48 \div 11 = 4\cdot363636 \dots$$
$$= 4\cdot\dot{3}\dot{6}$$

> The dots above the 3 and 6 tell us these digits repeat.

Practice Questions 2, 9, 14, 32, 35, 39

Integers

We can **add and subtract integers** using a number line.

Examples $-4 + 3 = -1$ $\qquad\qquad\qquad\qquad 2 - -1 = 3$

When we **multiply or divide two negative numbers** we get a positive answer.

Examples $-3 \times -5 = 15 \qquad -12 \div -3 = 4$

When we **multiply or divide a positive and a negative number** we get a negative answer.

Examples $-4 \times 7 = -28 \qquad -35 \div 7 = -5$

Practice Questions 1, 3, 4

HCF and LCM

The **HCF** (highest common factor) of two numbers is the largest factor common to both.

The **LCM** (lowest common multiple) of two numbers is the smallest number that is a multiple of both.

To find the HCF and LCM we first find prime factors.

Example $300 = 2^2 \times 3 \times 5^2$ $540 = 2^2 \times 3^3 \times 5$

We then use a Venn diagram to find the HCF and LCM.
HCF $= \mathbf{2 \times 2 \times 3 \times 5} = 60$
LCM $= \mathbf{2 \times 2 \times 3 \times 5} \times 5 \times 3 \times 3$
        $= 2700$

You can use a factor tree or table to find the prime factors.

300   540

Common prime factors

We can use the HCF to simplify fractions and the LCM when adding and subtracting fractions.

Practice Questions 15, 18

Powers

We write $5 \times 5 \times 5 \times 5$ as 5^4.
5^4 is read as '5 to the power of 4'.
The 4 in 5^4 is called an **index**.

The plural of index is indices.

On a calculator, squares are keyed using $\boxed{x^2}$,

cubes are keyed using $\boxed{x^3}$,

other indices are keyed using $\boxed{x^y}$.

Example $(^-3\cdot4)^3$ is keyed as $\boxed{(}\;\boxed{-}\;\boxed{3\cdot4}\;\boxed{)}\;\boxed{x^3}\;\boxed{=}$ to get $^-39\cdot304$.
        19^5 is keyed as $\boxed{19}\;\boxed{x^y}\;\boxed{5}\;\boxed{=}$ to get 2 476 099.

We must use brackets round a negative number.

Square roots and cube roots

$\sqrt{36}$ is read as 'the square root of 36'.
$^\pm\sqrt{36} = 6$ or $^-6$ since $6 \times 6 = 36$ and $^-6 \times ^-6 = 36$.
On a calculator we use $\boxed{\sqrt{}}$ to find square roots.
$\sqrt[3]{64}$ is read as 'the cube root of 64'.
$\sqrt[3]{64} = 4$ because $4 \times 4 \times 4 = 64$.
On a calculator we use $\boxed{\sqrt[3]{}}$ to find cube roots.

Usually we just want the positive square root.
$\sqrt{36} = 6$

Example $\sqrt[3]{96}$ is keyed as $\boxed{\sqrt[3]{}}\;\boxed{96}\;\boxed{=}$ to get $4\cdot6$ (1 d.p.).

Practice Questions 36, 53, 54, 60, 61, 62

Some calculators do not have $\boxed{\sqrt[3]{}}$. Then use the $\boxed{x^y}$ key.

Mental calculation

We use a range of strategies to **add and subtract** mentally.

These strategies can be used to **multiply and divide mentally**.

1 Place value

Example

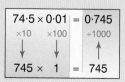

$74\cdot5 \times 0\cdot01$	$=$	$0\cdot745$
$\times 10$ $\times 100$	$\div 1000$	
$745 \times$	1	$= \; 745$

We multiplied by 1000 altogether so we must divide the answer by 1000.

2 Partitioning

Examples
$$14 \times 2 \cdot 4 = (10 \times 2 \cdot 4) + (4 \times 2 \cdot 4)$$
$$= 24 + 9 \cdot 6$$
$$= \mathbf{33 \cdot 6}$$

$$480 \div 5 = (400 \div 5) + (80 \div 5)$$
$$= 80 + 16$$
$$= \mathbf{96}$$

3 Factors

Examples
$$4 \cdot 5 \times 20 = 4 \cdot 5 \times 2 \times 10$$
$$= 9 \times 10$$
$$= \mathbf{90}$$

$$480 \div 16 = \mathbf{30} \text{ because } 480 \div 4 = 120$$
$$120 \div 4 = 30$$

4 Near tens

Examples
$$34 \times 11 = 34 \times 10 + 34$$
$$= 340 + 34$$
$$= \mathbf{374}$$

$$25 \times 39 = 25 \times (40 - 1)$$
$$= 25 \times 40 - 25$$
$$= 1000 - 25$$
$$= \mathbf{975}$$

5 Known facts

We know that $5 = 10 \div 2$, $\quad 25 = 100 \div 4$, $\quad 50 = 100 \div 2$ $\quad$ and $\quad 25 = 5 \times 5$.

Examples
$$324 \times 50 = 324 \times 100 \div 2$$
$$= 32\,400 \div 2$$
$$= \mathbf{16\,200}$$

$$625 \div 25 = 625 \div 5 \div 5$$
$$= 125 \div 5$$
$$= \mathbf{25}$$

6 Doubling and halving

$$\overset{\text{halve} \quad \text{double}}{2 \cdot 4 \times 3 \cdot 5} = 1 \cdot 2 \times 7$$

Examples
$$24 \times 16 = 24 \times 2 \times 2 \times 2 \times 2$$
$$= 48 \times 2 \times 2 \times 2$$
$$= 96 \times 2 \times 2$$
$$= 192 \times 2$$
$$= \mathbf{384}$$

$$2 \cdot 4 \times 3 \cdot 5 = 1 \cdot 2 \times 7$$
$$= 1 \times 7 + 0 \cdot 2 \times 7$$
$$= 7 + 1 \cdot 4$$
$$= \mathbf{8 \cdot 4}$$

Practice Questions 10, 11, 12, 17, 19, 22, 26c–m, 27, 31, 37, 38, 45

Order of operations

We work out brackets first, then do $\times$ and $\div$, then do $+$ and $-$.

Example
$$^-6 - 3 \times {}^-2 = {}^-6 - {}^-6$$
$$= 0$$

$$^-2({}^-5 + 2) = {}^-2 \times {}^-3$$
$$= 6$$

Remember BIDMAS.

Practice Questions 26a, b, n–r, 49

Estimating

To **estimate the answer to a calculation** we round the numbers so that we can do the calculation in our head.

Examples
$$492 \times 23 \approx 500 \times 20$$
$$= 10\,000$$

$$56 \cdot 37 \div 3 \cdot 4 \approx 60 \div 3$$
$$= 20$$

We try to round to '**nice numbers**' when estimating.

Example $\quad \frac{62}{7} \approx \frac{63}{7}$ $\quad$ We round to 63 rather than 60.
$\qquad\qquad\qquad$ 63 can be divided by 7.
$\qquad\qquad = 9$

Practice Questions 16, 55

Written calculations

When we **add and subtract decimals** using pencil and paper we line up the decimal points.

We can **multiply numbers** like this.

$58 \cdot 2 \times 6 \cdot 7$ is equivalent to $58 \cdot 2 \times 10 \times 6 \cdot 7 \times 10 \div 100$
or $582 \times 67 \div 100$.

$$
\begin{array}{r}
582 \\
\times\ 67 \\
\hline
34\ 920 \\
4074 \\
\hline
38\ 994 \\
\hline
\end{array}
\qquad
\begin{array}{l}
60 \times 582 \\
7 \times 582
\end{array}
$$

Answer $38\ 994 \div 100 = \mathbf{389 \cdot 94}$

When we **divide by a decimal** we do an equivalent calculation.

Example $47 \cdot 9 \div 3 \cdot 1$ is equivalent to $479 \div 31$.

$47 \cdot 9 \div 3 \cdot 1 \approx 48 \div 3 = 16$.

$$
\begin{array}{r}
31\,)\,479 \\
-310 \\
\hline
169 \\
-155 \\
\hline
14 \cdot 0 \\
-12 \cdot 4 \\
\hline
1 \cdot 60 \\
-1 \cdot 55 \\
\hline
0 \cdot 05 \\
\end{array}
\qquad
\begin{array}{l}
31 \times 10 \\
\\
31 \times 5 \\
\\
31 \times 0 \cdot 4 \\
\\
31 \times 0 \cdot 05 \\
\end{array}
$$

$15 \cdot 45$ R $0 \cdot 05 = \mathbf{15 \cdot 5}$ **(1 d.p.)**

Practice Questions 47, 50

Checking answers to calculations

We can **check an answer to a calculation** in one of these ways.

1 Check that the answer is sensible.

2 Check that the answer is the right order of magnitude by estimating.

Example   Marnie worked out the answer to $96 \times 2 \cdot 8$ as $268 \cdot 8$.
Estimate $96 \times 2 \cdot 8 \approx 100 \times 3 = 300$
So $268 \cdot 8$ is the right order of magnitude.

3 Check using inverse operations

Example $76 \cdot 8 \div 2 \cdot 4 = 32$
Check $32 \times 2 \cdot 4 = 76 \cdot 8$ Multiplying by 2·4 is the inverse of dividing by 2.4.

4 Check using an equivalent calculation.

Example $36 \times 12 = 432$ can be checked by calculating 72×6 or $72 \times 3 \times 2$ or
$36 \times 4 \times 3$ or $12 \times 3 \times 12$

5 Check the last digits.

Example $46 \times 2 \cdot 3 = 105 \cdot 6$ is wrong because $\mathbf{6} \times \mathbf{3} = 18$.
The last digit should be 8.

Practice Question 56

Using a calculator

Sometimes we use **brackets** on the calculator.

Example $\dfrac{14\cdot7 + 64\cdot8}{5\cdot8 - 1\cdot9}$

Key ((14·7 + 64·8) ÷ (5·8 − 1·9)) = to get 20·38 (2 d.p.)

Practice Question 46

Fractions, decimals and percentages

To **write a fraction or decimal as a percentage**, either

a write with a denominator of 100 or
b multiply by 100%.

Examples

$\dfrac{7}{20} = \dfrac{35}{100}$

$\qquad = 35\%$

$\dfrac{72}{96} = \dfrac{72}{96} \times 100\%$

$\qquad = 75\%$

$0\cdot35 \times 100\% = 35\%$

Key 72 ÷ 96 × 100 = to get 75

We **can compare fractions** by

1 converting them to decimals
2 writing them with a common denominator.

Examples $\dfrac{1}{3} < \dfrac{2}{5}$ because $0\cdot\dot3 < 0\cdot4$. $\qquad \dfrac{2}{3} > \dfrac{3}{5}$ because $\dfrac{10}{15} > \dfrac{9}{15}$.

When **comparing proportions** we convert them all to fractions, decimals or percentages.

To **write a percentage as a fraction or decimal**, first write the percentage as parts per hundred.

Examples $52\% = \dfrac{52}{100}$

$\qquad\qquad = 0\cdot52$

$\dfrac{52}{100}$ simplifies to $\dfrac{13}{25}$.

Fractions can be **added and subtracted** easily when they have the same denominator. If their denominators are different, we can use equivalent fractions.

Examples $\dfrac{3}{9} + \dfrac{4}{9} = \dfrac{3+4}{9}$

$\qquad\qquad = \dfrac{7}{9}$

$\dfrac{5}{12} + \dfrac{4}{6} = \dfrac{5}{12} + \dfrac{8}{12}$

$\qquad\qquad = \dfrac{13}{12}$

$\qquad\qquad = 1\dfrac{1}{12}$

$\dfrac{4}{6} = \dfrac{8}{12}$

12 is the LCM of 6 and 12.

To **multiply fractions** we

1 write whole numbers or mixed numbers as improper fractions
2 cancel if possible
3 multiply the numerators
4 multiply the denominators

Example $1\dfrac{2}{5} \times 2\dfrac{1}{4} \times 1\dfrac{1}{3} = \dfrac{7}{5} \times \dfrac{9^3}{4_1} \times \dfrac{4^1}{3_1}$

$\qquad\qquad\qquad = \dfrac{21}{5}$

$\qquad\qquad\qquad = 4\dfrac{1}{5}$

To **divide by a fraction**, we multiply by the inverse.

Example $1\dfrac{1}{4} \div \dfrac{3}{8} = \dfrac{5}{4_1} \times \dfrac{8^2}{3}$

$\qquad\qquad = \dfrac{10}{3}$

$\qquad\qquad = 3\dfrac{1}{3}$

Multiplying and dividing are inverse operations.

To find a **fraction of** or **percentage of** we multiply.

Examples $\frac{3}{5}$ of 55 g $= \frac{3}{5} \times 55$ g   15% of £45 $= \frac{15}{100} \times$ £45

$\frac{1}{5}$ of 55 g = 11 g   = £6·75

$\frac{3}{5}$ of 55 g = 3 × 11 g

= 33 g

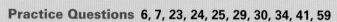

> To find 15% you could find 10% then 5% then add the answers.

Practice Questions 6, 7, 23, 24, 25, 29, 30, 34, 41, 59

Percentage change

We can calculate **percentage increase or decrease** using a single calculation.

To find an increase of 15% we find 115%.

100%	15%

115%

To find a decrease of 20% we find 80%.

80%	20%

100%

Example   Jarred bought a bike for £285.
He sold it for a 15% loss.
To find out how much he sold it for, find 85% of £285.
85% × 285 = 0·85 × 285   85% = 100% − 15%
= £242·25

Practice Questions 44, 48

Ratio and proportion

Ratio compares part to part.

Proportion compares part to whole.

Example   Ratio of blue to purple is 4 : 5.
Proportion of blue triangles is $\frac{4}{9}$.
Proportion of purple triangles is $\frac{5}{9}$.

We write ratios in their **simplest form by cancelling**.

Examples   12 : 27   20 min : 2 hours
÷ 3 (   ) ÷ 3   = 20 : 120
= 4 : 9   ÷ 20 (   ) ÷ 20
= 1 : 6

> All parts of a ratio must have the same units.

We can **solve problems using ratio and proportion**.

Example   9 packets of chips costs £3·50.
So 1 packet of chips costs $\frac{£3·50}{9}$.
So 15 packets of chips costs $\frac{£3·50}{9} \times 15 = £5·83$ (nearest penny).

We can **divide in a given ratio**.

Example   64 chocolates were shared out in the ratio 1 : 3 : 4.
There are 1 + 3 + 4 = 8 shares altogether.
One share is $\frac{64}{8} = 8$.
Three shares is 3 × 8 = 24.
Four shares is 4 × 8 = 32.

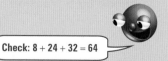

> Check: 8 + 24 + 32 = 64

64 chocolates divided in the ratio 1 : 3 : 4 is 8 : 24 : 32.

Practice Questions 20, 21, 28, 40, 42, 43, 51, 52, 57, 58

Practice Questions

1 I am an integer.
I am between ⁻6 and ⁻3.
I am less than ⁻4.
Which integer am I?

T

2 Use a copy of this.
Fill it in.

	Rounded to nearest 100	Rounded to nearest 1000	Rounded to nearest 10 000	Rounded to nearest 100 000
7 734 392				
3 848 312				
8 050 802				
775 500				

T

3 Use a copy of these.
What numbers go in the top boxes?

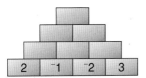 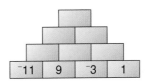

a In the red boxes, two numbers are added to get the number above.
b In the blue boxes, the number on the right is subtracted from the number on the left to get the number above.

4 Work these out.
a 7 + 4 + ⁻2 − 3   **b** 17 − 8 − ⁻4 + 2   **c** 0·6 + 0·8 + 1·3 − 0·2
d 3 + 4 × 2   **e** 4 × 3 + 6 × 5   **f** 7 + 2 × ⁻2

Remember BIDMAS

5 a 6·27 + 0·01   **b** 5·976 − 0·001   **c** 12·149 + 0·001   **d** 21·49 − 0·001

6 Copy these and fill in the first gap with a decimal and the second gap with a percentage.
a $\frac{3}{4}$ = ___ = ___%   **b** $\frac{2}{3}$ = ___ = ___%   **c** $\frac{7}{10}$ = ___ = ___%
d $\frac{17}{20}$ = ___ = ___%   **e** $\frac{12}{25}$ = ___ = ___%   **f** $\frac{3}{8}$ = ___ = ___%

7 Change each of these percentages to a fraction then to a decimal.
a 37%   **b** 45%   **c** 72%   **d** 135%

8 What goes in the box?
a 37 × ☐ = 0·37   **b** 0·1 × ☐ = 0·001   **c** 3·2 ÷ ☐ = 32   **d** 86 ÷ ☐ = 8600
e ☐ ÷ 0·1 = 50   **f** ☐ × 0·1 = 14   **g** ☐ ÷ 0·01 = 450   **h** ☐ × 0·01 = 2

9 Round these to the nearest whole number.
a 14·356   **b** 5·025   **c** 17·824   **d** 7·99

T

10 Use a copy of this.
Shade the answers to these on the diagram.
a 36 + 64   **b** 8 + ⁻2 + ⁻5   **c** 18 + 39
d 0·6 + 0·8   **e** 530 + 260   **f** 6·7 − 0·8
g 4600 − 2800   **h** 5·6 + 1·7   **i** 342 + 639
j 5·23 + 1·5   **k** 5·07 − 3·8   **l** 8·64 + 5·36
m 7·24 + 4·76   **n** 16 − 4·37   **o** 15·1 − 3·95
What shape does the shading make?

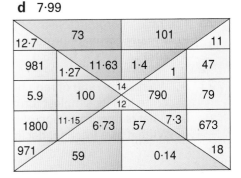

Number

11 Work these out mentally.
a $4·02 \times 0·1$	**b** 300×50	**c** $6·95 \div 0·1$	**d** $300 \div 600$	**e** $54\,000 \div 900$
f $0·45 \times 0·01$	**g** 180×4	**h** 324×4	**i** $0·16 \times 0·01$	**j** 12×18
k 18×25	**l** $0·7 \times 0·4$	**m** $1·5 \div 5$	**n** $740 \div 5$	**o** $572 \div 4$
p $96·6 \div 3$	**q** $450 \div 15$	**r** $0·54 \div 0·01$	***s** $^-11 \times 21$	***t** $^-19 \times {}^-34$

12 Decide which of these statements could be correct (✓) and which are definitely wrong (✗).
Justify your decision. Do not calculate the answer.
  a $5·02 \times 0·3 = 15·06$ **b** $2·75 \div 0·02 = 1·375$ **c** $6·24 \times 1·2 = 7·488$
  d $8·127 \div 1·4 = 58·05$ **e** $2·75 \times 1·1 < 2·75$ **f** $2·75 \div 0·8 > 2·75$

13 Jude has four number cards and a decimal point card.

Use the five cards to make a number that is as close as possible to 7·511.

14 Round these to 1 d.p. or 2 d.p. as indicated.
  a $16·357$ (2 d.p.)   **b** $0·155$ (1 d.p.)   **c** $0·306$ (2 d.p.)   **d** $9·015$ (1 d.p.)
  e $106·405$ (2 d.p.)   **f** $54·398$ (2 d.p.)   **g** $4·002$ (1 d.p.)   **h** $16·998$ (2 d.p.)

15 **a** Write the factors of 96 and 440 in index form.
  b Use a Venn diagram to find the HCF of 96 and 440.
  c Now find the LCM of 96 and 440.

16 Which is the best estimate for each of these?
  a $\dfrac{50·1 \times 16·2}{7·3}$ **A** 50   **B** 100   **C** 120   **D** 160
  b $3·2 \times (91·5 - 20·8)$ **A** 250   **B** 270   **C** 320   **D** 210
  c $\dfrac{22·3 \times 17·8}{4·2 + 1·9}$ **A** 100   **B** 40   **C** 60   **D** 400

17 Holly had these number cards.

She placed all the cards in the diagram so that each side added to the same total.
Make a copy of this.
Fill in the missing numbers on the cards.

18 **a** Write the factors of 224 and 308 in index form.
  b Use a Venn diagram to find the HCF of 224 and 308.
  c Now find the LCM of 224 and 308.

19 Use $+, -, \times, \div$ to make each calculation correct.
  a $4\,\square\,5 = 12\,\square\,3$ **b** $3\,\square\,1 = 12\,\square\,3$ **c** $5\,\square\,5 = 10\,\square\,10$

20 The ratio of protein to carbohydrate in a breakfast cereal is 2 : 15.
How many grams of protein are there if there is this much carbohydrate?
  a 30 g   **b** 45 g   **c** 150 g   **d** 75 g

21 **a** What is the ratio of red : blue : green in this shape?
    Write it in its simplest form.
  b What is the proportion of blue in the shape?
    Write it as a fraction, decimal and percentage.

22 Is it possible to place numbers in the squares which add across and down to the given numbers? If so, find possible answers. If it is not possible, explain why not.

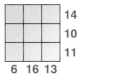

      a   **b**

23 Which is larger?
  a $1\frac{2}{3}$ or $11 \div 5$ **b** $\frac{59}{8}$ or $6\frac{5}{12}$

24 Find the answers to these mentally.
  a $\frac{2}{5}$ of 40 m   **b** 80% of 70 mℓ **c** $\frac{5}{9}$ of 45 g   **d** 30% of £160
  e $1\frac{1}{2}$ of 12 ℓ **f** 85% of 80 cm   **g** 35% of 140 g   **h** $1\frac{3}{4}$ of 60 m

25 A dog weighs 40 kg.
60% of its total mass is water.
What is the mass of this water?

26 Find the answers to these mentally. You may need jottings for some.
  a $11 + 3 \times 7$ **b** $27 - 2(4 + 3)$ **c** $4 + {}^{-}6$ **d** ${}^{-}4 - {}^{-}6$ **e** 7^2
  f $600 \times 0{\cdot}3$ **g** $\frac{600}{0{\cdot}3}$ **h** $\frac{5}{6}$ of 54   **i** $0{\cdot}64 \times 0.2$ **j** $\frac{0{\cdot}64}{0{\cdot}2}$
  k ${}^{-}5 \times {}^{-}2$ **l** $\frac{{}^{-}4 \times {}^{-}3}{{}^{-}2}$ **m** $8{\cdot}3^0$ **n** $15 \div (3 + 2)$
  o $(3 + 4 \times 8) \div 5$ **p** $\frac{37 + 43}{4 \times 5}$ **q** $\frac{35 + 14}{9 - 2}$ **r** $6 \times 2 \div (5 - 3)^2$

27 **a** How many seconds are there in three and three-quarter minutes?
   b Last year Tim grew from 1·58 m to 1·83 m. How much did he grow in centimetres?

28 Write these ratios in their simplest forms.
  a £2 : 90p   **b** 1·6 m : 480 cm   **c** 45 sec : 2 min

29 Each diagram below is drawn on a square grid.
  a Write what percentage of each diagram is shaded.
     i   **ii**

  b Explain how you know that $12\frac{1}{2}\%$ of the diagram below is shaded.

  c Copy this diagram.
    Shade $37\frac{1}{2}\%$ of it.

30 Three boys, Patrick, Brad and Harry, are in different classes for design and technology. They were given these marks.

Patrick
$\frac{16}{20}$
D and T

Brad
78%
D and T

Harry
$\frac{7}{10}$
D and T

Who got the highest mark? Explain how you know.

31 What number is halfway between
  a 4 and 10   **b** 7·5 and 12·5   **c** 2 and 6·8   **d** ${}^{-}3$ and 2?

32 Matt had these number cards.

[5] [0] [3] [8]

    a Make three numbers that round to 500 to the nearest 100.
    b Make three numbers that round to 4000 to the nearest 1000.

33 Which of <, > or = goes in the box?
    a 450 mm ☐ 4·05 m   **b** 7350 g ☐ 7·3 kg   **c** 7·3 km ☐ 7300 m   **d** 702 mm ☐ 72 cm

34 Use your calculator to find these. Give the answer to **d** and **e** to the nearest penny.
    a 8% of £28·50   **b** $12\frac{1}{2}$% of £108   **c** $17\frac{1}{2}$% of £118
    d 7% of £0.84   ✳**e** 18·75% of £1·75   ✳**f** $33\frac{1}{3}$% of £26·85

35 David rounded a whole number to the nearest 10 000 and got 80 000.
Alex rounded the same number to the nearest 1000 and got 75 000.
    a What is the smallest number it could have been?
    b What is the largest number it could have been?

36 Find these.
    a $\pm\sqrt{16}$ **b** $\pm\sqrt{49}$ **c** $\sqrt{64}$ **d** $\sqrt{144}$ **e** $\pm\sqrt{100}$

37 In design and technology, 22·5 m of braid was shared equally by 6 students.
How much did each get to the nearest metre?

38 Work this out mentally.
Sara needed 6 m of fabric to make a costume.
    a How much did it cost?
    b How much change did she get from £20?

Fabric
£2·20
per metre

39 A report on the number of police officers in 1995 said: **[SATs Paper 2 Level 6]**
'There were 119 000 police officers.
Almost 15% of them were women.'
    a The percentage was rounded to the nearest whole number, 15.
        What is the smallest value the percentage could have been, to one decimal place?
        Choose the correct answer below.
        14·1%   14·2%   14·3%   14·4%   14·5%
        14·6%   14·7%   14·8%   14·9%
    b What is the smallest number of women police officers that there might have been in
        1995?
        (Use your answer to part **a** to help you calculate this answer.)
        Show your working.

40 Amy poured 2 cartons of apple juice and 3 cartons of orange juice into a big jug.
    a What is the ratio of apple juice to orange juice in Amy's jug?
    b Tom pours 1 carton of apple juice and $2\frac{1}{2}$ cartons of orange juice into another big jug.
        What is the ratio of apple juice to orange juice in Tom's jug?
    c Chandri pours 2 cartons of apple juice and 2 cartons of orange juice into another big
        jug.
        But she wants only half as much apple juice as orange juice in her jug.
        What should Chandri pour into her jug now?

41 160 women gave birth at a hospital in July.
30% had twins.
The rest had just one baby.
How many babies were born in July at the hospital?

TWIN EXPLOSION
30% of women at local
hospital had twins in July.

42 Willie and Eleanor had a holiday gardening job.
Willie was paid £40 and Eleanor was paid £45
   a What is the ratio of the amounts Willie and
      Eleanor were paid?
   b What proportion of the total amount was Willie
      paid?
      Give your answer as a fraction.

43 Last year the amounts Mrs Booth spent on car insurance, road tax and petrol were in the
ratio 2 : 7 : 1.
She spent a total of £1650 on these three things.
How much did she spend on insurance?

44 Amelia bought some ski gear for £420.
The next year she sold it for a 35% loss.
How much did she sell it for?

45 A liquid is cooling under special conditions. Its temperature halves every hour.
   a If its temperature at the start was 360 °C what was its temperature after
      i 1 hour   **ii** 2 hours   **iii** 5 hours?
   b After how many hours will its temperature be less than 2 °C?

46 Use your calculator to find these. Round the answers to **b** and **c** to 2 d.p.
   a $15 \cdot 3 \times (25 \cdot 2 - 9 \cdot 8)$ **b** $\dfrac{25 \cdot 8 \times 17 \cdot 2}{9 \cdot 75 + 3 \cdot 26}$ **c** $\dfrac{21 \cdot 5(16 \cdot 2 + 23 \cdot 5)}{14 \cdot 2 \times 0 \cdot 875}$

47 A club wants to take 2000 people on a trip to London.
The club organiser says:
   'We can go in coaches.
   Each coach can carry 52 people.'
   a How many coaches do they need for the trip?
      Show your working.
   b Each coach costs £390.
      What is the total cost of the coaches?
   c How much is each person's share of the cost?

48

AUDIO
Discount Store
SALE
all CDs 20% off

Stereo Market
SALE
$\frac{1}{4}$ OFF all CDs

At both these shops a CD usually costs £8·80.
Sharne buys a CD in the sale at Stereo Market. Lucy buys a CD in the sale at Audio
Discount Store.
Who pays more, Sharne or Lucy? How much more?

49 **a** Write the answers.

$(6 + 3) \times 4 =$ ___ $6 + (3 \times 4) =$ ___

b Work out the answer to this.

$(3 + 5) \times (4 + 5 + 1)$

c Put brackets in the calculation to make the answer 64.

$3 + 2 + 3 \times 8$

d Now put brackets in the calculation to make the answer 43.

$3 + 2 + 3 \times 8$

50 Do these using a written method. Estimate first. Round to 2 d.p. if you need to round.

a $165 \cdot 072 + 29 \cdot 98$	**b** $15 \cdot 82 - 0 \cdot 709$	**c** $13 \cdot 65 \times 3 \cdot 7$
d $54 \cdot 35 \times 26$	**e** $32 \cdot 7 \times 5 \cdot 8$	**f** $567 \div 21$
g $677 \div 18$	**h** $325 \cdot 6 \div 4 \cdot 6$	**i** $762 \cdot 4 \div 9 \cdot 3$

51 Ben, Jason and Luke share £2560 in the ratio 1 : 4 : 3.
How much does each person get?

52 Paul is 14 years old. [SATs Paper 2 Level 6]

His sister is exactly 6 years younger, so she is 8 years old.

This year, the ratio of Paul's age to his sister's age is 14 : 8.

14 : 8 written as simply as possible is 7 : 4.

a When Paul is 21, what will be the ratio of Paul's age to his sister's age?
Write the ratio as simply as possible.

b When his sister is 36, what will be the ratio of Paul's age to his sister's age?
Write the ratio as simply as possible.

c Could the ratio of their ages ever be 7 : 7?
Explain how you know.

53 Use your calculator to find these. Give the answers to e, h and j to 2 d.p.

a $\sqrt{169}$	**b** $1 \cdot 3^2$	**c** 16^3	**d** $\sqrt[3]{2744}$	**e** $\sqrt{8 \cdot 7}$	**f** $15^3 - 11^2$
g 23^4	**h** $5 \cdot 6^5$	**i** $(^-5)^4$	**j** $(^-2 \cdot 3)^5$	**k** $(\frac{1}{4})^3$	**l** $7^2 \times 6^3$

54 Which is bigger, the square of 22 or the cube of 12?

55 Estimate the answers to these. Show how you found your estimate.

a 8×19	**b** $102 \div 9$	**c** 19×22	**d** $88 \div 31$	**e** $21 \cdot 9 \times 6 \cdot 2$
f $492 \div 4 \cdot 8$	**g** $\frac{3 \cdot 1 \times 4 \cdot 8}{5 \cdot 2}$	**h** $\frac{2 \cdot 6 + 6 \cdot 1}{7 \cdot 2 - 3 \cdot 8}$	**i** $8 \cdot 5 \div (7 \cdot 7 - 4 \cdot 8)$	

56 How can you tell this answer is wrong without doing the calculation?

$3 \cdot 64 \times 1 \cdot 7 = 6 \cdot 186$

57 Dressing is made by mixing oil and vinegar.

Recipe A says to mix 1 part oil with 2 parts vinegar.

Recipe B says to mix 1 part oil with 3 parts vinegar.

a In recipe A, how much oil should I mix with 250 mℓ of vinegar?

b In recipe B, how much oil should I mix with 450 mℓ of vinegar?

c How much oil and vinegar would I mix to make 600 mℓ of dressing in Recipe B?

d Is this correct?

50% of recipe A is oil.

Explain your answer.

58 These labels give the amount of fat in some muesli bars.

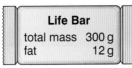

Energy Bar	
total mass	250 g
fat	8 g

Life Bar	
total mass	300 g
fat	12 g

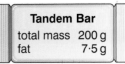

Tandem Bar	
total mass	200 g
fat	7·5 g

    a What is the ratio   *fat : total mass*   in each of the 3 bars?

    b Which bar has the highest percentage of fat?

    c What percentage of fat does it contain?

59 For a coursework task, Ben gained $\frac{5}{8}$ of the available marks while Natasha gained $\frac{2}{3}$.

    a Who got the better mark, Ben or Natasha?

    b If $a : b$ gives the ratio of Ben's mark to Natasha's, find possible values for a and b.

60 Show that $6 \cdot 547^2$ is about 43.

61 $9 \times x$ is a square number.
What is the smallest value of x?

62 $371 = 3^3 + 7^3 + 1^3$.
Find another three-digit number that is equal to the sum of the cubes of its digits.

1 Place Value, Ordering and Rounding

You need to know

✓ place value and powers of ten   page 1
✓ multiplying and dividing by powers of ten   page 1
✓ rounding   page 1

········ **Key vocabulary** ··

**billion,   exponent,   greater than or equal to ($\geqslant$),   index,
less than or equal to ($\leqslant$),   significant figures,
standard (index) form,   upper bound and lower bound**

 Planet X

The number system on Planet X has only these five number symbols.

0   1   2   3   4

On Planet X, when they get to 5 they have no more new symbols.

They write 5 as one lot of five and zero ones.
1 lot of five
and
0 ones

These are how some other numbers are written.

1 lot of five
and
3 ones
8

1 lot of five
and
1 one
6

2 lots of five
and
3 ones
13

Write these numbers in the Planet X number system.

7   18   24

How do you think 38 might be written?

Standard form

Discussion

● We know that:

$$10^{-1} = \frac{1}{10^1} = \frac{1}{10} = 0 \cdot 1$$

$$10^{-2} = \frac{1}{10^2} = \frac{1}{100} = 0 \cdot 01$$

$$10^{-3} = \frac{1}{10^3} = \frac{1}{1000} = 0 \cdot 001$$

You could check this by keying

$$\boxed{10} \ \boxed{x^y} \ \boxed{(-)} \ \boxed{2} \ \boxed{=} \ .$$

● $2^3 = 2 \times 2 \times 2$

So $\frac{1}{2^3} = \frac{1}{2 \times 2 \times 2}$

$$= \frac{1}{8}$$

$$= 0 \cdot 125$$

To find the value of 2^{-3} key $\boxed{2} \ \boxed{x^y} \ \boxed{(-)} \ \boxed{3} \ \boxed{=}$.

Compare the answer to 2^{-3} with $\frac{1}{2^3}$.

Now compare the answers to these.

2^{-4} and $\frac{1}{2^4}$ 2^{-5} and $\frac{1}{2^5}$ 2^{-2} and $\frac{1}{2^2}$

Make a statement about $\frac{1}{a^n}$ and a^{-n}. **Discuss**.

The area of the African Continent is about $30\ 000\ 000$ km^2.
The diameter of a human cell is about $0 \cdot 0000002$ m.

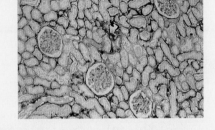

Very large and very small numbers can be difficult to work
with or write down because it is easy to make a mistake.

We often write very large or very small numbers in
standard form, sometimes called index notation.

Example $1 \cdot 86 \times 10^{-6}$ is written in standard form.

number from 1
up to but not
including 10

times 10 to
a power

$8 \cdot 735 \times 10^4$ and $7 \cdot 08 \times 10^{-3}$ are written in standard form.
$14 \cdot 73 \times 10^{-1}$ and $0 \cdot 0863 \times 10^4$ and $3 \cdot 86^4$ are *not* written in standard form.

14·73 is not from
1 up to 10.

0·0863 is not from
1 up to 10.

Discussion

Charlotte was asked to write the numbers $1 \cdot 57 \times 10^4$ and
$5 \cdot 27 \times 10^{-3}$ in decimal form.

'Decimal form' is sometimes
called 'ordinary form'.

She wrote $1.57 \times 10^4 = 1.57 \times 10\ 000$
$= 15\ 700$

$5.27 \times 10^{-3} = 5.27 \times \frac{1}{10^3}$
$= 5.27 \times \frac{1}{1000}$
$= \frac{5.27}{1000}$
$= 0.00527$

Is Charlotte correct? **Discuss**.

How could Charlotte write 53 800 in standard form?
What about 0·00538? **Discuss**.

Charlotte's friend asked her to write down some steps for changing numbers

    a from standard form to decimal form
    b from decimal form to standard form.

What might Charlotte write down? **Discuss**.

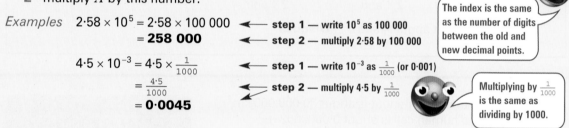

To write a number given in standard form ($A \times 10^n$) in decimal form,

1 write the 10^n part as a number not in index form,   *Example* $10^4 = 10\ 000$.
2 multiply A by this number.

Examples $2.58 \times 10^5 = 2.58 \times 100\ 000$ ←— **step 1** — write 10^5 as 100 000
                = 258 000   ←— **step 2** — multiply 2·58 by 100 000

The index is the same as the number of digits between the old and new decimal points.

$4.5 \times 10^{-3} = 4.5 \times \frac{1}{1000}$ ←— **step 1** — write 10^{-3} as $\frac{1}{1000}$ (or 0·001)
          $= \frac{4.5}{1000}$ ←— **step 2** — multiply 4·5 by $\frac{1}{1000}$
        = 0·0045

Multiplying by $\frac{1}{1000}$ is the same as dividing by 1000.

Exercise 1

1 Which of the following are in standard form?
    a 7.3×10^4 b 62.4×10^2 c 0.3×10^1 d 2.0×10^0
    e 3.49 f 8.2×10 g 3.05×10^{-17} h 80.1×10^{-3}
    i 7.6824×10^{92} j 0.305×10^4

2 Write these in decimal form.
    a 3.4×10^2 b 8.12×10^3 c 6.25×10^{-2} d 8.0×10^{-3}
    e 7.03×10^4 f 2.05×10^0 g 7.8×10^{-1} h 1.01×10^{-4}
    i 3.7×10^5 j 3.7×10^{-5} k 1.52×10^1 l 3.4×10^0
    m 4.81×10^{-3} n 8.0×10^1 o 2.61×10^{-5} p 6.0×10^{10}
    q 7.05×10^{-2} r 8.154×10^2 s 8.154×10^{-3} t 9.407×10^1
    u 9.407×10^{-1} v 6.0×10^4 w 6.0×10^{-4}

3 In the early 2000s, the population of the South-East of
   England was about 8.000064×10^6.
   Write this in decimal form.

4 About 1.2×10^9 credit card transactions take place in the
   UK each year.
   Write this in decimal form.

5 In a year, light travels about 9.46×10^{12} km.
Write this distance in decimal form.

6 The half-life of one of the polonium isotopes is about 3.0×10^{-7} seconds.
Write this in decimal form.

7 The wavelength of visible light is about 5.0×10^{-5} cm.
Write this in decimal form.

Review 1 Write these in decimal form.
a 2.3×10^{4} b 2.3×10^{0} c 2.3×10^{-4} d 3.0504×10^{-1}
e 9.01×10^{6} f 6.4×10^{-3} g 3.465×10^{2}

Review 2 About 3.4×10^{5} speeding tickets were issued in the UK last year.
Write this in decimal form.

Review 3 The diameter of a hydrogen atom is about 1.0×10^{-8} cm.
Write this in decimal form.

To write **a decimal number in standard form**,

   1 put the decimal point after the first non-zero digit, i.e. write the number as a number from 1 up to 10
   2 multiply by a power of 10, 10^{n}, so that your answer is equivalent to the original number.

Examples $5834 = 5.834 \times 10^{3}$

   Write as a number
   from 1 up to 10.

Multiply by a power of 10 so
that the number in standard
form is equal to 5834.

The index is the same
as the number of digits
between the old and
new decimal points.

$5 = 5 \times 10^{1}$ $0.0804 = 8.04 \times 10^{-2}$
$364.2 = 3.642 \times 10^{2}$ $0.004 = 4 \times 10^{-3}$

Exercise 2

1 Write the numbers in green as powers of 10.
  a The distance to the nearest star is about
    10 000 000 000 000 km.
  b The length of the Earth's orbit around the Sun is about
    1 000 000 000 km.
  c In a hydrogen bomb explosion, about **0·001 kg** of mass
    converts into energy.

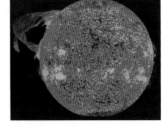

2 Find the missing index.
  a $36.4 = 3.64 \times 10-$ **b** $482.5 = 4.825 \times 10-$ **c** $7 = 7.0 \times 10-$
  d $17 = 1.7 \times 10-$ **e** $8478 = 8.478 \times 10-$ **f** $0.42 = 4.2 \times 10-$
  g $0.0591 = 5.91 \times 10-$ **h** $0.308 = 3.08 \times 10-$ **i** $0.008 = 8.0 \times 10-$
  j $2.6 = 2.6 \times 10-$ **k** $89 = 8.9 \times 10-$ **l** $0.05 = 5.0 \times 10-$
  m $0.6 = 6.0 \times 10-$ **n** $22.71 = 2.271 \times 10-$ **o** $0.00092 = 9.2 \times 10-$

3 Write these numbers in standard form.
  a 64   **b** 782   **c** 3640   **d** 55·2   **e** 7   **f** 1000
  g 34·2   **h** 555·61   **i** 72·4   **j** 0·8   **k** 0·04   **l** 0·0043
  m 0·804   **n** 0·91   **o** 2·4   **p** 0·24   **q** 24   **r** 0·0024
  s 240   **t** 9   **u** 90   **v** 0·09   **w** 0·9

4 Write the numbers in red in standard form.
    a The most distant objects yet observed are **15 000 000 000** light-years from the Earth.
    b The diameter of an atom is about **0·0000000001** mm.
    c The solubility of aluminium hydroxide is about **0·000000002** grams per litre.
    d The smallest insect, the fairy fly, is **0·02** mm long.
    e An estimate of the population in the UK in the year 2031 is **61·2 million**.

5 Put these in ascending order.
    a 3×10^{-2}, 4×10^{-1}, $4·5 \times 10^{-3}$, $3·8 \times 10^{-2}$, $2·9 \times 10^{-1}$
    b $8·1 \times 10^{-1}$, $9·4 \times 10^{-2}$, 8×10^{-2}, $7·3 \times 10^{-1}$, $9·9 \times 10^{-3}$

6 Write each of these probabilities as a decimal first and then in standard form.
    a The probability of dying in a road accident is 1 in 8000.
    b The probability of dying in the next ten years if you are aged 20 is about 1 in 950.
    c The probability of winning the National Lottery is 1 in 8 145 060.
    ***d** The probability of dying in a plane crash is 1 in 11 million.

Review 1   Write these in standard form.
a 52·7   **b** 16005   **c** 6   **d** 0·83   **e** 0·1   **f** 0·0002

Review 2   The Sun is about 0·000016 light-years from the Earth.
The centre of the Milky Way is about 26 000 light-years from the Earth.
Write these distances in standard form.

Review 3   Write each of these probabilities as a decimal first and then in standard form.
a The probability of a light bulb failing before 180 hours of use is 1 in 350.
* **b** The probability of living to be 116 is 1 in 2 billion.

Use 1 billion = 10^{12}.

Calculating with numbers in standard form

We use the **laws of indices** when we multiply or divide numbers written in standard form.
We use $a^x \times a^y = a^{x+y}$ and $a^x \div a^y = a^{x-y}$

Worked Example

Speed $= \dfrac{\text{distance}}{\text{time}}$

Calculate the time it would take for a spacecraft travelling
at an average speed of $2·6 \times 10^3$ km/h to reach the Moon
$3·844 \times 10^5$ km away.

Answer
Rearrange the equation to make time the subject.

$\text{Time} = \dfrac{\text{distance}}{\text{speed}}$

$= \dfrac{3·844 \times 10^5}{2·6 \times 10^3}$

$= \dfrac{3·844}{2·6} \times 10^{5-3}$ $\dfrac{3·844}{2·6} = 1·48$ **(2 d.p.)**

$= 1·48 \times 10^2$

$=$ **148 hours**

Discussion

What does 7.82×10^3 equal?

Key (7.82) (EXP) (3) (=).

What do you notice?

Discuss how to calculate these on a calculator.

$7.82 \times 10^3 \times 2.0 \times 10^{-4}$ $7.82 \times 10^{-3} \times 2.0 \times 10^{-4}$

To **key a number in standard form into the calculator** use the (EXP) key.

Example 5.7×10^{-2} is keyed as (5.7) (EXP) ((-)) (2).

If an answer is too large or too small to fit on your calculator screen, it will automatically be given in standard form.

Example   The answer to $87\,580\,000\,000 \times 5\,960\,000\,000$ is given as

5.219768 *20* ← This small number gives the power of 10.

See page 25 for significant figures.

The answer is **5.22×10^{20}** (3 s.f.).

To get the calculator to give answers to calculations such as 4.6×32.4 in standard form,

Key (MODE) (MODE) (MODE) (2).   ← This puts the calculator into scientific mode.

On the screen   *Sci 0~9?*   will appear.

Key in the number of significant figures you want the answer given to. If you key 4, the answer will be given to 4 significant figures.

Example 8.3657×95.6872 is keyed as (MODE) (MODE) (MODE) (2) (4) (8.3657) (×) (95.6872) (=) to

get   *8.005* *02*   which equals **8.005×10^2**.

To get the calculator out of scientific mode Key (MODE) (MODE) (MODE) (3).

On the screen   *Norm 1~2?*   will appear. Choose (1).

Exercise 3   **Except for questions 1 and 14.**

In this exercise, give all the answers in standard form.

1   Calculate.
   a $(3.7 \times 10^5) \times (2.0 \times 10^4)$ b $(2.4 \times 10^6) \times (2.0 \times 10^{-3})$ c $(4.12 \times 10^{-5}) \times (2.0 \times 10^{-3})$
   d $(4.24 \times 10^7) \div (2.0 \times 10^4)$ e $(6.9 \times 10^4) \div (3.0 \times 10^7)$ f $\frac{8.4 \times 10^6}{4.0 \times 10^{-2}}$

2 Calculate these. Round your answers sensibly.
  a $253\,000\,000\,000 \times 3\,640\,000\,000$ **b** $1\,300\,000\,000\,000 \times 28\,200\,000\,000$
  c $246\,000\,000 \div 134\,000\,000$ **d** $897\,200\,000\,000 \div 9\,982\,000\,000$
  e $0{\cdot}000000000082 \times 0{\cdot}00000000037$ **f** $0{\cdot}00000000789 \div 0{\cdot}000000000388$
  g $\dfrac{347\,000\,000\,000 \times 2\,440\,000\,000\,000}{2\,374\,000\,000}$ **h** $\dfrac{458\,900\,000\,000}{361\,000\,000 \times 783\,000\,000}$

3 A rectangle has length $2{\cdot}6 \times 10^3$ mm and width $1{\cdot}8 \times 10^2$ mm.
  a What is the area of this rectangle?
  b What is the perimeter of the rectangle?

4 **a** One oxygen atom has a mass of about $2{\cdot}7 \times 10^{-23}$ grams.
    How heavy would 5000 oxygen atoms be?
  b A hydrogen atom has a mass of about $1{\cdot}67 \times 10^{-24}$ grams.
    A water molecule is made up of two hydrogen atoms and one oxygen atom.
    What is the mass of one water molecule?

Link to science

5 The half-life of radium is about $1{\cdot}622 \times 10^3$ years.
  One year is about $8{\cdot}76 \times 10^3$ hours.
  How many hours is the half-life of radium?

6 One year about $9{\cdot}13 \times 10^4$ tonnes of shellfish
  and $5{\cdot}804 \times 10^5$ tonnes of other fish were caught.
  What total mass of fish was caught?

*7 The density of brass is $8{\cdot}5 \times 10^3$ kg/m³.
  What is the mass of $2{\cdot}4 \times 10^{-2}$
  cubic metres of brass?

Density $= \frac{mass}{volume}$

*8 A large stone has a mass of $5{\cdot}5 \times 10^2$ kg.
  It is known that the density of this stone is $2{\cdot}2 \times 10^3$ kg/m³.
  What is the volume of this stone?

Volume $= \frac{mass}{density}$

*9 The mass of a hydrogen atom is about $1{\cdot}67 \times 10^{-24}$ g while that of a uranium atom is
  about $3{\cdot}95 \times 10^{-22}$ g.
  About how many times heavier is a uranium atom than a hydrogen atom?
  (Round sensibly.)

*10 The nearest star other than the sun, Proxima Centauri, is about $4{\cdot}0 \times 10^{13}$ km from Earth.
  In a year, light travels about $9{\cdot}46 \times 10^{12}$ km.
  About how many light years is Proxima Centauri from Earth?

*11 **a** Which planet is the smallest?
  b Which planet is the largest?
  c Which planet has diameter about 10 times that of Venus?
  d Which planet has diameter about half that of Mercury?
  e One of the planets has diameter about 20 times larger than
    that of Pluto. Which planet?
  f The diameter of Venus is larger than that of Pluto.
    About how many times larger?
  g What is the ratio of the diameter of Pluto to the diameter of
    Saturn?

Planet	Diameter (km)
Mercury	$4{\cdot}9 \times 10^3$
Venus	$1{\cdot}2 \times 10^4$
Earth	$1{\cdot}3 \times 10^4$
Mars	$6{\cdot}8 \times 10^3$
Jupiter	$1{\cdot}4 \times 10^5$
Saturn	$1{\cdot}2 \times 10^5$
Uranus	$5{\cdot}2 \times 10^4$
Neptune	$4{\cdot}9 \times 10^4$
Pluto	$2{\cdot}4 \times 10^3$

*12 The mass of the Earth is about 5.97×10^{24} kg. How many tonnes is this?

*13 What goes in the box? The first one is done for you.

a $5 \times 10^n = 500 \times \boxed{10^{n-2}}$

b $0.5 \times 10^n = 500\,000 \times \boxed{}$

c $0.5 \times 10^n = 0.0005 \times \boxed{}$

d $5 \div 10^n = 0.0005 \times \boxed{}$

e $0.5 \div 10^n = 500 \times \boxed{}$

f $0.003 \div 10^n = 3 \times \boxed{}$

Remember:
$\div$ by 10^n is the same as $\times$ by 10^{-n}.

14 $\frac{1}{5000}$ is equal to 0.0002.

a Write 0.0002 in standard form.

b Write $\frac{1}{50\,000}$ in standard form.

c Work out $\frac{1}{5000} + \frac{1}{50\,000}$.
Show your working.

*15 This table shows information about some countries. **[SATs Paper 2 Level 8]**

Country	Population	Area (km²)
Canada	3.1×10^7	1.0×10^7
France	6.0×10^7	5.5×10^5
Gambia	1.4×10^6	1.1×10^4
India	1.0×10^9	3.3×10^6
United Kingdom	6.0×10^7	2.4×10^5
United States	2.8×10^8	9.3×10^6

a Use the table. What goes in the gaps?
  i The country with the largest population is _____.
  ii The country with the smallest area is _____.

b On average, how many **more** people per km² are there in the United Kingdom than in the United States?
Show your working.

*16 The star nearest the Earth (other than the Sun) is Proxima Centauri. **[SATs Paper 2 Level 8]**
Proxima Centauri is **4·22** light-years away.
(One light-year is 9.46×10^{12} kilometres.)
Suppose a spaceship could travel at **40 000 km per hour**.

Remember:
Speed $= \frac{distance}{time}$

a Write what the following calculations represent.
The first one is done for you.
  i $4.22 \times 9.46 \times 10^{12}$ **Number of km from Earth to Proxima Centauri**
  ii $\frac{4.22 \times 9.46 \times 10^{12}}{40\,000}$
  iii $\frac{4.22 \times 9.46 \times 10^{12}}{40\,000 \times 24 \times 365.25}$

b Work out $\frac{4.22 \times 9.46 \times 10^{12}}{40\,000 \times 24 \times 365.25}$.
Give your answer to the nearest thousand.

*17 A billion is a million million. A trillion is a million billion. A quadrillion is a million trillion.
Write, in standard form
a one trillion
b one quadrillion.

*18 1 micron is 10^{-4} centimetres.
a How many microns are there in 1 mm?
b 1 Angstrom is 10^{-8} centimetres. How many microns are there in 1 Angstrom?

Number

Review 1   Calculate.
a $(2\cdot4 \times 10^3) \times (4\cdot0 \times 10^2)$
b $(2\cdot0 \times 10^5) \times (5\cdot3 \times 10^0)$
c $(8\cdot4 \times 10^3) \div (2\cdot0 \times 10^{-2})$
d $(6\cdot5 \times 10^{-1}) \times (7\cdot8 \times 10^{-3})$
e $(2\cdot5 \times 10^{-3}) \div (5\cdot0 \times 10^{-2})$
f $(5\cdot3 \times 10^4) \times (3\cdot1 \times 10^{-5})$
g $(1\cdot64 \times 10^4) \div (2\cdot5 \times 10^{-1})$

Review 2   The UK value for one billion is $1\cdot0 \times 10^{12}$. The American value for one billion is $1\cdot0 \times 10^9$.
How much larger is the UK value for one billion than the American value?

Review 3   The Pacific Ocean covers an area of about $1\cdot65 \times 10^8$ km^2 while the Atlantic Ocean covers about $8\cdot22 \times 10^7$ km^2.
a   How many km^2 larger is the Pacific Ocean than the Atlantic Ocean?
b   What total area do these two oceans cover?

Review 4

City	Bombay	Cape Town	Darwin	London	Paris	Rome	Tokyo
Distance (in km) from Berlin	$6\cdot3 \times 10^3$	$9\cdot6 \times 10^3$	$1\cdot3 \times 10^4$	$9\cdot2 \times 10^2$	$8\cdot7 \times 10^2$	$1\cdot2 \times 10^3$	$8\cdot9 \times 10^3$

a   Which city is   i   closest to Berlin
                    ii   the greatest distance from Berlin
                    iii   about twice as far from Berlin as Bombay
                    iv   about 10 times closer to Berlin than Tokyo?
*b   What is the ratio of the distance between Cape Town and Berlin to the distance between London and Berlin?

*Review 5   1 hour = $3\cdot6 \times 10^3$ seconds
a   What is the velocity, in km/s, of an object which is travelling at $7\cdot2 \times 10^2$ km/h?
b   The velocity of light is about $3\cdot0 \times 10^5$ km/s. Give this velocity in km/h.

Upper and lower bounds

Discussion

● A report gave the size of the crowd at an athletics meeting as 8000. If the size of the crowd was given to the nearest thousand, how many people might have been at this meeting? **Discuss**.

What if the size of the crowd was given to the nearest hundred?

What if the size of the crowd was given to the nearest ten?

● 32 cm   28 cm   42 cm   34 cm   26 cm   40 cm   25 cm
  20 cm   35 cm   31·9 cm   24·8 cm   29·6 cm   30·4 cm   29·5 cm
  30·5 cm   29·4 cm   29·8 cm   30·2 cm

Which of the lengths in this list would be given as 30 cm, to the nearest 10 cm? **Discuss**.

Discuss some other lengths which would be given as 30 cm, to the nearest 10 cm.

What if the length was given as 30 cm, to the nearest cm?

What if the length was given as 30 cm, to the nearest $\frac{1}{2}$ cm?

- The size of the crowd given in the first bullet point above is discrete data. What does this mean? **Discuss**.
  What type of data are the lengths in the second bullet?

  > Jenni learnt in geography that the population of New Zealand, to the nearest 10 000 was 4 190 000.

  Link to handling Data.

  Is this discrete data?

  What are the greatest and least numbers for the population? **Discuss**.

Worked Example

1·9 million people to the nearest 100 000, visited the Tower of London last year.
What are the biggest and smallest number of people who could have visited?

Answer

This is discrete data as it can only take whole number values.
The **smallest** possible number of people is **1 850 000** because one fewer would be 1 849 999. This would have been rounded to 1·8 million, to the nearest 100 000.
The **biggest** possible number of people is **1 949 999** as one more would be 1 950 000 and this would have been rounded to 2·0 million, to the nearest 100 000.

We can write the **upper** and **lower bounds** for the number of people in the example above as:

1 850 000 $\leqslant p <$ 1 950 000

 This is called an inequality.

Worked Example

The height, to the nearest foot, of London Bridge Tower is 313 m.
What is the shortest and tallest height it could be?
Write your answer as an inequality.

Answer

The shortest it could be is 312·5 m.
The tallest it could be is right up to but not including 313·5 m.
We write this as:

312·5 $\leqslant h <$ 313·5

The height in the example above is continuous data. It can take *any* value in the given range.
The tallest height could be any value right up to but not including 313·5 m. If it was 313·5 m, this would have been rounded to the nearest metre as 314 m.

It *could* be 313·49999999 m.

Exercise 4

1 A construction firm estimated that it had employed 1700 people during the previous five years.
   a If this figure is correct to the nearest 10, find the biggest and smallest number of people employed by this firm during this time.
   b What goes in the gaps if you write your answer as an inequality?
   _____ $\leqslant$ number employed $<$ _____

2 It is projected that, to the nearest 1000, the population of Wales in 2021 will be 3 047 000.
   What is the greatest and the least number of people there could be?
   Write your answer as an inequality.
   _____ $\leqslant$ number of people $<$ _____

3 Carlton timber mills produce 3500 mm lengths of timber. Every fifth piece is measured to check it is the correct length to the nearest mm. Which of these pieces would be rejected?
- **a** 3500·7 mm
- **b** 3501 mm
- **c** 3495·7 mm
- **d** 3500·5 mm
- **e** 3500·2 mm
- **f** 3499·7 mm
- **g** 3499·3 mm

4 'Tour UK' advertised some day tours.
The distances were given to the nearest kilometre.
- **a** Write down the shortest and longest possible distances for each tour.
  Write your answers as inequalities.
- **b** Aunt Daisy went on the Lake District Luxury and Cotswold Delight one year.
  What is the shortest and longest distances altogether she could have travelled on these tours?

Lake District Luxury — 65 km
Midlands Rural Tour — 78 km
Cotswold Delight — 49 km
Norfolk Novelty — 104 km

5 a Give the shortest and longest lengths for these.
Write each answer as an inequality.
- **i** The length of the Nile, to the nearest 10 kilometres, is 6670 km.
- **ii** The length of the Amazon, to the nearest kilometre, is 6570 km.
- **iii** The length of the Zaire (Congo), to the nearest kilometre, is 4630 km.

b What is the longest and shortest total length of these rivers if all of the distances are given to the nearest kilometre?

Review 1 The audience at an outdoor concert was estimated at 18 000 to the nearest thousand.
a What is the greatest and least number of people that could have attended the concert?
b Write your answer as an inequality.   _____ $\leqslant$ no. of people $<$ _____.

Review 2
a Write down the shortest and longest lengths of the sides of this rectangle.
Assume the measurements are correct to the nearest cm (0·01 m).
b Calculate the longest and shortest perimeter and biggest and smallest area of the rectangle.
c Write each answer in inequality form. Round sensibly.

4·82 m

8·36 m

Rounding

We sometimes **round** to enable us to make an **estimate**.

Discussion

Ammoniacal nitrogen is toxic in fresh water if it exceeds one part per million (1 mg per m^3 of water).
The flow in a river is 4200 litres per second.
A waste water discharge from a local town contains 38 parts per million of ammoniacal nitrogen.
At how many litres per second can the town discharge into the stream for it not to be toxic?

Discuss how you could use rounding to *estimate* the answer.

Rounding to decimal places and significant figures

Remember

Follow these steps to **round to a given number of decimal places**. For example, to round 4·06992 to 3 decimal places.

1 Keep the number of digits you want after the decimal point. **4·06992**

— keep

2 Before discarding the rest, look at the next decimal place. 4·06992

— look at this digit

3 If this digit is 5 or greater, increase the last digit you are keeping by 1. Otherwise leave it as it is.
Discard the unwanted digits.

4·069|92

— This digit is greater than 5 so add 1 to the 9

4·07**0**

adding 1 to 9 makes 10 so 4·069 becomes 4·070

We keep this zero to show the number has been rounded to 3 d.p. not 2 d.p.

We sometimes round to a number of **significant figures**.

3·6482 rounded to 1 significant figure is 4.

This is the highest significant figure.

3·6482 rounded to 2 significant figures is 3·6.
3·6482 rounded to 3 significant figures 3·65.
3·6482 rounded to 4 significant figures 3·648.

Follow these steps to **round to a given number of significant figures**. For example, to round 0·2465 to 2 significant figures.

1 Count from the left and keep the number of digits asked for, from the first *non-zero* digit.

0·2465

— keep

2 Before deleting the rest, look at the next decimal place.

0·2465

— look at this digit

3 If this digit is 5 or greater, increase the last digit you are keeping by 1. Otherwise leave it as it is.
Delete the unwanted digits.

0·2465

— this digit is greater than 5 so add 1 to the 4

0·25

The words 'significant figures' are often abbreviated to s.f.

Worked Example
Round these to 3 s.f.
a 19·753   b 9·832.

Answer
a 1 19·753

— keep

2 19·753

— Look at this digit. It is 5 or greater so 7 becomes 8.

19·753 to 3 s.f. is **19·8**.

b 1 9·832

— keep

2 9·832

— Look at this digit. It is less than 5 so 3 remains as 3.

9·832 to 3 s.f. is **9·83**.

Number

When we **round to s.f.** we count the number of figures from the first non-zero figure.

Worked Example
Write 0·0504715 to 4 s.f.

Answer
1   0·0504715
    — keep the first 4 non zero digits
2   0·0504715
    — Look at this digit. It is less than 5 so 7 stays as 7.
    0·0504715 to 4 s.f. is 0·05047

The first zero after the ·
is *not* 'significant'.

The second zero is 'significant' since
it comes after the first non-zero digit.

When we **round large numbers to s.f.** we may have to insert zeros so the size of the
number is unchanged.

For example, 34 592 to 1 s.f. is written as 30 000, not as 3.

These zeros are not 'significant'.
They are place-holder zeros.

Exercise 5

1 Round these to 3 d.p.
   a   28·9872   b   3·4179   c   3·6004   d   8·0327   e   31·2496
   f   40·0801   g   7·96381   h   62·1522   i   12·7399   j   4·1595
   k   29·7094   l   3·0099   m   49·9999

2 Round these to 1 s.f.
   a   34   b   43   c   407   d   970   e   750
   f   2348   g   2843   h   3099   i   2415   j   4512
   k   84 030   l   253·8   m   76·821   n   954·3

3 Round these to 3 s.f.
   a   0·04826   b   0·00090218   c   0·060071   d   0·34052   e   0·92471
   f   0·0098888   g   0·0088148   h   0·080008   i   0·099952

4 Hayley used her calculator for some calculations. The calculator displays are shown.
   Hayley rounded the answers to 2 significant figures. What answers did she give?

   a   34.84   b   2521   c   356.09
   d   1.005   e   0.00628   f   9.88
   g   836241   h   0.0005555

5 What answers would Hayley have given if she had rounded the calculator displays to
   1 significant figure?

6 Round to the number of significant figures or decimal places given in the brackets.
   a   3·405 (2 d.p.)   b   27·4 (1 s.f.)   c   243 (2 s.f.)   d   18·996 (3 s.f.)
   e   0·0238 (2 s.f.)   f   0·79 (1 d.p.)   g   1405 (1 s.f.)   h   17·63 (0 d.p.)
   i   149 (2 s.f.)   j   0·0598 (3 d.p.)   k   0·00864 (1 s.f.)   l   8·0995 (3 d.p.)

7 Evaluate $\frac{1}{960}$ to **a** 1 d.p. **b** 3 s.f.

8 A reporter estimated that there were 37 455 spectators at a football game.
Approximate this number to
a the nearest ten
b the nearest hundred
c the nearest thousand
d the nearest ten thousand
e one significant figure
f two significant figures.

9 Use your calculator for these, giving the answer to 2 significant figures.
   a $19 \div 6$ **b** $23 \div 13$ **c** $4 \div 71$ **d** 2600×88
   e 540×290 **f** $0.2 \div 21$ **g** $0.0055 \div 0.18$ **h** $\frac{2}{3}$ of 140
   i 17.5% of 9 **j** $\sqrt{5}$

10 About 55% of all new cars in Great Britain are imported. Of the 213 000 new cars registered in one month, about how many would be imported?
(Answer to 3 s.f.)

11 The film *E.T.* made approximately $210 million (US). The film *Gone with the Wind* made approximately 37% of that.
How much did *Gone with the Wind* make? Give the answer in US dollars to two significant figures.

12 Jeff found the answer to $\frac{597 \cdot 3}{4 \cdot 4 \times 1 \cdot 38}$ like this.

He keyed 4·4 × 1·38 = to get 6·072.
He rounded this to the nearest whole number, 6.
Then he keyed 597·3 ÷ 6 = to get 99·55.
He rounded this to 2 significant figures to get 100.
Marie found the answer like this.
She keyed 597·3 ÷ (4·4 × 1·38) = to get 98·36956522.
She rounded this to 2 significant figures to get 98.
Who is correct? Explain why.

13 Find the answer to these. Round your answer to 3 s.f.
   a $\frac{5 \cdot 72 \times 34 \cdot 61}{6 \cdot 3}$ **b** $\frac{483 \cdot 6}{5 \cdot 1 \times 2 \cdot 73}$ **c** $\frac{0 \cdot 642}{19 \cdot 63 + 18 \cdot 27}$
   d $68 \cdot 4 + \frac{53 \cdot 6 \times 4 \cdot 2}{3 \times 6 \cdot 3^2}$ **e** $112^2(\frac{0 \cdot 0427}{5 \cdot 68 \times 3 \cdot 27} + 8 \cdot 96^2)$
   f $\frac{6 \cdot 2 \times 10^5}{3 \cdot 6 \times 10^7 \times 4 \cdot 1 \times 10^2}$

> Remember not to round till the final answer.

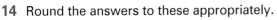

14 Round the answers to these appropriately.
Say what you have rounded to.
   a The volume of a sphere is found using the formula $V = \frac{4}{3}\pi r^3$.
      Find the volume of a sphere of radius 8·64 m.
   b The area of a circle is given by $A = \pi r^2$, where r is the radius.
      Aidan measured the radius of a circular base he made in design and technology as 0·68 m.
      Find the area of the circular base.
   c If $a = 6 \cdot 4$ and $b = 0 \cdot 83$ find the value of
      i ab **ii** $a^2 b$ **iii** $\frac{a}{b}$ **iv** $\frac{a+b}{a^2}$

*15 Use rounding to *estimate* the answer to this.

The water quality guideline for copper is that it is toxic when there is greater than 5 parts per billion (5 mg per m^3 of water).
A factory discharges 10·3 kg per day in 2100 m^3 of water through an ocean outfall.

The billion used is the American billion, 1 000 000 000.

What dilution rate is needed to make it safe?

Review 1   Round these to the given number of significant figures or decimal places.
a   0·0294 (3 d.p.)   b   68·7 (2 s.f.)   c   0·0079 (1 s.f.)
d   3069 (3 s.f.)   e   98·601 (2 d.p.)   f   398·6 (2 s.f.)

Review 2   The mass of a bus is given as 4527 kg.
Approximate this mass to
a   the nearest 1000 kg   b   two significant figures   c   the nearest 10 kg
d   one significant figure   e   the nearest 100 kg.

Review 3   Jenny wanted to make a square vegetable garden which would only take up 5 m^2.
She worked out the length she needed to make each side as
$\sqrt{5} = 2\cdot23607$ m (6 s.f.).

Why might this number of significant figures not be appropriate?
What might have been better?

Review 4   Find the answer to these. Round your answers to 3 s.f.
a $\dfrac{3\cdot87 \times 5\cdot68}{4\cdot92}$ b $\dfrac{2\cdot46}{32\cdot98 + 16\cdot84}$ c $17^3(\frac{4\cdot83}{0\cdot25} + 7\cdot92^2)$

Review 5   Find the correct answer using a calculator. Say what you rounded the answer to.
a   The surface area of a sphere is given by the formula $A = 4\pi r^2$.
    What is the surface area if $r = 5\cdot72$ cm?
b   The volume of a cylinder is $V = \pi r^2 h$.
    What is the volume if $r = 3\cdot62$ cm and $h = 9\cdot81$ cm?

Investigation

Significant figures

$5\cdot5 \times 5\cdot5 = 30\cdot25$
$5\cdot25 \times 5\cdot25 = 27\cdot5625$
$5\cdot255 \times 5\cdot255 = 27\cdot615\,025$

Investigate the following statements. They may be true or false.

1   If we multiply a 2 s.f. number by a 2 s.f. number we always get a 4 s.f. answer.
2   If we multiply a 3 s.f. number by a 3 s.f. number we always get a 6 s.f. answer.
3   If we multiply an n s.f. number by an n s.f. number we always get a $2n$ s.f. answer.
4   If we multiply an n s.f. number by an m s.f. number we always get an $n + m$ s.f. answer.

Try to find a counter-example to show a statement is false.

What if you divide instead of multiplying?

Summary of key points

 A number is written in **standard form** when it is written as $A \times 10^n$, where $1 \leqslant A < 10$ and n is an integer.

Example 6.35×10^4 and 5.21×10^{-3} are written in standard form.

If a number is given in standard form, we can write it in decimal form by

    1 writing 10^n as a number not in index form

    2 multiplying A by this number.

Examples $\begin{aligned} 1.52 \times 10^3 &= 1.52 \times 1000 \\ &= 1520 \end{aligned}$ $\qquad \begin{aligned} 4.6 \times 10^{-2} &= 4.6 \times \frac{1}{10^2} \\ &= 4.6 \times \frac{1}{100} \\ &= \frac{4.6}{100} \\ &= 0.046 \end{aligned}$

If a number is given in decimal form we can write it in standard form by

    1 putting the decimal point after the first non-zero digit

    2 multiplying by a power of 10 so that your answer is equal to the original decimal number.

Examples $6245 = 6.245 \times 10^3$

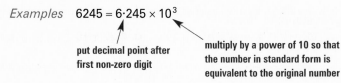

put decimal point after first non-zero digit

multiply by a power of 10 so that the number in standard form is equivalent to the original number

 We can **calculate with numbers in standard form**.

We may need to use these laws of indices.

    $a^x \times a^y = a^{x+y} \qquad a^x \div a^y = a^{x-y}$

To key a number given in standard form into your calculator use the $\boxed{\text{EXP}}$ key.

Example 6.5×10^{-4} is keyed as $\boxed{6.5}\ \boxed{\text{EXP}}\ \boxed{(-)}\ \boxed{4}$.

See page 19 for how to get your calculator to give answers in standard form.

 When numbers are rounded, we sometimes give **upper and lower** bounds for the rounded number.

Example   56 000 people, to the nearest 1000, went to a rock concert.

    The smallest number of people who could have been there is 55 500 and the biggest number is 56 499.

 Number of people is discrete data. It can only have whole-number values.

    This can be written as $55\ 500 \leqslant p < 56\ 500$.

Example   On his cycling holiday, Tim cycled 320 km, to the nearest 10 km.

    The shortest distance he could have cycled is 315 km and the longest distance is up to but not including 325 km. This can be written as the inequality $315 \leqslant d < 325$.

 Distance is continuous data. It can take any value in the range. The longest distance could be 324.9999 ... km

D Follow these steps to **round to a particular number of significant figures**.

1 Count from the left from the first *non-zero* digit and keep the number of digits asked for.

2 Delete all the following digits.

If the first digit to be deleted is 5 or greater, increase the last digit kept by 1.

Examples   24·679 to 3 s.f. is 24·7.

0·060502 to 3 s.f. is 0·0605

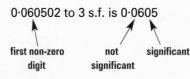

first non-zero digit   not significant   significant

Test yourself

1 Which of these are written in standard form?   **A**
  a $3\cdot4 \times 10^4$ **b** $5\cdot6^7$ **c** 80×10^{-1} **d** $5\cdot65 \times 10^6$ **e** $4\cdot27 \times 10^{-2}$

2 Write these numbers in decimal form.   **A**
  a $3\cdot24 \times 10^3$ **b** $6\cdot04 \times 10^4$ **c** $8\cdot2 \times 10^{-2}$ **d** $9\cdot0 \times 10^1$
  e $7\cdot5 \times 10^{-1}$ **f** $4\cdot0 \times 10^5$ **g** $5\cdot05 \times 10^{-6}$ **h** $5\cdot126 \times 10^0$

3 Write these numbers in standard form.   **A**
  a 52   **b** 654   **c** 32·6   **d** 3   **e** 642·35
  f 0·7   **g** 0·36   **h** 6   **i** 0·063   **j** 2006

4 Write the numbers in red in decimal form.   **A**
  a The distance of the Sun from Earth is $\mathbf{1\cdot5 \times 10^6}$ km.
  b The diameter of a white blood cell is $\mathbf{1\cdot243 \times 10^{-3}}$ cm.

5 Put these in ascending order.   **A**
  a $6\cdot4 \times 10^3$, $9\cdot6 \times 10^2$, $1\cdot8 \times 10^4$, $7\cdot1 \times 10^3$, $2\cdot9 \times 10^2$
  b $6\cdot9 \times 10^{-4}$, $7\cdot5 \times 10^{-3}$, $1\cdot6 \times 10^{-5}$, 9×10^{-3}, $2\cdot3 \times 10^{-5}$

6 Calculate these. Give your answers in standard form.   **B**
  a $(4\cdot3 \times 10^6) \times (2\cdot0 \times 10^2)$ **b** $(3\cdot14 \times 10^{-2}) \times (2\cdot0 \times 10^{-4})$
  c $(8\cdot64 \times 10^6) \div (2\cdot0 \times 10^4)$ **d** $(9\cdot63 \times 10^2) \div (3\cdot0 \times 10^{-3})$

7 Calculate these. Round your answers sensibly.   **A B**
  Give your answers in standard form.
  a $42\,000\,000 \times 21\,500\,000$ **b** $0\cdot0000056 \times 0\cdot00000063$
  c $254\,300\,000 \div 86\,400\,000$ **d** $0\cdot000000654 \div 0\cdot00000258$
  e $\dfrac{615\,200\,000\,000}{14\,500\,000 \times 7\,120\,000}$

8 $\frac{1}{2500}$ is equal to 0·0004.   [SATs Paper 1 Level 8]
   a Write 0·0004 in standard form.
   b Write $\frac{1}{25\,000}$ in standard form.
   c Work out $\frac{1}{2500} + \frac{1}{25\,000}$.
      Show your working and write your answer in standard form.

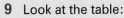 9 Look at the table: [SATs Paper 2 Level 8]

	Earth	Mercury
Mass (kg)	$5 \cdot 98 \times 10^{24}$	$3 \cdot 59 \times 10^{23}$
Atmospheric pressure (N/m²)		2×10^{-8}

   a The atmospheric pressure on Earth is $5 \cdot 05 \times 10^{12}$ times
      as great as the atmospheric pressure on Mercury.
      Calculate the atmospheric pressure on Earth.
   b What is the ratio of the mass of Earth to the mass of
      Mercury?
      Write your answer in the form $x : 1$.
   c The approximate volume, V, of a planet with radius r is given by $V = \frac{4}{3}\pi r^3$.
      Assume the radius of Mercury is 2400 km.
      Calculate the volume of Mercury.
      Give your answer, to 1 significant figure, in standard form.

10 6×10^{23} carbon dioxide molecules have a mass of about 44 g.
    How heavy would 4 million carbon dioxide molecules be?

11 In 2001 the population of India was about $1 \cdot 03 \times 10^9$ while that of France was about
    $6 \cdot 07 \times 10^7$.
    The area of India is about $3 \cdot 3 \times 10^6$ km^2 while that of France is about $5 \cdot 5 \times 10^5$ km^2.
    Find
    a the population density (number of people per square kilometre) of India in 2001
    b the population density of France in 2001.

12 The population of the world is about 5300 million.
    The approximate population of the four largest metropolitan areas in the world are
         Tokyo 28 million   New York 20.1 million
         Mexico City 18.1 million   Mumbai, India 18 million
    The tenth most heavily populated area is Buenos Aires with 12.5 million.
    Estimate the percentage of the world's population which lives in the ten largest
    metropolitan areas.

13 a Jessie read that there were 15 000, to the nearest 1000, at the cricket match.
       i   What is the smallest and biggest number of people that could have been there?
       ii Give your answer as an inequality.
    b Jessie heard on the radio that there were 15 200, to the nearest 100, at the match.
       i   What are the smallest and biggest numbers of people that could be there now?
       ii Give your answer as an inequality.

T

14 Use a copy of this.
Fill it in.

Number	Nearest whole number	To 1 d.p.	To 2 d.p.
46·075			
0·625			
16·995			

15 a Give the smallest and biggest volumes of acids these students made up in a
chemistry experiment. Give your answers as inequalities.
  i Jody used 210 mℓ to the nearest 10 mℓ.
  ii Marcel used 200 mℓ to the nearest 10 mℓ.
  iii Sam used 180 mℓ to the nearest 10 mℓ.

  *b What is the smallest and biggest total volume used by Jody, Marcel and Sam if all the
  volumes are given to the nearest 1 mℓ?

16 Two satellites circle the Earth.   **[SATs Paper 2 Level 8]**
Their distance from the centre of the Earth is:
    $1·53 \times 10^7$ m   Satellite A
    $9·48 \times 10^6$ m   Satellite B
  a What is the minimum distance apart the satellites
  could be?
  Show your working and give your answer in
  standard form.
  b What is the maximum distance apart the satellites
  could be?
  Show your working and give your answer in standard form.

17 Round each of these to the stated number of significant figures.
  a 54·6 (1 s.f.)   **b** 362 (2 s.f.)   **c** 8·399 (2 s.f.)
  d 0·024 (1 s.f.)   **e** 3925 (1 s.f.)   **f** 639 (2 s.f.)
  g 9·0992 (3 s.f.)   **h** 0·00572 (1 s.f.) **i** 0·06042 (3 s.f.)

18 A university enrolled 18 465 full-time students one year.
Approximate this number to
  a the nearest 10   **b** the nearest 100
  c the nearest 1000   **d** the nearest 10 000
  e one significant figure **f** two significant figures
  g three significant figures.

19 The export earnings of a country were £21 565 000. About 22% of this was earned
from timber.
How much was earned from timber? (Answer to 3 s.f.)

20 Round the answer to these sensibly. Say what you
have rounded them to.
  a A 6 kg bag of fruit and nuts was shared equally by
  14 people. How much did each person get?
  b The area of a rectangular mat is 36·5 m^2.
  One side of the mat is 7·5 m.
  What is the length of the other side?

2 Integers, Powers and Roots

You need to know

✓ integers   page 1

✓ prime numbers, prime factors, HCF, LCM   page 2

✓ powers   page 2

             – squares and cubes

             – square roots and cube roots

·· Key vocabulary ·······

common factor, cube, cubed, cube root, highest common factor (HCF), index, index law, indices, lowest common multiple (LCM), prime factor decomposition, square, squared, square root, take out common factor

⏵⏵ Shaping Numbers

1

  1   5   12   22

These diagrams show the first four pentagonal numbers.
Why are 1, 5, 12, 22 called pentagonal numbers?
Draw a diagram to find the fifth pentagonal number.
What is this fifth pentagonal number?

2

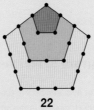

  1   6   15

These diagrams show the first three hexagonal numbers.
Why are 1, 6, 15 called hexagonal numbers?
Draw a diagram to show the fourth hexagonal number.
Write down the first four hexagonal numbers.

Using prime factor decomposition

Remember

The **HCF (highest common factor)** of two numbers is the largest factor common to both.

Example   8 is the HCF of 24 and 32.   Factors of 24   1, 2, 3, 4, 6, **8**, 12, 24
Factors of 32   1, 2, 4, **8**, 16, 32

See page 2 for more on HCF and LCM.

The **LCM (lowest common multiple)** of two numbers is the smallest number that is a multiple of both.

Example   24 is the LCM of 6 and 8.   6, 12, 18, **24**, 30, 36
8, 16, **24**, 32

We can use **prime factors** to find the HCF and LCM.

Example   To find the LCM and HCF of 700 and 1440, we find the prime factors of 700 and 1440 using a table or factor tree.

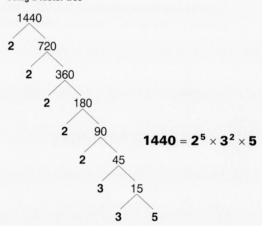

Using a table

2	700
2	350
5	175
5	35
	7

Using a factor tree

1440 → 2, 720 → 2, 360 → 2, 180 → 2, 90 → 2, 45 → 3, 15 → 3, 5

$$1440 = 2^5 \times 3^2 \times 5$$

$$700 = 2^2 \times 5^2 \times 7$$

This is 700 expressed as a product of prime factors in index notation.

Now use a **Venn diagram** to find the HCF and LCM.

Put the common prime factors of both numbers in the middle where the ovals overlap.

Put the other prime factors for each number in the other part of the oval for that number.

$HCF = 2 \times 2 \times 5 = 20$
$LCM = 2 \times 2 \times 2 \times 2 \times 2 \times 3 \times 3 \times 5 \times 5 \times 7$
$= 50\,400$

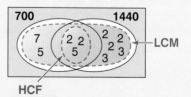

We use the **HCF when cancelling fractions** and the **LCM when adding and subtracting fractions**.

Example $\dfrac{700}{1440} = \dfrac{35}{72}$ dividing both numerator and denominator by 20, the HCF of 700 and 1440

Example $\dfrac{17}{700} - \dfrac{23}{1440} = \dfrac{1224}{50\,400} - \dfrac{805}{50\,400}$ common denominator, 50 400, is the LCM of 700 and 1440

$= \dfrac{1224 - 805}{50\,400}$

$= \dfrac{419}{50\,400}$

Discussion

A number is a multiple of 3 and 4. It has 3 digits.
What is the smallest number it could be? **Discuss**.
What if the numbers were 12 and 21?

Exercise 1

1   Express each of these numbers as a product of prime factors in index notation.
   a   450   b   630   c   3168   d   2160

2   Find the HCF and LCM of these.
   You could draw Venn diagrams to help.
   a   240 and 180   b   280 and 392   c   112 and 154   d   525 and 1250

> See page 2 for finding HCF and LCM.

3   Cancel these fractions by finding the HCF of the numerator and denominator.
   a $\frac{45}{72}$ b $\frac{180}{240}$ c $\frac{280}{392}$
   d $\frac{112}{154}$ e $\frac{525}{1250}$ f $\frac{2160}{5760}$

> Use your answers from questions **1** and **2**.

4   Add or subtract these fractions by finding the LCM of the denominators.
   a $\frac{7}{45} + \frac{9}{72}$ b $\frac{29}{180} - \frac{37}{240}$ c $\frac{11}{280} + \frac{53}{392}$

5   A pair of numbers have an HCF of 8 and a LCM of 480.
   What might they be?
   Find all the answers.

6   Use each of the digits in the box once to make two three-digit
   numbers with an HCF of 24.

$$\begin{array}{ccc} 1 & 5 & 0 \\ 8 & 6 & 4 \end{array}$$

7   $\frac{\boxed{A}}{\boxed{B}} + \frac{\boxed{C}}{\boxed{D}} = \frac{17}{12}$

   The HCF of A and B is 4.
   The LCM of B and D is 24.
   What numbers might A, B, C and D be if they are all different?

*8   Jack thinks of a number. He works out that all except two of the numbers from 1 up to 10
   are factors of this number.
   The two numbers that are not factors are consecutive.
   What is the smallest number Jack could be thinking of?

*9   Three bells are being rung, the first every 40 seconds, the
   second every 45 seconds and the third every 60 seconds. If they
   ring together at 11 a.m., when will they next ring together?

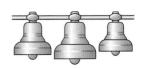

Review 1   Cancel these fractions by finding the HCF of the numerator and the denominator.
a $\frac{154}{330}$ b $\frac{525}{2310}$ c $\frac{1980}{4752}$

Review 2   Add or subtract these fractions by finding the LCM of the two denominators.
a $\frac{5}{54} + \frac{11}{90}$ b $\frac{37}{240} - \frac{17}{192}$

*Review 3   The HCF of three numbers is 10 and the LCM is 2000.
Two of the numbers are 80 and 100. What is the third number?

Common factors of algebraic expressions

In algebra, letters stand for numbers.

We can find **common factors of algebraic expressions**.

Example $2n = 2 \times n$
$3n = 3 \times n$
Both $2n$ and $3n$ can be divided exactly by n.
n is a common factor of $2n$ and $3n$.

Example $2xy^2z$ and $3wxy$ can both be divided exactly by x and by y.
xy is a common factor of $2xy^2z$ and $3wxy$.

This is linked to factorising in algebra, page 163.

We can find the HCF and LCM of two expressions using a Venn diagram.

Example   To find the HCF and LCM of $3m^4n^3$ and m^6n^4 draw a Venn diagram.

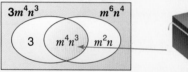

This section has the highest power of each letter that both expressions have.

$\mathbf{HCF} = \boldsymbol{m^4n^3}$ $\mathbf{LCM} = \mathbf{3} \times \boldsymbol{m^4n^3} \times \boldsymbol{m^2n}$ $m^4 \times m^2 = m^6$ by laws of indices
                                    $= \boldsymbol{3m^6n^4}$

Exercise 2

1   Find the highest common factor of each of these.
   a $5n$ and $8n$ b $3m$ and m c $5b$ and b^2 d $2n$ and $2n^2$
   e $3p$ and p f n and $2n$ g $8x$ and 2 h $3a$ and $6a$
   i $5m$ and $10m$ j $5xy$ and $3yz$ k $4ab$ and $8bc$ l $5a^2b$ and ab
   m $3x^2y$ and x^2y n $5pq^2$ and $4pq$ o $8a^2bc$ and b^2c p $12xy^2z$ and $4xy$

2   Use a copy of this.
    Fill in the Venn diagrams for the expressions given.
    Use this to find the HCF and LCM of the expressions.

a
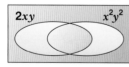

HCF = _____
LCM = _____

b
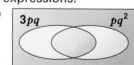

HCF = _____
LCM = _____

c
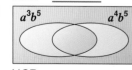

HCF = _____
LCM = _____

d

12m^4n^4 3m^6n^2

HCF = _____
LCM = _____

3 Find the HCF and LCM of these.
   a cd and c^2d **b** $8y^2$ and $4x^2y$ **c** mn^3p^2 and m^2n^4
   d $24a^4b^3c^2$ and $8a^4b^2c$ ***e** $25a^2b^3$ and $15a^3c^4$ ***f** $48pqr$ and $32p^2q^4$

4 $x^2 + 5x$ can be factorised as $x(x + 5)$.
   a What can $x^2 + 3x$ be factorised as?
   b What is the common factor of $x(x + 5)$ and $x(x + 3)$?
   c What is the common factor of these?
      i $x(x - 2)$ and $x(x + 4)$ **ii** $y(y + 8)$ and $y(y - 3)$ **iii** $a^2(a - b)$ and $a^2(2a + b)$
      iv $m(n + p)$ and $m^2(n - p)$ **v** $p^2q(r + 2)$ and $pq(3r - 4)$ **vi** $s^2t(r - q)$ and $st^2(r + q)$

5 Mario thinks that $(y - 3)$ is the HCF of these expressions.
   $(y - 3)(5y + 1)$ and $(y - 3)^2(5y - 1)^3$
   a Is Mario correct? Explain.
  *b What if the expressions were
      $(y - 3)(5y + 1)$ and $(y - 3)^2(5y + 1)^3$?

The HCF is $y - 3$.

*6 Write true or false for these.
   a The HCF of $(p - 2)^2$ and $(p - 2)$ is $(p - 2)^2$.
   b The HCF of $(m + 3)^2$ and $(m + 3)$ is $(m + 3)$.
   c The HCF of $(a - 2)(a + 2)$ and $(a - 2)^2$ is $(a - 2)$.
   d The HCF of $(b - 3)^2(2b + 1)$ and $(b - 3)(2b - 1)$ is $(b - 3)(2b + 1)$.
   e The HCF of $(h + 1)^2(2h - 3)^2$ and $(h + 1)(2h - 3)$ is $(h + 1)(2h - 3)$.

*7 Find the HCF of these.
   a $(x - 2)(3x + 1)$ and $(x - 2)^2(2x - 4)$ **b** $(a + 3)^2(2a + 1)$ and $(a + 3)(2a - 1)$
   c $(b + 2)^2(3b - 2)$ and $(b + 2)^2(3b + 2)^2$ **d** $(y + 3)(y - 3)$ and $(y + 3)^2(y - 3)^2$

Review 1 Write down the common factor of these.
a $5a$ and $7a$ **b** $3x^2$ and x **c** $5n$ and 10 **d** $8pq$ and $16q$ **e** $4a^2b$ and $6ab^2$

Review 2 Use a copy of this box. Write the letter beside each question above its answer in the box.

$\overline{7xz}$	$\overline{24ab^3c}$	$\overline{9x^3yz^2}$	$\overline{54x^5y^2z^3}$

$\overline{a^2b^3c^4}$	$\overline{48x^3y^2z}$	$\overline{144a^3b^5c^2}$	$\overline{9x^3yz^2}$	$\overline{54x^5y^2z^3}$	$\overline{48x^3y^2z}$	$\overline{a^2b^2}$	$\overline{abc}$	$\overline{9x^3yz^2}$

A

$\overline{a^2b^2}$	$\overline{24ab^3c}$	$\overline{210x^2yz^2}$	$\overline{54x^5y^2z^3}$	$\overline{ab}$	$\overline{48x^3y^2z}$	$\overline{210x^2yz^2}$

A

$\overline{24a^2b}$	$\overline{48x^3y^2z}$	$\overline{9x^3yz^2}$	$\overline{8x^2y^2}$		$\overline{9x^3yz^2}$	$\overline{a^2b^2}$	$\overline{ab}$	$\overline{a^2b^3c^4}$ $\overline{2ab}$ $\overline{9x^3yz^2}$

Find the HCF and LCM of these.

	HCF	LCM		HCF	LCM
a^2b and ab^2	**A** $= ab$	**C**	$35xyz$ and $42x^2z^2$	**M**	**N**
$6ab$ and $8a^2b$	**E**	**F**	$48a^3b^5c$ and $72ab^3c^2$	**O**	**P**
$16x^3y^2$ and $24x^2y^2z$	**H**	**I**	$18x^5y^2z^3$ and $27x^3yz^2$	**S**	**T**
ab^3c^4 and a^2bc	**K**	**L**			

Review 3 Match the HCF of the expressions in Box 1 with the answers given in box 2.

    Box 1

    1 $(2x - 1)^2(3x + 2)$ and $(2x - 1)(3x - 2)^2$
    2 $(2x - 1)^2(3x - 2)$ and $(2x - 1)(3x - 2)^2$
    3 $(2x - 1)^2(3x + 2)$ and $(2x + 1)(3x + 2)^2$
    4 $(2x - 1)(2x + 1)(3x + 2)$ and $(2x + 1)(3x + 2)(3x - 2)$

    Box 2

    A $(2x + 1)(3x + 2)$
    B $(3x + 2)$
    C $(2x - 1)$
    D $(2x - 1)(3x - 2)$

T

Powers and roots

Remember

8^6 is read as '8 to the power of 6'.

The 6 in 8^6 is called an **index**. The plural is **indices**.

On a calculator, squares are keyed using $\boxed{x^2}$,

cubes are keyed using $\boxed{x^3}$,

other powers are keyed using $\boxed{\wedge}$.

> We need the brackets or else we get the answer to $^-4 \cdot 3^2$.

Examples $(^-4 \cdot 3)^2$ is keyed as $\boxed{(}\ \boxed{(-)}\ \boxed{4 \cdot 3}\ \boxed{)}\ \boxed{\wedge}\ \boxed{=}$ to get 18·49.

$5 \cdot 6^7$ is keyed as $\boxed{5 \cdot 6}\ \boxed{\wedge}\ \boxed{7}\ \boxed{=}$ to get 172709·485.

Note Any number to the power of zero equals 1.

$x^0 = 1$

$0 \cdot 8 \times 0 \cdot 8 = 0 \cdot 64$ **and** $^-0 \cdot 8 \times ^-0 \cdot 8 = 0 \cdot 64$

$\sqrt{0 \cdot 64}$ has two roots, $0 \cdot 8$ and $^-0 \cdot 8$.

We usually only give the positive square root unless

– the notation $^\pm\sqrt{0 \cdot 64}$ is given

– we are finding $\sqrt{0 \cdot 64}$ as part of a problem.

On a calculator to find

$\sqrt{}$ we key $\boxed{\sqrt{}}$,

$\sqrt[3]{}$ we key $\boxed{\sqrt[3]{}}$,

other roots we use the $\boxed{\sqrt[x]{y}}$ key.

Examples $\sqrt[3]{^-8 \cdot 4}$ is keyed as $\boxed{\sqrt[3]{}}\ \boxed{(-)}\ \boxed{8 \cdot 4}\ \boxed{=}$ to get $^-2 \cdot 03$ (2 d.p.).

$\sqrt[5]{5 \cdot 7}$ is keyed as $\boxed{5}\ \boxed{\text{Shift}}\ \boxed{\sqrt[x]{y}}\ \boxed{5 \cdot 7}\ \boxed{=}$ to get $1 \cdot 42$ (2 d.p.).

Discussion

- Chelsea keyed the number 6·31 into her calculator, and cubed it.

  What must she do to get back to the starting number?

  Why, on some calculators, is it not possible to get back to the exact starting number? **Discuss**.

- Is it possible to find the square root of a negative number? **Discuss**.

  What about the cube root of a negative number? Explain.

- What can you always say about the

  cube root of a positive number

  cube root of a negative number?

Exercise 3

1 $28 = 1^3 + 3^3$
Write 35 as the sum of two cubes.

2 $10^2 - 8^2 = 36$
  a The difference of the squares of two other consecutive even numbers is 44.
       What are these even numbers?
  b The difference of the squares of two consecutive even numbers less than 30 is a square
       number. What are these even numbers?

3 $8^3 = 512$
The units digit is 2.
Without using a calculator, work out the units digit of 8^{12}.
Explain how you did this.

4 Use your calculator to find these.
  Give the answers to 2 d.p.
  a $\sqrt{304}$ **b** $\sqrt[3]{7}$ **c** $\sqrt[4]{842}$ **d** $\sqrt[5]{367}$ **e** $\sqrt{187 - 46}$
  f $\sqrt{26^2 - 8^2}$ **g** $\sqrt[3]{32^2 - 17^2}$ **h** $\sqrt{27} + \sqrt{39}$ * **i** $\sqrt[4]{3} + \sqrt[3]{5}$

5 Which of these are true?
  a $\sqrt{25} + \sqrt{9} = \sqrt{25 + 9}$ **b** $\sqrt{25} - \sqrt{9} = \sqrt{25 - 9}$ **c** $\sqrt{5} \times \sqrt{5} = \sqrt{25}$
  d $\sqrt{4} \times \sqrt{9} = \sqrt{4 \times 9}$ **e** $\sqrt{a} + \sqrt{b} = \sqrt{a + b}$

6 $\sqrt[3]{46\ 656} = 36$.   Explain how you know this is wrong without doing the calculation.

7 **a** A 3-digit square number has a units digit of 1.
       __ __ 1
       What could the units digit of the square root of this number be?
  b A 6-digit square number has a units digit of 4.
       __ __ __ __ __ 4
       Could its square root be a prime number?
       Explain your answer.

8 **a** Annalyn wrote this.

       √19 is between 5 and 6.
          try 5·5   5·5 × 5·5 = 30·25 which is too big.
          try 5·2   5·2 × 5·2 = 27·04 which is too small

       Annalyn was using trial and improvement to find $\sqrt{29}$. Continue Annalyn's working to
       find $\sqrt{29}$ correct to 2 decimal places.
  b Find $\sqrt[3]{70}$ to 3 d.p. using trial and improvement. Show your working.

9 Find the value of x.
  a $2^x = 16$ * **b** $3^x = 729$

10 Write the following in ascending order.
       2^{19} 3^9 6^7

11 Katie found three integers, a, b and c, so that $c^2 = a^2 + b^2$.
    What might Katie's integers be, if each is less than 100?
    Find all the possible answers.

This is linked to
Pythagoras.

Number

12 a Look at these numbers. [SATs Paper 1 Level 7]

i Which is the **largest**? **ii** Which is equal to 9^2?
b Which **two** of the numbers below are **not** square numbers?

13 Use your calculator to find these.
Write the answer as a decimal and as a fraction.
a 3^{-1} **b** 2^{-3}

14 $\sqrt[3]{n} = {}^-m$
m is a positive number.
Is n positive or negative?

15 y^2 represents a square number; y is an integer. [SATs Paper 2 Level 8]

a Think about the expression $9 + y^2$
Explain how you know there are values of y for which this expression does **not**
represent a square number.
b Explain why the expression $16y^2$ **must** represent a square number.

*16 Write Always, Sometimes or Never for these
a (negative number)$^{\text{odd number}}$ = negative odd number
b (negative number)^{zero} = 1

*17 Keely wanted to find an accurate estimate for $\sqrt{7}$.
She said 'I know it is between 2 and 3 because $\sqrt{4} < \sqrt{7} < \sqrt{9}$.
I will start by estimating the answer as 2.'
She then multiplied this number by a number to give the answer 7.
To find this number she drew a rectangle.
A rectangle with sides 2 and $\frac{7}{2}$ has area 7.

Keely then used her graphical calculator to find the mean, x,
of 2 and $\frac{7}{2}$ to get a **better** estimate for $\sqrt{7}$.
a Explain why finding the mean of 2 and $\frac{7}{2}$ gives a better
estimate for $\sqrt{7}$ than 2.
b Explain why continuing to find the mean of the new
value, x, and a value y, such that $xy = 7$ gives a
better approximation for $\sqrt{7}$ each time.

$\frac{7}{2}$

| 2 | rectangle |

```
2
                    2
(Ans+7÷Ans)÷2
                 2.75
```

Hint. You could plot a
graph of y versus x.

Review 1 $\sqrt{21}$ lies between 4 and 5. Explain how you know this.
$4 \cdot 5^2 = 20 \cdot 25$. Use trial and improvement to find $\sqrt{21}$ to 2 d.p.

Review 2
a A 4-digit square number has a units digit of 6. _ _ _ _ 6
(i) What could the square root of this number end in?
(ii) Could this square root be an odd number? Explain your answer.
b A 5-digit square number has a units digit of 9. _ _ _ _ 9
Will the square root of this number be even or odd? Explain.

*Review 3 Find the value of n. **a** $8^4 = 2^n$ **b** $27^2 = 3^n$

Investigation

Squares of sums

Sofia thinks that adding two numbers and then squaring is the same as squaring each number and then adding.

Investigate if this is never true, sometimes true or always true.

Index laws

Remember

The **index laws** are:

    Indices are added when **multiplying**. $a^m \times a^n = a^{m+n}$

    Indices are subtracted when **dividing**. $a^m \div a^n = a^{m-n}$

Examples

$4^2 \times 4^6 = 4^{2+6}$ because

$\qquad = 4^8 \qquad 4 \times 4 \quad \times \quad 4 \times 4 \times 4 \times 4 \times 4 \times 4 = 4^8$

$\dfrac{5^9}{5^4} = 5^{9-4}$

$\qquad = 5^5$

because $\dfrac{5 \times 5 \times 5 \times 5 \times 5 \times \cancel{5} \times \cancel{5} \times \cancel{5} \times \cancel{5}}{\cancel{5} \times \cancel{5} \times \cancel{5} \times \cancel{5}} = 5^5$

Discussion

- $2^3 = 2 \times 2 \times 2$ $(2^3)^2 = 8^2$

    $= 8$ $= 64$

  Write 64 as a power of 2.

  Compare this with $(2^3)^2$. **Discuss**.

  What if the indices were 5 and 2 rather than 3 and 2?

  What if the indices were 2 and 4 rather than 3 and 2?

  What if the base was 4 instead of 2?

  What if ...

  How would this rule be completed? **Discuss**. $(a^m)^n =$ _____

- The answer is 2^6.

  Think of some expressions that could be simplified to 2^6. **Discuss**.

Example $(7^5)^3 = 7^{5 \times 3}$

             $= 7^{15}$ **because** $(7 \times 7 \times 7 \times 7 \times 7) \times (7 \times 7 \times 7 \times 7 \times 7) \times (7 \times 7 \times 7 \times 7 \times 7) = 7^{15}$

Worked Example

Write $\dfrac{8^2 \times 2^5}{4^4}$ as a power of 2.

Answer

$8 = 2^3$ and $4 = 2^2$, $\dfrac{8^2 \times 2^5}{4^4} = \dfrac{(2^3)^2 \times 2^5}{(2^2)^4}$

                               $= \dfrac{2^6 \times 2^5}{2^8}$

                               $= \mathbf{2^3}$

The **index laws apply to algebraic expressions**.

Worked Example
Simplify these.
a $x^2 \times x^3$ b $a^4 \times a^3 \times a^2$ c $\frac{x^6}{x^3}$ d $(x^4)^3$

Answer
a $x^2 \times x^3 = x^{2+3}$ b $a^4 \times a^3 \times a^2 = a^{4+3+2}$ c $\frac{x^6}{x^3} = x^{6-3}$ d $(x^4)^3 = x^{4 \times 3}$
$\quad\quad = \mathbf{x^5}$ $= \mathbf{a^9}$ $= \mathbf{x^3}$ $= \mathbf{x^{12}}$

Worked Example
Write the following without brackets.
a $(b^3h^2)^4$ b $(2a^2b)^3$

Answer
a $(b^3h^2)^4 = b^3h^2 \times b^3h^2 \times b^3h^2 \times b^3h^2$ b $(2a^2b)^3 = 2a^2b \times 2a^2b \times 2a^2b$
$\quad\quad = b^{3+3+3+3} \times h^{2+2+2+2}$ $= 2 \times 2 \times 2 \times a^{2+2+2} \times b \times b \times b$
$\quad\quad = \mathbf{b^{12}h^8}$ $= \mathbf{8a^6b^3}$

Exercise 4

1 Write these as a single power of 8.
   a $8^7 \times 8^3$ b $8^2 \times 8^9$ c $8^1 \times 8^5$ d $8^6 \div 8^4$ e $\frac{8^5}{8^2}$
   f $8^{12} \div 8^4$ g $(8^3)^2$ h $(8^4)^5$ i $(8^4)^7$ j $(8^2)^8$

2 Which of these are true?
   a $a^6 \div a^2 = a^3$ b $a^6 \times a^2 = a^8$ c $(x^2)^3 = x^6$ d $x^3 \times x^4 \times x = x^8$
   e $(p^5)^3 = p^{15}$ f $a^6 - a^2 = a^4$ g $a^7 + a^2 = a^9$ h $(2a^3)^2 = 2a^6$

3 Simplify these.
   a $x^4 \times x^5$ b $a^3 \times a^9$ c $p^{12} \div p^4$ d $x^{16} \div x^2$ e $b^a \times b^x$ f $b^a \times b^c$
   g $a^x \div a^b$ h $(a^x)^y$ i $(x^{2a})^3$ j $(p^{3x})^4$ k $x^p \times x^p$ l $b^a \times b^a$
   m $\frac{a^7 \times a^4}{a^3}$ n $\frac{x^4 \times x^5}{(x^3)^2}$ *o $(y^{\frac{2}{3}})^3$ *p $(m^{\frac{3}{2}})^2$ *q $x^{1.5} \times x^{0.5}$

4 a Write the values of k and m.   **[SATs Paper 1 Level 7]**
        $64 = 8^2 = 4^k = 2^m$ **a and b only**
   b Copy and complete the following.   *c Copy and complete.
        $2^{15} = 32\ 768$ $2^{14} = \underline{\quad\quad}$
        $2^{14} = \underline{\quad\quad}$ $2^{12} = \underline{\quad\quad}$
                                             $2^7 = \underline{\quad\quad}$

5 Find y if
   a $4^3 \times 4^y = 2^{10}$ b $2^5 \times 2^y = 4^6$.

6 Write these without brackets.
   a $(x^2y^3)^2$ b $(a^3b)^4$ c $(x^2yz^3)^5$ d $(2a^2b^3)^4$ e $(3p^3q^5)^3$ f $(4a^5x^4)^2$ *g $(2xy^3z^3)^5$

7 a Write $\frac{9^2 \times 27}{3^5}$ as a power of 3.   b Write $\frac{125 \times 5^{10}}{25^2}$ as a power of 5.
   c Write $\frac{16^3 \times 8^3}{2^9 \times 4^5}$ as a power of 2.

8 a What are the values of x and y?   i $2^x = 32$ ii $3^x = 27$ iii $2^{24} = 4^x = 8^y$
   b Does $2^{x+1} = 2^x \times 2^1$?
   c Does $3^{3(x+1)} = 3^{3x+3}$ or $(3^3)^{x+1}$ or both?
   d What is the value of x?
       i $2^{x-1} = 16$ ii $3^{3(x+1)} = 729$

9 Look at the table.

[SATs Paper 1 Level 8]

$7^0 =$	1
$7^1 =$	7
$7^2 =$	49
$7^3 =$	343
$7^4 =$	2401
$7^5 =$	16 807
$7^6 =$	117 649
$7^7 =$	823 543
$7^8 =$	5 764 801

a Explain how the table shows that $49 \times 343 = 16\ 807$.

b Use the table to help you work out the value of $\frac{5\ 764\ 801}{823\ 543}$.

c Use the table to help you work out the value of $\frac{117\ 649}{2\ 401}$.

d The units digit of 7^6 is 9.
   What is the units digit of 7^{12}?

Review 1 Which of the following statements are correct?

a $4^5 + 4^7 = 4^{12}$ b $(3^2)^5 = 3^7$ c $7^7 - 7^2 = 7^5$

d $a^5 \times a^3 \times a^2 = a^{10}$ e $(2a^3)^4 = 16a^7$ f $x^8 \div x^2 = x^4$

Review 2 Use the rules of indices to simplify these.

a $\frac{a^{14}}{a^7}$ b $p^5 \times p^7$ c $a^2 \times a^3 \times a$ d $\frac{y^7 \times y^2}{y^5}$

e $(x^3)^5$ f $(3a^4)^2$ g $(2x^3y^4)^3$ *h $(x^{\frac{2}{5}})^5$

Review 3 Write $\frac{2^8 \times 4^3}{16 \times 8^2}$ as a power of 2.

Review 4 Use the information in this box to answer the questions. Do not use a calculator.

$3^0 = 1$	$3^5 = 243$	$3^9 = 19\ 683$	$3^{13} = 1\ 594\ 323$
$3^1 = 3$	$3^6 = 729$	$3^{10} = 59\ 049$	$3^{14} = 4\ 782\ 969$
$3^2 = 9$	$3^7 = 2187$	$3^{11} = 177\ 147$	$3^{15} = 14\ 348\ 907$
$3^3 = 27$	$3^8 = 6561$	$3^{12} = 531\ 441$	
$3^4 = 81$			

a Explain why $2187 \times 243 = 531\ 441$.

b Find the values of these.

   i $81 \times 19\ 683$ ii $\frac{531\ 441}{6561}$ iii $243 \times 59\ 049$ *iv $\frac{4\ 782\ 969 \times 59\ 049}{531\ 441}$

Negative and fractional indices

Discussion

What is the value of $\sqrt{16}$?

To find $16^{\frac{1}{2}}$, **key** (16) (x^y) (((1) ($a^{b/c}$) (2))) (=).

Compare your answers for $\sqrt{16}$ and $16^{\frac{1}{2}}$. **Discuss**.
What if 16 was replaced with 49?
What if 16 was replaced with 25?
What if 16 was replaced with 7?
What can you say about $a^{\frac{1}{2}}$ and $\sqrt{a}$? **Discuss**.

What is $\sqrt[3]{8}$?

To find $8^{\frac{1}{3}}$, key ⬢8⬢ ⬢∧⬢ ⬢(⬢ ⬢1⬢ ⬢a^{b/c}⬢ ⬢3⬢ ⬢)⬢ ⬢=⬢ .

Compare your answers for $\sqrt[3]{8}$ and $8^{\frac{1}{3}}$. **Discuss**.

What if 8 was replaced by 125?
What if 8 was replaced by 64?
What if 8 was replaced by 1000?

Could we have keyed $8^{\frac{1}{2}}$ as ⬢8⬢ ⬢∧⬢ ⬢0·5⬢ ⬢=⬢?

What if the index is $\frac{1}{3}, \frac{1}{4}, \frac{1}{5}, \frac{1}{6}, \frac{1}{7} \dots$?

What can you say about $a^{\frac{1}{3}}$ and $\sqrt[3]{a}$? **Discuss**.

Compare the values of $\sqrt[4]{16}$ and $16^{\frac{1}{4}}$.

What does $\frac{2^3}{2^5}$ equal as a decimal?

Find 2^{-2} by keying ⬢2⬢ ⬢∧⬢ ⬢(-)⬢ ⬢2⬢ ⬢=⬢ .

Compare $\frac{1}{2^2}$ and 2^{-2}.

Use your calculator to compare these. **Discuss** the results.

3^{-2} and $\frac{1}{3^2}$ $\qquad$ 4^{-3} and $\frac{1}{4^3}$ $\qquad$ 5^{-2} and $\frac{1}{5^2}$.

Two more rules of indices are:

$$a^{-n} = \frac{1}{a^n} \qquad a^{\frac{1}{n}} = \sqrt[n]{a}$$

Worked Example
Without using the calculator find these.
a $\quad 16^{\frac{1}{4}}$ $\hspace{5cm}$ b $\quad 2^{-4}$

Answer
a $\quad 16^{\frac{1}{4}} = \sqrt[4]{16}$

$\qquad = \pm 2$

$$\sqrt[\text{even number}]{\text{positive number}} = \pm \text{ answer}$$

b $\quad 2^{-4} = \frac{1}{2^4}$

$\qquad = \frac{1}{16}$

Exercise 5

1 What is the missing index?
 a $\quad \sqrt{36} = 36^{\cdots}$ $\quad$ b $\quad \sqrt{28} = 28^{\cdots}$ $\quad$ c $\quad \sqrt{100} = 100^{\cdots}$ $\quad$ d $\quad \sqrt[3]{64} = 64^{\cdots}$
 e $\quad \sqrt[3]{10} = 10^{\cdots}$ $\quad$ f $\quad \sqrt[4]{18} = 18^{\cdots}$

2 Copy and fill in the bottom row of this table with fractions or whole numbers.

2^{-3}	2^{-2}	2^{-1}	2^0	2^1	2^2	2^3
						8

3 Without using the calculator, find these.
 a $\quad 4^{\frac{1}{2}}$ $\quad$ b $\quad 16^{\frac{1}{2}}$ $\quad$ c $\quad 9^{\frac{1}{2}}$ $\quad$ d $\quad 64^{\frac{1}{2}}$ $\quad$ e $\quad 100^{\frac{1}{2}}$ $\quad$ f $\quad 8^{\frac{1}{3}}$
 g $\quad 64^{\frac{1}{6}}$ $\quad$ h $\quad 125^{\frac{1}{3}}$ $\quad$ i $\quad 16^{\frac{1}{4}}$ $\quad$ j $\quad 81^{\frac{1}{4}}$ $\quad$ k $\quad 32^{\frac{1}{5}}$

4 Without using the calculator, find these.
 a $\quad 2^4$ $\quad$ b $\quad 4^{-1}$ $\quad$ c $\quad (\frac{1}{2})^4$ $\quad$ d $\quad 3^{-3}$ $\quad$ e $\quad 5^0$ $\quad$ f $\quad 2^{-3}$ $\quad$ *g $\quad (0·04)^{\frac{1}{2}}$

*5 Write in the form x^-.

a $(x^{-4})^{\frac{1}{2}}$ b $\dfrac{x^2}{x^3}$ c $\dfrac{x^{-2}}{x^3}$ d $\dfrac{x^2 \times x^{0.5}}{x^{-1}}$ e $\dfrac{x^{1.5}}{x^{0.5}}$

T

Review   Use a copy of this box. Write the letter beside each question above its answer in the box.

$\frac{1}{2}$	$\frac{1}{32}$	9	4	1	8	$\frac{1}{16}$	4		$\frac{1}{9}$	5	0.3	8
$\bar{9}$	$\frac{1}{2}$	$\frac{1}{3}$	$\bar{8}$		$\bar{9}$	$\frac{1}{2}$	$\frac{1}{32}$	$\frac{1}{32}$	$\bar{3}$			

Without using the calculator find these.

A $25^{\frac{1}{2}}$ B $81^{\frac{1}{2}}$ D $27^{\frac{1}{3}}$
E 2^3 H 3^{-2} L $4^{-\frac{1}{2}}$
O 2^{-5} R $(\frac{1}{4})^2$ S $64^{\frac{1}{3}}$
T 100^0 U $(\frac{1}{27})^{\frac{1}{3}}$ *V $(0.09)^{\frac{1}{2}}$

Surds

The value of $\sqrt{5}$ cannot be found exactly. It cannot be written as a fraction. The decimal form is not terminating or repeating.

The only **exact** value of $\sqrt{5}$ is $\sqrt{5}$.
$\sqrt{5}$ is called a **surd**. A surd is a square root that does not have an exact value.

Discussion

What does $\sqrt{5} \times \sqrt{5}$ equal? **Discuss**.
What does $\sqrt{2} \times \sqrt{6}$ equal? Compare this to the value of $\sqrt{12}$.
How else could you write $\sqrt{5} \times \sqrt{5} \times \sqrt{5}$ as an exact value? **Discuss**.

$\sqrt{a} \times \sqrt{a} = a$ $\sqrt{b} \times \sqrt{b} \times \sqrt{b} = b\sqrt{b}$ $\sqrt{ab} = \sqrt{a} \times \sqrt{b}$

Worked Example

Write these as exact values without any surds.
a $\sqrt{6} \times \sqrt{6}$ b $\sqrt{8} \times \sqrt{2}$ c $\sqrt{32} \times \sqrt{2}$

Answer

a $\sqrt{6} \times \sqrt{6} = \mathbf{6}$ b $\sqrt{8} \times \sqrt{2} = \sqrt{8 \times 2}$ c $\sqrt{32} \times \sqrt{2} = \sqrt{32 \times 2}$
                                                     $= \sqrt{16}$ $= \sqrt{64}$
                                                     $= \mathbf{4}$ $= \mathbf{8}$

We can **simplify surds** so they are in simplified surd form. This means that the number under the square root has no factors which are square numbers.

Worked Example

Write these in simplified surd form, e.g. $2\sqrt{3}$.

a $\sqrt{50}$ **b** $\sqrt{48}$ **c** $2\sqrt{12}$ *d $\sqrt{25\,200}$

Answer

a $\sqrt{50} = \sqrt{25 \times 2}$ Write the number as a multiple   **b** $\sqrt{48} = \sqrt{16 \times 3}$

$\qquad = \sqrt{25} \times \sqrt{2}$ of a square number if possible. $= \sqrt{16} \times \sqrt{3}$

$\qquad = \mathbf{5\sqrt{2}}$ $= \mathbf{4\sqrt{3}}$

c $2\sqrt{12} = \sqrt{4} \times \sqrt{12}$ *d $\sqrt{25\,200} = \sqrt{2^4 \times 3^2 \times 5^2 \times 7}$ divide even powers by 2.

$\qquad = \sqrt{48}$ $= 2^2 \times 3 \times 5 \sqrt{7}$

$\qquad = \sqrt{16} \times \sqrt{3}$ $= \mathbf{60\sqrt{7}}$

$\qquad = \mathbf{4\sqrt{3}}$

2	25200
2	12600
2	6300
2	3150
3	1575
3	525
5	175
5	35
	7

Exercise 6

1 Write these in simplified surd form.

     a $\sqrt{7} \times \sqrt{7}$ **b** $\sqrt{9} \times \sqrt{9}$ **c** $\sqrt{2} \times \sqrt{2} \times \sqrt{2}$ **d** $\sqrt{14} \times \sqrt{14} \times \sqrt{14}$ **e** $\sqrt{5} \times \sqrt{3}$

     f $\sqrt{2} \times \sqrt{7}$ **g** $\sqrt{12} \times \sqrt{3}$ **h** $\sqrt{50} \times \sqrt{2}$ **i** $\sqrt{8} \times \sqrt{18}$

2 Simplify these surds.

     a $\sqrt{20}$ **b** $\sqrt{18}$ **c** $\sqrt{72}$ **d** $\sqrt{80}$ **e** $\sqrt{8}$ **f** $\sqrt{75}$

     g $\sqrt{60}$ **h** $\sqrt{24}$ **i** $\sqrt{27}$ *j $\sqrt{432}$ *k $\sqrt{245}$ *l $\sqrt{56\,700}$

     *m $\sqrt{19\,404}$ *n $\sqrt{26\,325}$ *o $\sqrt{11\,200}$ *p $\sqrt{33\,075}$

3 Write each of these as a single surd.

     a $3\sqrt{4}$ **b** $5\sqrt{7}$ **c** $3\sqrt{6}$ **d** $10\sqrt{2}$ **e** $8\sqrt{5}$

*4 Simplify these expressions by writing them as simplified surds.

     a $5\sqrt{8}$ **b** $3\sqrt{8}$ **c** $4\sqrt{12}$ **d** $2\sqrt{75}$ **e** $\sqrt{8} \times \sqrt{3}$

     f $\sqrt{6} \times \sqrt{2}$ **g** $\sqrt{12} \times \sqrt{18}$ **h** $\sqrt{72} \times \sqrt{5}$

*5 Simplify these by looking for common factors.

     a $\dfrac{\sqrt{5}}{\sqrt{20}}$ **b** $\dfrac{2\sqrt{8}}{3\sqrt{18}}$ **c** $\dfrac{\sqrt{125}}{3\sqrt{10}}$ **d** $\dfrac{5\sqrt{6}}{2\sqrt{10}}$

*6 Can a square have an exact area of 48 m²?

     If so, what is its exact perimeter?

7 In this question, you should **not** use a calculator.   **[SATs EP]**

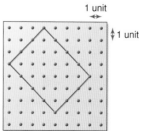

     a An elastic band is fixed on four pins on a pinboard. Show that the total length of the band in this position is $14\sqrt{2}$ units.

     b What is the length of the band in this new position? Write your answer in its simplest form using roots.

T c Use a copy of this. Draw a square on the pinboard that has a perimeter of $4\sqrt{29}$. Show your working.

T d Use a copy of this. Now draw a trapezium on the pinboard that has a perimeter of $6 + 4\sqrt{2}$. Show your working.

T **Review 1** Use a copy of this box. Write the letter beside each question above its answer in the box.

10	$4\sqrt{3}$	$\sqrt{125}$	$4\sqrt{6}$		$\sqrt{800}$	11	$10\sqrt{5}$	$\sqrt{63}$	$\sqrt{63}$	$2\sqrt{3}$	$3\sqrt{3}$	$4\sqrt{3}$	$15\sqrt{2}$	$4\sqrt{6}$

Write in their simplest form.

A $\sqrt{11} \times \sqrt{11}$ B $\sqrt{3} \times \sqrt{3} \times \sqrt{3}$ D $\sqrt{5} \times \sqrt{20}$

Write these as a single surd.

E $3\sqrt{7}$ G $5\sqrt{5}$ H $10\sqrt{8}$

Write these in simplified surd form.

L $\sqrt{12}$ O $\sqrt{48}$ S $\sqrt{96}$ V $5\sqrt{20}$ *W $3\sqrt{50}$

* **Review 2** Find the perimeter and area of these two rectangles, formed by bands put around pegs on a pinboard. Simplify your answers.

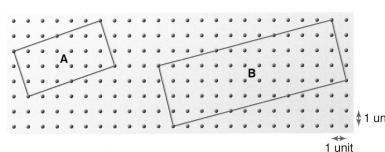

1 unit

1 unit

Summary of key points

A We can use the **HCF when cancelling fractions** and the **LCM when adding and subtracting fractions**. See page 2 for how to find the HCF and LCM of a number.

Example $\frac{360}{840} = \frac{3}{7}$ dividing numerator and denominator by 120, the HCF of 360 and 840

Example $\frac{19}{360} + \frac{31}{840} = \frac{133}{2520} + \frac{93}{2520}$ the common denominator is the LCM of 360 and 840

$$= \frac{226}{2520}$$

$$= \frac{113}{1260}$$

B We can find **common factors of algebraic expressions**.

Examples x is a common factor of $3x$ and $4x$. x is the HCF of $3x$ and $4x$.

          $2y$ is a common factor of $4y^2$ and $2y$. $2y$ is the HCF of $4y^2$ and $2y$.

          $3xyz$ is a common factor of $3x^2yz$ and $6xy^2z^2$. $3xyz$ is the HCF of $3x^2yz$ and $6xy^2z^2$.

We can use a Venn diagram to find the HCF and LCM of algebraic expressions.

Example

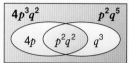

HCF $= p^2 q^2$

LCM $= 4p^3 q^5$

 On a calculator **squares** are keyed using $\boxed{x^2}$,

cubes are keyed using $\boxed{x^3}$,

other powers are keyed using $\boxed{\wedge}$,

square roots are keyed using $\boxed{\sqrt{}}$,

cube roots are keyed using $\boxed{\sqrt[3]{}}$,

other roots are keyed using $\boxed{\sqrt[x]{y}}$.

Examples $(^-6\cdot2)^3$ is keyed as $\boxed{(}\ \boxed{(-)}\ \boxed{6\cdot2}\ \boxed{)}\ \boxed{x^3}\ \boxed{=}$ to get $^-238\cdot328$.

$(3\cdot7)^6$ is keyed as $\boxed{3\cdot7}\ \boxed{\wedge}\ \boxed{6}\ \boxed{=}$ to get $2565\cdot726409$.

$\sqrt[3]{340}$ is keyed as $\boxed{\sqrt[3]{}}\ \boxed{340}\ \boxed{=}$ to get $6\cdot979532047$.

$\sqrt[4]{964}$ is keyed as $\boxed{4}\ \boxed{\text{Shift}}\ \boxed{\sqrt[x]{}}\ \boxed{964}\ \boxed{=}$ to get $5\cdot572104575$.

 The **index laws** can be used to simplify algebraic expressions.

$$a^m \times a^n = a^{m+n}$$

$$a^m \div a^n = a^{m-n}$$

$$(a^m)^n = a^{m \times n}$$

Examples $(a^5)^4 = a^{5 \times 4}$ $\qquad \dfrac{x^2 \times x^4}{x^3} = \dfrac{x^{2+4}}{x^3}$ $\qquad (x^2 y^3)^3 = x^{2 \times 3} y^{3 \times 3}$

$\qquad\qquad\qquad = a^{20}$ $\qquad\qquad\qquad = \dfrac{x^6}{x^3}$ $\qquad\qquad\qquad = x^6 y^9$

$\qquad\qquad\qquad\qquad\qquad\qquad = x^3$

 Two more **rules of indices** are:

$$a^{-n} = \frac{1}{a^n} \qquad\qquad a^{\frac{1}{n}} = \sqrt[n]{a}$$

Examples **a** $81^{\frac{1}{4}} = \sqrt[4]{81}$ $\qquad$ **b** $5^{-3} = \dfrac{1}{5^3}$

$\qquad\qquad\qquad = 3$ $\qquad\qquad\qquad = \dfrac{1}{125}$

 Sometimes a square root cannot be found exactly.

$\sqrt{3}$ cannot be found exactly. The only exact value of $\sqrt{3}$ is $\sqrt{3}$.

$\sqrt{3}$ is called a **surd**.

$$\sqrt{x} \times \sqrt{x} = x$$
$$\sqrt{y} \times \sqrt{y} \times \sqrt{y} = y\sqrt{y}$$
$$\sqrt{xy} = \sqrt{x} \times \sqrt{y}$$

We write surds so that the number under the square root sign has no factors which are square numbers.

Examples $\sqrt{72}$ in its simplest surd form is $\sqrt{36 \times 2}$

$\qquad\qquad\qquad\qquad\qquad\qquad = \sqrt{36} \times \sqrt{2}$

$\qquad\qquad\qquad\qquad\qquad\qquad = 6\sqrt{2}$.

Test yourself **Except for questions 5 and 8b.**

1 a Express these numbers as a product of prime factors in index notation.
   i 504   **ii** 8800
  b Find the HCF and LCM of 504 and 8800 using a Venn diagram to help.

2 Cancel these fractions by finding the HCF of the numerator and denominator.
  a $\frac{360}{540}$ **b** $\frac{112}{192}$ **c** $\frac{180}{288}$

3 Add or subtract these fractions by finding the LCM of the denominator.
  a $\frac{17}{240} + \frac{37}{720}$ **b** $\frac{372}{560} - \frac{84}{196}$

4 Find the HCF and LCM of these.
  Use Venn diagrams to help.
  a $m^2 n$ and mn^2 **b** $6a^2bc^3$ and $18ab^3c^2$

5 Give the answers to these to 2 d.p.
  a $\sqrt{366} + 86$ **b** $\sqrt[3]{4^2} + 38$ **c** $\sqrt[5]{116} + 205$ **d** $\sqrt[3]{24^2 - 7^2}$

6 Write the following in ascending order.
   3^{16} 7^8 2^{20}

7 Simplify these.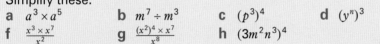
  a $a^3 \times a^5$ **b** $m^7 \div m^3$ **c** $(p^3)^4$ **d** $(y^n)^3$ **e** $m^p \times m^q$
  f $\frac{x^3 \times x^7}{x^2}$ **g** $\frac{(x^2)^4 \times x^7}{x^8}$ **h** $(3m^2n^3)^4$

8 a $625 = 25^2 = 5^k$
   What does k equal?
  b $3^{16} = 43\,046\,721$
   What does 3^{15} equal?

9 Work these out without using a calculator.
  a $25^{\frac{1}{2}}$ **b** $27^{\frac{1}{3}}$ **c** $32^{\frac{1}{5}}$ **d** 2^5 **e** 3^{-1}
  f 6^0 **g** $(\frac{1}{3})^3$ **h** $(0.16)^{\frac{1}{2}}$ **i** 6^{-1}

10 Simplify these.
  a $(x^{\frac{2}{5}})^5$ **b** $\frac{x^2}{x^5}$ **c** $\frac{m^3 \times m^{1.5}}{m^{-2}}$

11 Write these in simplified surd form.
  a $\sqrt{6} \times \sqrt{6}$ **b** $\sqrt{3} \times \sqrt{3} \times \sqrt{3}$ **c** $\sqrt{18} \times \sqrt{18}$ **d** $\sqrt{5 \times 7}$
  e $2\sqrt{45}$ **f** $\sqrt{8} \times \sqrt{5}$ **g** $\sqrt{8} \times \sqrt{2}$ **h** $\sqrt{63} \times \sqrt{4}$

12 Simplify these surds.
  a $\sqrt{54}$ **b** $\sqrt{405}$ * **c** $\sqrt{2205}$

3 Calculation

You need to know

✓ mental calculation   page 2
     – add and subtract mentally
     – multiply and divide mentally

✓ order of operations   page 3

✓ estimating   page 3

✓ written calculations   page 4

✓ checking answers to calculations   page 4

✓ using a calculator   page 5

Key vocabulary

best estimate,   **order of operations**,   **quotient**,   **reciprocal**

Mystery Numbers

1
```
    **
 ×  **
   ─────
   *2*
```
What digit does * stand for?

2 **a** A number when divided by 5 leaves no remainder.
When divided by 6 there is a remainder of 5.
It is between 20 and 50
What could it be?

   b A number is between 0 and 70.
When divided by 3 or 7 there is no remainder.
When divided by 4 there is remainder 2.
What number is it?

   c A number is a multiple of 4.
It leaves remainder 4 when divided by 5.
What could the number be?
Is there more than one answer?

   d A number has a digit sum of 9.
It is divisible by 8.
What number could it be?
Is there more than one answer?

17 2
6085 10
6 502
475 123
87
7·2

50

Mental calculation

Remember
We carry out operations in this order, working from left to right.
Brackets
Indices (powers)
Division and **M**ultiplication
Addition and **S**ubtraction

Use **BIDMAS** to help you remember.

You can **add and subtract mentally** using

complements	partitioning
counting up	nearly numbers
compensation	facts you already know.

You can **multiply and divide mentally** using

place value knowledge	partitioning
factors	near tens
known facts	doubling and halving.

See pages 2–3 for examples of these.

Try to use the most efficient method.

Discussion

● **Discuss** the most efficient method to calculate these mentally.
$(\frac{5}{2})^2$ $55{\cdot}3 \times 0{\cdot}01$ $3{\cdot}4 \times 4{\cdot}5$ $8{\cdot}6 \times 4{\cdot}2$

● Find the length of tube needed to build this frame.
The frame is covered to make a closed box.
What is the volume of the box?

1·5 cm
3 cm
5·5 cm

● **Discuss** whether these statements are true or false.
When we multiply a positive number, n, by 0·6 we get an answer less than n.

When we divide a positive number, m, by 0·6, we get an answer less than m.

Exercise 1   **This exercise is to be done mentally.**

1   Answer these questions as quickly as possible.
  a   What is the area of this triangle?
  b   A fair spinner has eight equal sections.
     Five are red and three are blue.
     What is the probability I will spin red?
  c   The scale on a map is 2 centimetres to 1 kilometre.
     On the map, the distance to the airport is 12 centimetres.
     How many kilometres is it to the airport?
  d   What is the volume of a cuboid measuring 3 cm by 5 cm by 4 cm?
  e   How many pairs of parallel sides does a parallelogram have?
  f   In a test I got eighteen out of twenty questions correct.
     What percentage did I get correct?
  g a, b and c are equal angles.
     What is the size of angle a?

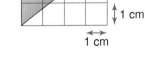

1 cm
1 cm

h Multiply $2y$ by $4y$. Give your answer in its simplest form.
i A bird flies at an average speed of 30 kilometres per hour.
How far will it fly in 12 minutes at this speed?
j Find the HCF of 24 and 36.
k What is the value of x if $x + 28 = 104$?
l $39 = 50 - x$
What is the value of x?
m $x^2 = 25$
What are the two possible values of x?
n What is the next number in this sequence?
8, 3, ⁻2, ⁻7, ...
o What is one-half of one point five five?
p $p = 3q - 8$
What is the value of p when $q = 2$?
q How much must you add to 3087 to get 5000?
r Find the LCM of 15 and 20.

2 Answer these as quickly as possible.
a Subtract three from minus six.
b $y = x^2 + 1$.
What does y equal if $x = 5$?
c What is seven point nine divided by two?
d John measured the length of a room as ten metres, to the nearest metre.
What is the shortest length the room could be?
e What are the coordinates of point D?
f What is the area of ABCD in square units?
g How many zero point fives are there in eight?
h Write $\frac{160}{240}$ in its simplest form.
i What is the area of a square of side 2·5 cm?
j Jane's height was 160 cm when she was 20.
By the time she was 75, she had shrunk by 5%.
How tall was she at 75?

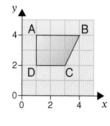

k A trailer measures two metres by two point five metres and is 30 cm deep.
Which of these calculations gives the volume of soil that will fit on the trailer if it is level with the top.
A $2 \times 2·5 \times 30$ **B** $2 \times 2·5 + 30$ **C** $2 \times 2·5 + 0·3$ **D** $2 \times 2·5 \times 0·3$
l Expand these brackets. $^-2(3x - 4)$
m Estimate the answer to $\frac{49·85 \times 30·6}{5·03}$.
n The teacher is going to choose a pupil at random to answer a question.
The probability it will be a boy is two-fifths.
There are twelve boys in the class. How many girls are there?
o What 3-D shape has 4 vertices?
p What would be the last digit of one hundred and sixty-two to the power of four?
q The mean of two numbers is seven.
One of the numbers is negative two.
What is the other number?
r 4·2 multiplied by what is the same as $8·4 \times 4·5$?
s I write all the integers from one to seventy.
How many of these integers contain the digit three?

3 Kezia did this calculation. $3·06 \times 0·4 = 12·24$
Explain how you can tell, without doing it, that it is wrong.

4 Which of these answers will be less than 4·6?
   a $4·6 \div 0·1$ b $4·6 \times 0·35$ c $4·6 \div 1·7$ d $4·6 \times 8·3$
   e $4·6 \times 0·83$ f $4·6 \div 1·003$ g $4·6 \div 0·006$ h $4·6 \times 0·006$

5 Write down the letters that are beside correct answers.
   The letters, when unjumbled, make the name of a famous Irish lead singer.
   N $3^2 \times 5^2 = 225$ D $(12 + 8 - 17)^2 = {}^-269$ B $\sqrt{12 + 13} = 5$
   O $\sqrt[3]{57 + 68} = 5$ A $(2 - 5)^2 = 25$ O $(22 - 15 + 4 - 7)^3 = 64$

6 a $(2 + {}^-1)^3$ b $(5 - {}^-4)^2$ c $({}^-4)^2 + {}^-3$ d $({}^-11)^2 - 3(5 - {}^-2)$
   e $(25 - 17 + 2 + {}^-5)^3$ f $4(5 - 3)^2 + {}^-2$ g $18 - 3 \times 2^2$ h $(\frac{4}{3})^2$
   i $24 \div (4 + 8) - 6 + 3 \times (10 \div 2)^2$
   j $\frac{(3 + 2)^2}{8 - 3}$ k $\frac{(3 + 1)^2}{(3 - 1)^2}$ l $\frac{5 \times 6^2}{5 \times 2}$ m $\frac{(5 \times 6)^2}{5 \times 4}$ *n $\frac{{}^-8^2 + 4}{7^2 + 1}$

7 Here are six number cards.

   Arrange the six cards to make these calculations.
   a $579 = \boxed{}\boxed{}\boxed{} + \boxed{}\boxed{}\boxed{}$
   b $1146 = \boxed{}\boxed{}\boxed{} + \boxed{}\boxed{}\boxed{}$
   c $660 = \boxed{}\boxed{}\boxed{} + \boxed{}\boxed{}\boxed{}$
   d $324 = \boxed{}\boxed{}\boxed{} - \boxed{}\boxed{}\boxed{}$

8 The number 4 is halfway between 2·5 and 5·5.
   a What goes in the gap?

      The number 4 is halfway between 1·8 and ____.
      The number 4 is halfway between ⁻10 and ____.
   b Work out the number that is halfway between 37×46 and 43×46.

9 Which of these are true for *all* positive values of p and q?
   a $p > 1$ and $q > 1$ so $pq > 1$ b $p > 1$ and $q > 1$ so $\frac{p}{q} > 1$
   c $p > 1$ and $0 < q < 1$ so $pq < p$ d $p > 1$ and $0 < q < 1$ so $\frac{p}{q} > p$

10 On my map 2 cm represents 5000 m. If the distance from Cambridge to Huntingdon is approximately 11 cm, how far are the towns apart, in kilometres?

11 Jess was going to visit her aunt in Australia. She was given £40 to spend. How many Australian dollars would she have if £100 is approximately equal to AU$250?

12 Some boys and girls have £30 between them.
   Each boy has £3 and each girl has £5.
   How many girls are there?

13 Arnie wrote these jottings to help him work out the volume of this square-ended cuboid.

   $3·5 \times 3·5 \times 8 = \frac{7}{2} \times \frac{7}{2} \times 8$

   a Copy and finish Arnie's jottings to find the answer.
   b Show another way you could work out the volume mentally.

14 Calculate these.
   a 1.3×14 **b** 6.2×21 **c** 8.2×18 **d** 23×11 **e** 65×29

15 Calculate these.
   a $30 \div 0.5$ **b** $16 \div 0.8$ **c** $300 \div 0.6$ **d** $180 \div 0.6$ **e** $30 \div 0.06$ **f** $45 \div 0.09$

16 What goes in the box?
   a $0.01 \times \square = 1.5$ **b** $\square \div 0.01 = 2.4$

17 Look at these number cards.

| 0·4 | 4 | 10 | 0·1 | 0·05 | 1 |

a Which two cards multiplied together will give the smallest possible answer?
   What is this answer?

$\square \times \square = \_\_$

b Choose two of the cards to give the answer 100.

$\square \div \square = 100$

18 What might go in the boxes?
   Find at least four different ways to fill them.

$\square \times \square \times \square = 0.08$

19 Given that $46 \times 45 = 2070$, find these.
   a 46×44 **b** 45^2 **c** 92×45

20 **a** Roslyn takes 12 minutes to run the 1·5 km to her friend's house.
         Calculate her average speed in km/h.

$\text{Speed} = \frac{\text{distance}}{\text{time}}$

   b £1 = 1·65 euros and £1 = AU$2·50
       How many euros would you get for AU$10?
   c A boat travels 240 km on 60 litres of fuel.
       How many litres will it use to travel 96 km?

21 Put $+, -, \times$ or $\div$ into these calculations to make the answers correct.
   a $(3 \_\_\_ 8) \_\_\_ 9 \_\_\_ 3 = 30$ **b** $\frac{5 \_\_ 7}{3^2 \_\_ 3} = 2$ **c** $5(8 \_\_\_ 4) - (2 \_\_\_ 3)^2 = {}^-5$

22 $\frac{8+8}{8} + 8 = 10$
   Use seven 8s and any of $+, -, \times, \div$ to make 1000.

23 I am between 40 and 50.
   I am made from the numbers $^-4$, 7, $^-2$ and the operations $+$, $(\ )$, ^{2}.
   What number am I?

24 The product of two whole numbers is 4 times their sum.
   The sum of their squares is 180.
   What are the two numbers?

25 Put the digits 1, 2, 3, 4, 5 and 6 in place of boxes to make this true.

$\square\square \times \square = \square\square\square$

Review 1   Answer these as quickly as possible.
a What is the square of the sum of seven and three?
b What is the difference between five and negative six?
c What are the next two numbers in the sequence 25, 18, 11, 4, ... ?
d What is the value of ab^2 if $a = 2$ and $b = 3$?
e Find x if $48 = x + 75$.

f Four positive whole numbers have a mean of 5. Write down three possible sets of numbers, assuming all the numbers are different.

g Lisa's height is 172 cm to the nearest cm.
  i What is her minimum possible height?
  ***ii** If the height was to the nearest 0·5 cm, what would her minimum possible height have been?

h i Gavin writes all the numbers from 1 to 60. How many times does he write the digit 2?
  ***ii** He now writes all the numbers from 1 to 600. How many times does he write the digit 2?

1,2,3,4

Review 2 Find a and b if
a the sum of a and $b = {}^-4$, product = $^-96$ **b** the sum of a and $b = 12$, product = $^-48$.

Review 3 A number of years ago, it cost 30 pence for an airmail letter and 24 pence to send a letter by surface mail. Jenni posted 18 letters altogether at a total cost of £4·74. How many letters did she send by airmail?

Puzzle

Find the value of each letter.

a A·B × B·A = B·CB   **b** L·M × L·L = ML·LM

Calculating mentally with fractions, decimals and percentages

Remember
Fractions, decimals and percentages are all ways of expressing **proportions**.

$0·1 = \frac{1}{10} = 10\%$ $0·01 = \frac{1}{100} = 1\%$

$0·2 = \frac{1}{5} = 20\%$ $0·02 = \frac{1}{50} = 2\%$

$\frac{1}{8} = 0·125 = 12\frac{1}{2}\%$ $\frac{1}{3} = 0·\dot{3} = 33\frac{1}{3}\%$ $\frac{2}{3} = 0·\dot{6} = 66\frac{2}{3}\%$

Discussion

Discuss how to find $17\frac{1}{2}\%$ of £165 mentally.
What about $32\frac{1}{2}\%$ of 650 g?

Number

Exercise 2 **This exercise is to be done mentally.**

1 a What is $\frac{3}{8}$ as a decimal?
b Convert $7\frac{2}{3}$ to an improper fraction.
c Convert $\frac{38}{5}$ to a mixed number.
d Express 0·625 as a percentage.
e Express 10·5 as a percentage.
f Simplify $\frac{85}{100}$.
g Simplify $\frac{450}{630}$.
h Convert 0·45 to a fraction.
i Convert $\frac{4}{25}$ to a decimal.
j Convert 1·2 to a percentage.
k Convert 165% to a decimal.
l Increase 200 by 10%.
m 25% of a number is 8. What is the number?
n 20% of a number is 12. What is the number?
o Increase 500 ℓ by 12%.
p Decrease 200 mm by 15%.
q Increase 360 m by 30%.
r Increase 7 by 150%.
∗s Find $\frac{5}{8}$ of 20.
∗t Find $17\frac{1}{2}$% of 3200 kg.
∗u Find 85% of 32 mm.
∗v Find $2\frac{7}{12}$ of £1200.

2 a In a sale a pair of shoes was reduced by 30%.
The original price was £85, what was the sale price?
∗b In the same sale, Victoria bought a coat for £112.
How much did the coat cost before the sale?

3 I start with a fraction $\frac{a}{b}$.
I add 1 to the numerator and 1 to the denominator.
The new fraction is equivalent to $\frac{3}{4}$.
Write down three possible fractions I could have started with.

∗4 How much water must be added to 12 litres of an 80% acid solution to make a 60% acid solution?

Review 1
a Simplify $\frac{360}{450}$.
b Write $\frac{1}{2}$% as a decimal.
c Increase 300 by 8%.
d Decrease 2000 by 15%.
e If $\frac{1}{16}$ = 0·0625, find $\frac{5}{16}$ as a decimal.
f Convert 0·064 to a fraction.
g What is three point seven five divided by 2?
h Three-quarters of a class took part in a maths competition and $\frac{1}{6}$ of the participants won prizes.
  i What fraction of the class won prizes?
  ii If 4 students won prizes, how many students were in the class?

Review 2 In a sale, a camera which cost £175 is reduced by 30%. How much would I save if I bought it in the sale?

Review 3 Jane bought a box of books for £5 per book. At a fair she sold them to make a profit of 40% on each book. She made a total profit of £480. How many books were in the box?

Review 4 A school's staff have instant coffee at morning break. The coffee manufacturer increased the size of its large 750 g tin by $\frac{2}{5}$ to make a giant sized tin.
a How much coffee is in the giant sized tin?
∗b If there are 15 staff members and they take 20 days to empty the large tin, how long will it take a staff of 20 to empty the giant tin?

Estimating answers to calculations

Remember

Guidelines for estimating

1 Look for 'nice' numbers.
   Example $208 \div 5{\cdot}6 \approx 200 \div 5$ rather than $200 \div 6$

2 Look for numbers that will cancel.
   Example $\frac{29}{4{\cdot}2 \times 7{\cdot}1} \approx \frac{28}{4 \times 7}$ rather than $\frac{30}{4 \times 7}$

3 When **multiplying** two numbers, try to **round one up and one down**.
   Example   It is better to estimate $4{\cdot}5 \times 3{\cdot}5$ as 5×3 or 4×4 rather than 5×4.

4 When **dividing** two numbers, try to **round both numbers up** or **both numbers down**.
   Example   It is better to estimate $\frac{79{\cdot}3}{8{\cdot}5}$ as $\frac{81}{9}$ rather than $\frac{80}{8}$.

   both up   one up
                  one down

When **estimating** answers to calculations we often **round** all the numbers **to one significant figure** (1 s.f.).

Worked Example
Estimate the answer to 231×38.

Answer
To 1 s.f. 231 is 200.
To 1 s.f. 38 is 40.
$200 \times 40 = 8000$
so $231 \times 38 \approx \mathbf{8000}$

Example $\frac{3{\cdot}68 \times 29{\cdot}05}{6{\cdot}83} \approx \frac{4 \times 30}{7}$ rounding each number to 1 s.f.

$\approx \frac{4 \times 28}{7}$ changing 30 to a number that cancels with 7

$= 16$

so $\frac{3{\cdot}68 \times 29{\cdot}05}{6{\cdot}83} \approx \mathbf{16}$

Exercise 3

1 Estimate the answers to these calculations by rounding the numbers to 1 s.f.
   a 409×28 b $596 \div 23$ c $683 \times 62{\cdot}9$ d £$774{\cdot}82 \div 37$
   e £$821{\cdot}76 \div 48$ f $350 \times 281{\cdot}5$ g $854{\cdot}7 \div 62$ h $1087 \times 56{\cdot}7$
   i $487 \times 321{\cdot}7$ j $981 \div 38{\cdot}4$ k $5{\cdot}98 \times 7052$ l $4{\cdot}3 \div 0{\cdot}82$
   m $57{\cdot}4 \times 39{\cdot}1$ n $4862{\cdot}9 \div 3{\cdot}7$ o $89{\cdot}2 \div 4{\cdot}29$ p $48{\cdot}6^2$

2 Estimate the answers to these. Justify your estimates.
   a $\frac{4{\cdot}98 \times 32{\cdot}41}{6{\cdot}2}$ b $\frac{597{\cdot}3}{4{\cdot}3 \times 1{\cdot}94}$ c $\frac{214{\cdot}7 \times 300{\cdot}9}{58{\cdot}72}$
   d $\frac{3{\cdot}74 \times 209{\cdot}9}{16{\cdot}59}$ e $\frac{52{\cdot}38 \times 64{\cdot}7}{5{\cdot}07 \times 3{\cdot}5}$ f $68{\cdot}3 + \frac{27{\cdot}9 \times 3{\cdot}4}{5{\cdot}28}$
   g $43{\cdot}2(87{\cdot}94 - 5{\cdot}2)$ h $39{\cdot}8(52{\cdot}6 + 187{\cdot}4)$ i $\frac{89{\cdot}6 + 42{\cdot}7 \times 321{\cdot}6}{42{\cdot}9}$
   j $82{\cdot}68 + 4{\cdot}7(32{\cdot}1 + 89{\cdot}4)$ k $4{\cdot}75 \times (4{\cdot}69 - 0{\cdot}86^2)$ l $\frac{108 \times 0{\cdot}67}{\sqrt{195}}$

3 Use a calculator for these calculations. Give your answers to 3 s.f.
Check your answers by estimating.
  a 47.23×89.64 **b** $8.34 \div 1.24$ **c** $162.93 \div 32.6$ **d** $\frac{53.7 \times 8.49}{5.92}$ **e** $\frac{4.53 \times 4.1}{6.3}$

4 **a** Identify the **best** estimate of the answer to $72.34 \div 8.91$.   [SATs Paper 1 Level 7]
          6   7   8   9   10   11
  b Identify the **best** estimate of the answer to 32.7×0.48.
          1·2   1·6   12   16   120   160
  c Estimate the answer to $\frac{8.62 + 22.1}{5.23}$.
     Give your answer to **1 significant figure**.
  d **Estimate** the answer to $\frac{28.6 \times 24.4}{5.67 \times 4.02}$.

5 Decide who **you** think is right.
Give a reason for your answer.
You could choose either as
long as you give a reason for
your choice.

Penn   **Nadia**

6 **Estimate**, then use your calculator, to find the answers to the following.
  a A hiking club took 27 people on a weekend hike.
     The total cost was £635·85.
     How much did each pay?
  b Nazir bought 47 boxes of fruit, each costing £12·85. The fruit was shared equally
     between 19 families. What was the cost to each family?
  c A wooden walkway has 4·65 m platforms at each end and 38 joined sections
     each 3·2 m long.
     What is the total length of the walkway?
  *d This is the formula for finding the surface area of a cylinder.
     $A = 2\pi r^2 + 2\pi rh$
     What is the surface area of a cylinder with radius 4·2 cm and height 7·8 cm?

*7 A rectangular field is 102·4 m by 293·6 m. Estimate the number of fence posts needed to
fence this field if the posts are 4·8 m apart.

*8 Mason estimated the answer to $\frac{31.3}{5.27 - 3.14}$ as $\frac{30}{2} = 15$.
Write down four more calculations that could have an estimated answer of 15.
Use at least three of the operations $+$, $-$, $\times$, $\div$ and squaring in each calculation.

*9 Explain why, when estimating a multiplication, it is better to round one number up and the
other down but when estimating a division, it is better to round both numbers up or both
down. Give examples.

Review 1   Estimate the answers to these calculations.
a £832·78 ÷ 18   **b** 34.96×89.7 **c** $\frac{3.09 \times 64.2}{8.7}$ **d** $\frac{5.99 \times 31.7}{6.32 \times 28.4}$ **e** $52.3(6.7 + 8.9)$

Review 2   Use a calculator for these calculations. Give your answers to 3 s.f.
Check your answers by estimating.
a 72.8×96.3 **b** $527.3 \div 82.1$ **c** $\frac{23.01 \times 3.8}{9.72}$

Review 3 The Sun is 1 392 530 kilometres in diameter.
The Earth is 12 756·6 kilometres in diameter.
Estimate how many Earths would fit across the diameter of the Sun.

*__Review 4__ A car travels for 6 hours 12 minutes at an average speed of 82 km/h. The fuel
consumption is 8·12 ℓ/100 km. The cost of fuel is 91 pence per litre.
Estimate the approximate cost of the journey.

 Practical

Estimate how many barrels of oil must be used each day to provide enough
petrol for all the cars in Britain.

Written calculation

Example A vet needs to give a dog 0·035 cℓ of antibiotic per kilogram of mass.
The dog has a mass of 19·26 kg.
The vet must give the dog 19·26 × 0·035 cℓ.
$19 \cdot 26 \times 0 \cdot 035 \approx 20 \times 0 \cdot 04 = 0 \cdot 8$
19·26 × 0·035 is equivalent to 1926 ÷ 100 × 35 ÷ 1000
= 1926 × 35 ÷ 100 000.
First multiply the integers.

```
    1926
  ×   35
   57780     1926 × 30
    9630     1926 × 5
   67410
```

$19 \cdot 26 \times 0 \cdot 035 = 67\,410 \div 100\,000 = 0 \cdot 67$ cℓ to 2 s.f.

We choose to round to 2 s.f. because
the amount of antibiotic per kg is given
to 2 s.f. and this seems appropriate.

Remember
To divide by a decimal we do an equivalent calculation.

Worked Example
Calculate 0·059 ÷ 0·29 to 3 significant figures.

Answer
$0 \cdot 059 \div 0 \cdot 29 \approx 0 \cdot 06 \div 0 \cdot 3 = 0 \cdot 2$ and is equivalent to $\frac{59 \div 1000}{29 \div 100} = (59 \div 29) \div 10$

```
29 ) 59
   −58        29 × 2
    1·00
  − 0·87      29 × 0·03
    0·130
  − 0·116     29 × 0·004
    0·014
```

Answer 2·034 R 0·014 = 2·03 correct to 3 s.f.

So 0·059 ÷ 0·29 = 2·03 ÷ 10 = **0·203** (3 s.f.)
When we check with our estimate of 0·2, the answer is the right order of magnitude.

Number

> **Remember**
> Check your answer using one or more of these ways.
>
> 1 Check the answer is sensible.
> 2 Estimate to check the answer is the right order of magnitude.
> 3 Check by working the problem backwards (using inverse operations).
> 4 Check using an equivalent calculation.
> 5 Check the last digits.
>
>
>
> See page 4 for examples of each of these.

Exercise 4

1 **a** Mark pays £16·80 to travel to school each week.
       He goes to school 38 weeks each year.
       How much does he pay to travel to school each year?
       Show your working.
   b Mark could buy a season ticket that would let him travel for all 38 weeks.
       It would cost £532.
       How much is that per week?
   c Show how you could check your answer to **a**.

2 Find the answers to these, to 4 d.p.
   a $8·63 \times 0·842$ **b** $0·00724 \times 0·035$ **c** $0·183 \times 0·029$ **d** $0·0804 \times 0·00603$
   e $18·2 \div 4·3$ **f** $94·5 \div 90·1$ **g** $0·0834 \div 0·84$ **h** $0·0684 \div 0·35$

3 What is the cost of 0·45 kg of peanuts at £0·89 per kilogram?

4 Which is the better buy?
        15 pieces of pizza for £12·60
     or 25 pieces of pizza for £20·50
   Show how you could check your answer.

5 Find the answers to these to 3 s.f.
   a $0·00826 \times 0·045$ **b** $4·81 \times 7·62$ **c** $0·0827 \div 0·39$

6 Card used to make storage boxes weighs
   300 grams per square metre. The net for
   one of these boxes is shown.
   a Calculate the area of card used to make
       one box.
   b A company makes about 28 000 of the
       boxes. Calculate the approximate total
       mass of card used. Round your answer
       appropriately.

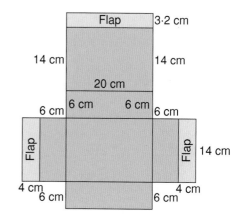

*7 Your body mass index is found by dividing your mass by the square of your height. Mass is measured in kilograms, height is measured in metres.

    a Paul is 1·75 m tall and weighs 69 kg. What is Paul's body mass index?

    b Helen is 1·6 m tall and weighs 70 kg. Helen's body mass index is within the healthy range. What is the minimum age Helen could be?

    c Andrew is 25 and is 1·69 m tall. What weight range would give Andrew a healthy body mass index?

Body mass index changes with age

Age	Body mass index healthy range
19 – 34	20 – 25
35 – 44	20 – 26
45 – 54	20 – 27
55 – 64	20 – 28
greater than 65	20 – 29

*8 Heat is lost from a room at approximately the rate given in the table below.

    a Calculate the amount of heat lost from a room with the following dimensions and window sizes. The ceiling is insulated, the walls are insulated, the windows all have curtains and the floor is concrete.

          Dimensions of room
             3·2 m by 4·1 m by 2·4 m
          Windows (two in total)
             1·8 m by 1·3 m in the 4·1 by 2·4 wall
             1·2 m by 1·3 m in the 3·2 by 2·4 wall

    b To heat a room a heater must output at least 50% more watts than the room is losing. Heaters are available with the following power outputs.
          1 kW   1·5 kW   2 kW
    Which heater would you choose? Explain your choice.

Heat Loss

Source of heat loss	Watts lost
Ceiling – insulated	$20 \times$ area in m^2
Floor – concrete	$10 \times$ area in m^2
Walls – insulated	$25 \times$ area in m^2
– not insulated	$65 \times$ area in m^2
Windows – with curtains	$40 \times$ area in m^2
– without curtains	$80 \times$ area in m^2

Review 1   Find the answers. Round your answers to 4 d.p. where appropriate.
a 7.45×0.00056 **b** $0.00532 \div 2.9$

Review 2   Which is the best buy? Justify your answer.
     A 2 litres of washing-up liquid £4·89
     B 500 mℓ of washing-up liquid £1·25
     C 750 mℓ of washing-up liquid £1·82

Review 3   Find the answers to 3 s.f.
a $43\,600 \times 280$ **b** $0.000564 \div 0.013$ **c** 0.0256^3

Review 4   Sheets of paper have a g.s.m rating where g.s.m stands for grams per square metre. A booklet contains two types of paper. The cover uses 85 gsm paper and folds out to measure 30 cm by 20 cm. The booklet contains 4 sheets of paper inside. These are all 70 g.s.m paper and each sheet folds out to measure the same as the cover.
a What is the mass of one cover?
b What is the mass of one booklet?
c 2500 booklets are ordered. What is the mass, in kilograms, of a package containing the booklets? Ignore the mass of the packaging.

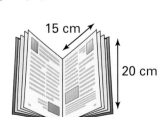
15 cm
20 cm

Reciprocals

Discussion

What is the relationship between the following pairs of numbers? **Discuss**.

$$2, \frac{1}{2} \qquad 4, \frac{1}{4} \qquad \frac{1}{7}, 7 \qquad \frac{2}{3}, \frac{3}{2} \qquad \frac{5}{4}, \frac{4}{5} \qquad 2\frac{1}{2}, \frac{2}{5}$$

Write down some other pairs of numbers with this relationship. **Discuss**.

To find the **reciprocal** of a number express the number as a fraction and invert.
The reciprocal of $\frac{3}{5}$ is $\frac{5}{3}$; the reciprocal of $\frac{a}{b}$ is $\frac{b}{a}$; the reciprocal of a is $\frac{1}{a}$.
A number multiplied by its reciprocal equals 1.

Worked Example
Find the reciprocals of these.   **a** $\frac{2}{5}$ **b** 5   **c** 0·8   **d** $\frac{3}{x}$ **a** $3x$

Answer
a The reciprocal of $\frac{2}{5}$ is $\frac{5}{2}$.   **b** Since $5 = \frac{5}{1}$, the reciprocal of 5 is $\frac{1}{5}$.

c $0·8 = \frac{8}{10}$ or $\frac{4}{5}$. The reciprocal of 0·8 is $\frac{5}{4}$ or 1·25.   **d** The reciprocal of $\frac{3}{x}$ is $\frac{x}{3}$.

e Since $3x = \frac{3x}{1}$, the reciprocal of $3x$ is $\frac{1}{3x}$.

The key on a calculator is used to find the reciprocal of a number.

Example   The reciprocal of 2 is found by keying $\boxed{2}\boxed{x^{-1}}\boxed{=}$ to get 0·5.

Example   The reciprocal of 0·6 is found by keying $\boxed{0·6}\boxed{x^{-1}}\boxed{=}$ to get 1·$\dot{6}$ or 1·67 (2 d.p.).
          The answer is not exact.

Investigation

Dividing by zero

Start with the number 200 on your screen.
Divide 200 by 10, 9, 8, 7, ... 1
Now divide it by 0·9, 0·8, 0·7, 0·6, ...
Now divide it by 0·09, 0·08, 0·07, 0·06, ...
Now divide ...

> Start with 200 again for each division.

> The number you divide by is the divisor.

What happens to the answer as the divisor gets smaller and smaller? **Investigate**.
Does dividing by zero have any meaning?

Exercise 5   Except for questions 4, 5 and 6.

1 Without using a calculator, find the reciprocals of these.

   a $\frac{2}{3}$ **b** $\frac{3}{4}$ **c** $\frac{3}{10}$ **d** $\frac{8}{7}$ **e** $\frac{5}{4}$

   f $\frac{1}{6}$ **g** 7   **h** 8   **i** 0·7   **j** 1·3

2 Find the reciprocals of these.

a $\frac{c}{a}$ b $\frac{a}{c}$ c $\frac{z}{x}$ d $\frac{x}{z}$ e x f $2z$ g $\frac{x}{4}$

3 Without doing the calculations, work out what these equal.

a $1\frac{1}{2} \times \frac{2}{3}$ b $2\frac{3}{4} \times \frac{4}{11}$ c $0.4 \times 2\frac{1}{2}$

4 Use a calculator to find the reciprocals of these. Give the answers to 2 significant figures where rounding is necessary.

a 0·14   b 7·2   c 0·3   d 10   e 25   f 54

5 ▲, ∗ and ● stand for missing digits.

a   The reciprocal of a whole number between 0 and 50 is 0·02 ▲ 83 to four significant figures.
Find the number and the missing digit, ▲.

b   The reciprocal of a whole number between 200 and 1000 is 0·0027 ∗ ● 7 to five significant figures.
Find the number and the missing digits, ● and ∗.

∗6 Investigate the sequences given by these functions.

a $x \longrightarrow \frac{1}{x-2}$ b $x \longrightarrow \frac{1}{x^2}$ c $x \longrightarrow \frac{x-1}{x-3}$

> Investigate what happens as x gets very large or very small.

Review 1 Give the reciprocals of these as whole numbers or fractions.

a $\frac{3}{4}$ b 9   c 0·1   d k e $3k$ f $\frac{a}{3k}$

Review 2 Find the reciprocals of these, giving answers to 2 d.p. if rounding is necessary.

a 0·6   b 20   c 1·16

Review 3 The reciprocal of a whole number between 50 and 100 is 0·016 ∗ 93 to 5 s.f. Find the number and the missing digit, ∗.

Discussion

● **Discuss** the following statements which may be true or false. For those that are false, make a similar correct statement.

Statement 1   There is just one whole number that has no reciprocal.
Statement 2   Dividing by a number is the same as multiplying by the reciprocal of that number.
Statement 3   When a number is multiplied by its reciprocal the answer is 0.
Statement 4   Negative numbers have no reciprocals.

● Adding and subtracting, multiplying and dividing, squaring and taking the square root are all inverse operations. What is the inverse operation for 'taking the reciprocal'? **Discuss**.

Sometimes it is useful to 'take the reciprocal' when **solving equations**.

Worked Example
Solve $\frac{3}{x} = 5$.

Answer

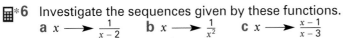

Begin with x → | Take the reciprocal | → $\frac{1}{x}$ → | ×3 | → $\frac{3}{x}$

$x = \frac{3}{5}$ **or 0·6** $\frac{3}{5}$ ← | Take the reciprocal | ← $\frac{5}{3}$ ← | ÷3 | ← Begin with 5

Number

Worked Example

Solve $\dfrac{2}{x+1} = 5$.

Answer

$\dfrac{2}{x+1} = 5$

$\dfrac{x+1}{2} = \dfrac{1}{5}$ **taking the reciprocal of both sides**

$x + 1 = 2 \times \dfrac{1}{5}$ **multiplying both sides by 2**

$x + 1 = \dfrac{2}{5}$

$x = \dfrac{2}{5} - 1$ **subtracting 1 from both sides**

$x = \dfrac{^-3}{5}$ **or** $^-0 \cdot 6$

Note If the equation is $3 + \dfrac{2}{x+1} = 8$, we must subtract 3 from both sides *first* before taking the reciprocal. $3 + \dfrac{2}{x+1} = 8 \neq \dfrac{1}{3} + \dfrac{x+1}{2} = \dfrac{1}{8}$

Exercise 6   **Use 'taking the reciprocal' for all questions in this exercise.**

1 Solve these equations.

   a $\dfrac{2}{x} = 9$ **b** $\dfrac{4}{x} = 10$ **c** $\dfrac{5}{x} = 1 \cdot 8$ **d** $\dfrac{2}{x-1} = 4$ **e** $\dfrac{5}{x+3} = 1$

   f $\dfrac{1}{2x-1} = 3$ **g** $\dfrac{2}{1+3x} = 5$ **h** $\dfrac{4}{5x+2} = 5$ **i** $1 + \dfrac{5}{x-2} = 3$

2 $s = \dfrac{d}{t}$ is a formula which gives the average speed, s, if a distance, d, is covered in time t.

   Use this formula to find the time taken if:

   a $s = 50$ km/h, $d = 240$ km   **b** $s = 120$ km/h, $d = 500$ km

   c $s = 10$ m/s, $d = 44$ m   **d** $s = 20$ m/s, $d = 550$ m

This exercise has lots of links to Science.

3 $d = \dfrac{m}{V}$ gives the density, d, of an object of mass, m, and volume V.

   Find the volume of an object if:

   a $d = 0 \cdot 25$ g/cm^3, $m = 150$ g   **b** $d = 1 \cdot 5$ g/cm^3, $m = 60$ g

*4   If just the hot tap is turned on, a bath can be filled in h minutes; if just the cold tap is turned on, a bath can be filled in c minutes. If both taps are turned on, a bath can be filled in t minutes where t is given by $\dfrac{1}{t} = \dfrac{1}{h} + \dfrac{1}{c}$. For Angela's bath, $h = 8$ minutes and $c = 5$ minutes. Find the time taken to fill Angela's bath if both taps are turned on. (Answer to the nearest minute.)

*5

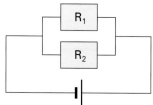

   The total resistance R in this circuit, with resistors R_1 and R_2 in parallel, is given by $\dfrac{1}{R} = \dfrac{1}{R_1} + \dfrac{1}{R_2}$.

   Find the total resistance, R, if $R_1 = 3 \cdot 5$ ohms and $R_2 = 4 \cdot 5$ ohms.

Review 1   Solve these equations.   **a** $\dfrac{3}{x} = 8$ **b** $\dfrac{5}{x+2} = 2$ **c* $1 + \dfrac{2}{3x-1} = 4$

Review 2   Ohm's Law, $\frac{V}{I} = R$,   gives the relationship between voltage V, current I and resistance R. V is measured in volts, I is measured in amps and R is measured in ohms. Find the current if the voltage is 12 volts and the resistance is 2·4 ohms.

Using a calculator

On a calculator we use

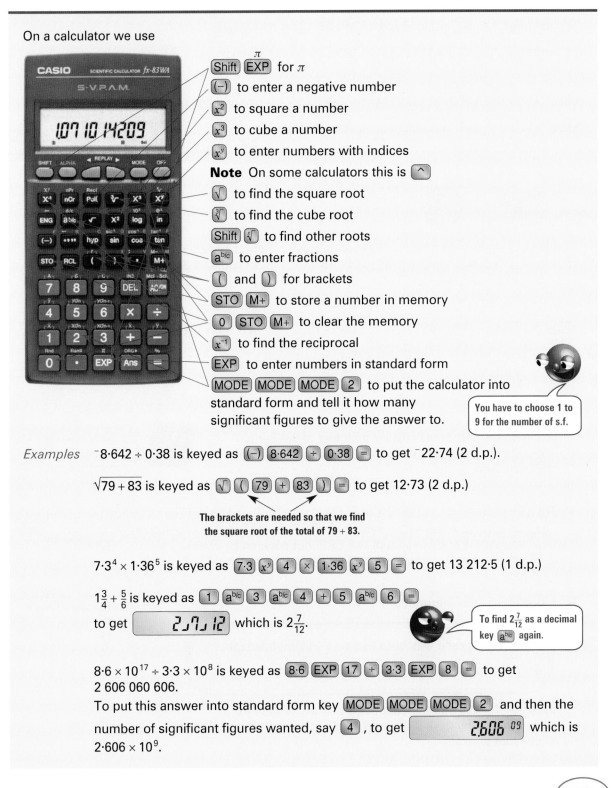

Shift EXP for π

(−) to enter a negative number

x^2 to square a number

x^3 to cube a number

x^y to enter numbers with indices

Note On some calculators this is ^

√ to find the square root

∛ to find the cube root

Shift √ to find other roots

$a^{b/c}$ to enter fractions

(and) for brackets

STO M+ to store a number in memory

0 STO M+ to clear the memory

x^{-1} to find the reciprocal

EXP to enter numbers in standard form

MODE MODE MODE 2 to put the calculator into standard form and tell it how many significant figures to give the answer to.

You have to choose 1 to 9 for the number of s.f.

Examples $^-8·642 \div 0·38$ is keyed as (−) 8·642 ÷ 0·38 = to get $^-22·74$ (2 d.p.).

$\sqrt{79 + 83}$ is keyed as √ (79 + 83) = to get 12·73 (2 d.p.)

The brackets are needed so that we find the square root of the total of 79 + 83.

$7·3^4 \times 1·36^5$ is keyed as 7·3 x^y 4 × 1·36 x^y 5 = to get 13 212·5 (1 d.p.)

$1\frac{3}{4} + \frac{5}{6}$ is keyed as 1 $a^{b/c}$ 3 $a^{b/c}$ 4 + 5 $a^{b/c}$ 6 = to get [2⌐7⌐12] which is $2\frac{7}{12}$.

To find $2\frac{7}{12}$ as a decimal key $a^{b/c}$ again.

$8·6 \times 10^{17} \div 3·3 \times 10^8$ is keyed as 8·6 EXP 17 ÷ 3·3 EXP 8 = to get 2 606 060 606.

To put this answer into standard form key MODE MODE MODE 2 and then the number of significant figures wanted, say 4, to get [2606 09] which is $2·606 \times 10^9$.

Number

Worked Example

Find $\dfrac{(5\cdot8 - 4\cdot3)^3 - 2}{3 \times 6 + (3\cdot5 - 0\cdot42)^2}$ using a calculator, to 3 significant figures.

Answer

Key [MODE] [MODE] [MODE] [2] [3] so that the answer is given in standard form to 3 s.f.

The whole numerator must be divided by the whole denominator.
To find the denominator, key this

[0] [STO] [M+] [3] [×] [6] [+] [(] [(] [3·5] [−] [0·42] [)] [x²] [=] [STO] [M+]

  to clear   to store in
  memory   memory

To find the numerator and then divide it by the denominator, key

[(] [5·8] [−] [4·3] [)] [x³] [−] [2] [=] [÷] [RCL] [M+] [=] to get $0\cdot0500$ to 3 significant figures.

Note There are other possible keying sequences.

Exercise 7

1. Find the answers to these using your calculator. Round sensibly.

   a $^-4\cdot32 + 8\cdot74 - {}^-3\cdot86$
   b $^-3\cdot42 \times 5\cdot37$
   c $86\cdot47 \div 3\cdot28$
   d $8\cdot3^2 \times 0\cdot21^3$

   e $\dfrac{3\cdot65^4}{2\cdot16^3}$
   f $\sqrt{8\cdot4^4 + 3\cdot1^2}$
   g $\sqrt{7\cdot1^4 - 8\cdot3^3}$
   h $5\frac{1}{2} + 3\frac{2}{7}$

   i $8\frac{3}{11} - 5\frac{7}{8}$
   j $4\frac{1}{2} \times 3\frac{3}{4}$
   k $8\frac{1}{3} \div 2\frac{3}{4}$
   l $\sqrt{\dfrac{82\cdot3}{\pi}}$

   m $\sqrt[4]{1296}$
   n $\dfrac{(3\cdot6 - 2\cdot1)^2 - 3}{4 \times 5 \div (8\cdot7 - 6\cdot2) - 2^2}$
   o $^-(231 \times 3 + 432) + 4 \times 5\frac{1}{2} \div 1\frac{3}{4}$

   p $\dfrac{8\frac{3}{4}(4\frac{1}{3} \div 1\frac{1}{2})}{5\frac{7}{8}}$
   q $\dfrac{3 \times 10^4 \div 8 \times 10^{-3}}{1\cdot4 \times 10^{-5}}$
   r $\sqrt{8\cdot7 \times 10^4}$

2. The mass of the Earth is $5\cdot94783 \times 10^{21}$ tonnes. The mass
of the Moon is 81 times smaller than that of the Earth.
Work out the mass of the Moon, giving your answer in
standard form.

Link to science and geography.

3. The surface area of the Earth is about $3\cdot17 \times 10^8$ square kilometres.
The surface area covered by water is about $2\cdot25 \times 10^8$ square kilometres.
   a Calculate the surface area of the Earth not covered by water.
   b Calculate the percentage of the Earth's surface covered by water.

4. A book gives this information:
                                                                         [SATs Paper 2 Level 7]

> A baby giraffe was born that was $1\cdot58$ metres high.
>
> It grew at a rate of $1\cdot3$ centimetres **every hour**.

Suppose the baby giraffe continued to grow at this rate.
About how many days old would it be when it was **6 metres** high?
Show your working.

5 Carbon dioxide (CO_2) is produced by burning fossil fuels such as coal and petrol. CO_2, along with methane, is a major cause of the greenhouse effect, which causes global warming.
One litre of petrol produces about 1000 litres of CO_2.
   a There are about 24 000 000 cars registered in the UK. If each uses, on average, about 25 litres of petrol a week, calculate the volume of CO_2 produced in a week. Give your answer in standard form.
   b 1000 litres of CO_2 weighs about 1·96 kg. What mass of CO_2 is produced by the cars registered in the UK each week? Give your answer in standard form.

Trees use and store CO_2 in photosynthesis. A medium-sized tree can use up to 2 tonnes of CO_2 in one year.
   c How many medium-sized trees are needed to use up the CO_2 produced by the cars registered in the UK?

6 A science and technology exhibition committee are trying to decide what admission price to charge. From previous experience they know that if they charge £8, about 1500 people will attend. If the price is increased by £1, the number of people attending will drop by about 10%.

Admission price	Approximate number of people expected to attend	Total received from admissions

If the price drops by £1, the number of people attending increases by about 10%.
What price should they charge to get the maximum amount from admissions? Assume they want to charge a whole number of pounds.
You may find it useful to copy and fill in the table above.
(**Note** Round to the nearest person.)

Review 1 The Earth's crust was formed about $4·5 \times 10^9$ years ago.
The first land plants grew about $4·5 \times 10^8$ years ago.
The first flowering plants grew about $1·35 \times 10^8$ years ago.
   a How many years after the Earth's crust was formed did the first land plants grow?
   b How long before the first flowering plants grew did the first land plants grow?

Review 2 To raise funds for a new local sports centre, the council is planning a Celebrities Debate with wine and cheese in the school hall. Three celebrities will debate with three locals. The celebrities' fees are £750 each, the locals will donate their time. The school hall can seat 800 guests and the rent will be £300 for the night. The committee estimates the cost price of the wine and cheese will be £5 per guest. The local supermarket has offered a 20% discount on the wine and cheese if the order exceeds £3000. A survey showed there would be a full house if the ticket price was £12·50 per person. For every £2·50 increase in price the number attending would drop by 120.
   a The fixed costs for the evening will be £2550. Explain.
   b Copy and complete the table to find the most profitable ticket price.

Ticket price	Guests	Income	Wine and cheese (minus discount)	Profit
£12·50	800	£10 000	£3200	£4250
£15				
£17·50				
£20				
£22·50				

Number

? Puzzle

1 This calculation uses the digits 2, 3, 4, 5 and 6.

$$23 \times 45 \times 6$$

What multiplication using these same digits, has the smallest answer?
Explain your answer.

2 a The sum of two whole numbers is 19.
What is the greatest product they can have?

b What if there are three whole numbers?

c What if there are four whole numbers?

Summary of key points

 Mental Calculation

We can **add and subtract mentally** using

complements	partitioning
counting up	nearly numbers
compensation	facts we already know.

We can **multiply and divide mentally** using

place value knowledge	partitioning
factors	near tens
known facts	doubling and halving.

See pages 2–3 for examples.

We carry out **operations in this order**, working from left to right.

Brackets

Indices

Division and **M**ultiplication

Addition and **S**ubtraction

Remember BIDMAS.

 We can **convert between fractions, decimals and percentages mentally**.

We can also do simple fraction, decimal and percentage calculations mentally.

You need to know these.

$\frac{1}{8} = 0 \cdot 125 = 12\frac{1}{2}\%$ $\frac{1}{3} = 0 \cdot \dot{3} = 33\frac{1}{3}\%$ $\frac{2}{3} = 0 \cdot \dot{6} = 66\frac{2}{3}\%$

Examples $\frac{2}{3}$ of $45 \cdot 3 = \frac{2}{3} \times 45 \cdot 3$

$\frac{1}{3} \times 45 \cdot 3 = 45 \cdot 3 \div 3$

$= 15 \cdot 1$

$\frac{2}{3} \times 45 \cdot 3 = 2 \times 15 \cdot 1$

$= 30 \cdot 2$

To find 17·5% of 96

10% of 96 = 9·6

5% of 96 = 4·8   (half of 10%)

2·5% of 96 = 2·4   (half of 5%)

so 17·5% of 96 = 9·6 + 4·8 + 2·4

$= 16 \cdot 8$

Estimating

When estimating answers we use these guidelines.

    1 Approximate to '**nice numbers**' that are easy to work with.

    2 Approximate to **numbers that will cancel**.

    3 When **multiplying**, try to round **one number up and one number down**.

    4 When **dividing**, try to round **both numbers up or both numbers down**.

When **estimating** answers we often round all numbers to one significant figure (1 s.f.).

Example $\dfrac{4 \cdot 38 \times 28 \cdot 2}{5 \cdot 78} \approx \dfrac{5 \times 30}{6}$

                      $= 25$

 We use **written methods** with the four operations and with whole numbers and decimals.

To **divide by a decimal** we do an equivalent calculation so that we divide by a whole number.

Worked Example   Calculate $0 \cdot 0248 \div 0 \cdot 46$.

Answer $0 \cdot 0248 \div 0 \cdot 46 \approx 0 \cdot 02 \div 0 \cdot 5 = 0 \cdot 04$ and is equivalent to

    $\dfrac{248 \div 10\,000}{46 \div 100} = (248 \div 46) \div 100$

$$
\begin{array}{ll}
46\,)\,248 & \\
\quad\underline{230} & 46 \times \mathbf{5} \\
\quad 18 \cdot 0 & \\
\quad\underline{13 \cdot 8} & 46 \times \mathbf{0 \cdot 3} \\
\quad\ 4 \cdot 20 & \\
\quad\underline{4 \cdot 14} & 46 \times \mathbf{0 \cdot 09} \\
\quad\ 0 \cdot 06 &
\end{array}
$$

Answer 5·39 R 0·06 = 5·39 to 3 s.f.

So $0 \cdot 0248 \div 0 \cdot 46 = 5 \cdot 39 \div 100 = \mathbf{0 \cdot 0539}$ (3 s.f.)

When we check with our estimate of 0·04, the answer is the right order of magnitude.

 The **reciprocal** of $\frac{3}{8}$ is $\frac{8}{3}$; the **reciprocal** of $\frac{x}{y}$ is $\frac{y}{x}$; the **reciprocal** of 7 is $\frac{1}{7}$.

A number multiplied by its reciprocal equals 1.

We use the $\boxed{x^{-1}}$ key on a calculator to find the reciprocal.

Example   To find the reciprocal of 0·8 we key $\boxed{0 \cdot 8}\ \boxed{x^{-1}}\ \boxed{=}$ to get 1·25.

 See page 65 for calculator keys.

Example   To find $4 \cdot 62^{4} \times 3 \cdot 1^{5}$ to 4 s.f. key

    $\boxed{\text{MODE}}\ \boxed{\text{MODE}}\ \boxed{\text{MODE}}\ \boxed{2}\ \boxed{4}\ \boxed{4.62}\ \boxed{x^{y}}\ \boxed{4}\ \boxed{\times}\ \boxed{3.1}\ \boxed{x^{y}}\ \boxed{5}\ \boxed{=}$

                                 4 s.f.

to get $1 \cdot 304 \times 10^{5}$ (4 s.f.)

Example   To find $2\frac{1}{3} + 4\frac{5}{8}$ key $\boxed{2}\ \boxed{a^{b/c}}\ \boxed{1}\ \boxed{a^{b/c}}\ \boxed{3}\ \boxed{+}\ \boxed{4}\ \boxed{a^{b/c}}\ \boxed{5}\ \boxed{a^{b/c}}\ \boxed{8}\ \boxed{=}$

                    to get $6\frac{23}{24}$.

Number

1 Answer these as quickly as possible.
  a The scale on a plan is 1 cm to 2 m.
     On the plan the length of a hall is 8 cm.
     How long is the hall in real life?
  b Mary had £25. She spent £5. What percentage did she spend?
  c How many $\frac{1}{2}$s in 6?
  d Find the HCF of 20 and 36.
  e A car travels at an average speed of 75 km/h.
     How far will it travel in 24 minutes at this speed?
  f What is the area of this trapezium? ⟶ 10 cm
  g Multiply $3x$ by x^2.
  h Solve $x - 14 = 88$.
  i What is the next number in the sequence ⁻1, 5, 11, 17, ...?
  j What is the value of x when $y = 4$? $x = 7 - 3y$
  k What is 11·5 divided by 2?
  l What is the volume of a cuboid 5 cm by 4 cm by 6 cm?
  m Estimate $\frac{62 \cdot 1 \times 19 \cdot 8}{5 \cdot 23}$.

2 Which of the following calculations will have an answer greater than 0·34?
  a $0·34 \div 1·5$ **b** $0·34 \times 0·04$ **c** $0·34 \times 12·76$ **d** $\frac{0·34}{0·76}$

3 Find the answer to these mentally.
  a $\frac{3 + 4 \times 8}{7 - 2}$ **b** $\frac{3 \times 4^2}{6 \times {}^-2}$ **c** $\frac{(2 + 4)^2}{(5 - 3)^2}$ **d** $\sqrt{54 - 18}$ **e** $\sqrt[3]{6^2 - 9}$

4 Choose two numbers out of the box so that
  a they multiply to give the biggest answer
  b they divide to give the biggest answer
  c they multiply to give the answer closest to 1.

5 Find three consecutive even numbers with a sum of 114.

6 Two numbers are added to get the number above.
   Find out a possible way to fill in this diagram.

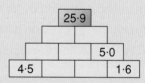

7 Thirty-four pieces of protein food are to be fed to 8 animals.
   The dogs get 5 pieces each, the cats get 3 pieces each and the mice get 1 piece each.
   How many of each animal are fed the protein food?

8 Use the digits 3, 4, 5 and 6, the operations +, ×, −, ² and two sets of brackets to get the answer 32.
   Use each digit and operation just once.

9 Answer these mentally.

    a Write $\frac{2}{5}$ as a percentage.   **b** Write 0·6 as a fraction.

    c Write $\frac{3}{8}$ as a decimal.   **d** Write 0·625 as a percentage.

    e Find $\frac{3}{5}$ of 45.   **f** Find $2\frac{1}{4}$ of 24.

    g Find 120% of 40.   ***h** Find 35% of 25.

10 Find the answer mentally.
Shonagh makes silver necklaces.
A market stall sells them for Shonagh.
The necklaces sell for £6·20 each.
The stall keeps 25% of the sale.
How much does Shonagh get for each necklace sold?

11 **a** Choose the **best** estimate of the answer to 72·36 ÷ 8·92.
        A 6   **B** 7   **C** 8   **D** 9   **E** 10   **F** 11

    b Choose the **best** estimate of the answer to 32·6 × 0·47.
        A 1·2   **B** 1·6   **C** 12   **D** 16   **E** 120   **F** 160

    c Estimate the answer to $\frac{9\cdot31 + 22\cdot4}{5\cdot26}$.

      Give your answer to 1 s.f.

    d Estimate the answer to $\frac{29\cdot3 \times 24\cdot6}{6\cdot12 \times 4\cdot08}$.

12 **a** Estimate the perimeter and area of this shape.
    b Calculate the perimeter and area, using your
       estimates as a check.

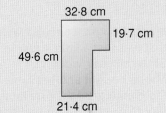

32·8 cm
19·7 cm
49·6 cm
21·4 cm

13 Give a calculation which would be equivalent to the one given.
The divisor should not be a decimal number.
    a 4·52 ÷ 0·08   **b** 0·0694 ÷ 0·007

14 Give the answers to these to 3 d.p.
    a 7·62 × 0·045   **b** 0·0925 × 0·0402   **c** 0·587 ÷ 1·6   **d** 0·00429 ÷ 0·28

15 A Year 11 class decided to bake 100 cakes to raise money for their school trip.
The class surveyed parents to find out how many they would buy at various prices.
The results showed that the cakes would all sell if they were priced at £2·50 each.
For each 50p increase in price, the number of cakes sold would drop by 12%.
Calculate what the class should price the cakes at to make the maximum profit.

16 Find the reciprocals of these.
    a $\frac{2}{5}$ **b** $\frac{10}{7}$ **c** $\frac{1}{8}$ **d** 6   **e** 0·4
    f 1·7   **g** $\frac{x}{y}$ **h** $\frac{m}{3}$ **i** p **j** $3q$

 17 Use your calculator to find the reciprocals of these.
Give your answers to 2 significant figures if you need to round.
    a 0·25   **b** 6·2   **c** 0·17   **d** 23   **e** 87

18 Solve these equations by taking the reciprocal.
    a $\frac{3}{x} = 12$ **b** $\frac{7}{x+2} = 1$ **c** $\frac{2}{3x-2} = 5$

19 A formula in physics is $\frac{1}{f} = \frac{1}{u} + \frac{1}{v}$.
   Calculate the value of f when $u = 8$ and $v = 14$.
   Give the answer to 4 s.f.

20 Use your calculator to find the answers to these.
   Round your answers to 3 s.f. if you need to round.
   Estimate first.

   a $\pi \times 4 \cdot 7^2$
   b $\sqrt{14^2 - 6^2}$
   c $(^-5 \cdot 1)^3 + 6 \cdot 1^2$
   d $4\frac{2}{5} + 3\frac{6}{7}$
   e $6\frac{2}{3} \div 3\frac{1}{4}$
   f $\frac{6 \cdot 24^4}{3 \cdot 7^3}$
   g $\frac{(7-4)^2 + 2 \cdot 4^3}{6 \times 2 - 3 \cdot 6}$
   h $\frac{6 \times 10^4 + 9 \times 10^{-6}}{2 \cdot 3 \times 10^{-7}}$

21

PATIO UMBRELLAS

Circumference 3·5 m, 4·0 m, 4·5 m, 5·0 m, 5·5 m
£89 for 3·5 m CIRCUMFERENCE
others priced proportionately according to circumference
10% OFF 4·5 m, 5·0 m and 5·5 m WHILE STOCKS LAST

a Copy and complete this table.

Circumference	3·5 m	4·0 m	4·5 m	5·0 m	5·5 m
Price (£)	89				

b Marita bought a circular table of diameter 1·2 m for her patio. She wanted an umbrella that was just a little larger in area than the table. Which umbrella should Marita buy? Show all your working.

c Marita bought another table for the patio. The diameter of this table is 30 cm more than the diameter of the first table. How much greater is the area of the larger table?

d Marita wanted an umbrella for her second table as well. Which size should she choose? Give a reason for your answer.

4 Fractions, Decimals and Percentages

You need to know

✓ converting between fractions, decimals and percentages   page 5

✓ comparing and ordering fractions   page 5

✓ finding fraction of and percentage of   page 6

Key vocabulary

⩽ **less than or equal to,** ⩾ **greater than or equal to**

 Line Up

This is called a **nomogram**.
It converts fractions to percentages.

1 To use it join the numerator and denominator of the fraction with a line.

2 Extend the line to the percentage scale.

3 Where the line crosses the percentage scale gives the percentage equivalent to the fraction.

Example $\frac{40}{50} = 80\%$ is shown with the purple line.

Use the nomogram to answer these questions.

1 Write these as percentages.

   a $\frac{5}{20}$ **b** $\frac{27}{50}$ **c** $\frac{53}{92}$

2 What score out of 60 is needed to get 75%?

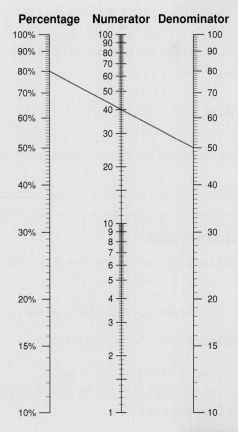

Fractions, decimals and percentages

Fractions, decimals and percentages are all ways of expressing **proportions**.

We can **convert** between them.

Remember

To write a **fraction as a percentage**

either **a Write with a denominator of 100**

Example $\frac{9}{25}$ ⤻×4 $= \frac{36}{100}$

$= 36\%$

or **b multiply by 100%**

Example $\frac{450}{784} = \frac{450}{784} \times 100\%$ **Key** [450] [÷] [784] [×] [100] to get **57·39795918**

$= 57\%$ to the nearest per cent.

For notes and practice at converting between fractions, decimals and percentages see pages 5–6.

We can **compare fractions** by

1   converting them to decimals   *Example* $\frac{2}{3} > \frac{5}{8}$ because $0 \cdot \dot{6} > 0 \cdot 625$.

2   writing them with a common denominator   *Example* $\frac{3}{4} > \frac{2}{3}$ because $\frac{9}{12} > \frac{8}{12}$.

We can compare proportions by converting them all to either fractions, decimals or percentages.

Example

This table gives the number of ewes on three different farms and the number of lambs they produced.

To find which farm had the greatest lambing success, we need to change the fractions to decimals or percentages.

	Farm		
	Upton	Dale	Tandor
Ewes	364	132	247
Lambs	412	186	324

Upton $\frac{412}{364} = 1 \cdot 13$ (2 d.p.)   **Dale** $\frac{186}{132} = 1 \cdot 41$ (2 d.p.)

  $= 113\%$ $= 141\%$

Tandor $\frac{324}{247} = 1 \cdot 31$ (2 d.p.)

  $= 131\%$

Dale Farm had the highest lambing percentage.

Discussion

Discuss how to find the answer to these.

  What fraction of 80 is 60?
  What fraction of 60 is 80?
  What fraction of 2 metres is 340 cm?

Hint: The number after 'of' is the denominator.

Discuss how to compare these.

a $\frac{7}{20}$ and $\frac{3}{10}$

b 6 out of 13 and 13 out of 28

c 135 muffins cost £116·10 and 35 scones cost £29·75. Is a muffin or a scone cheaper?

d $0 \cdot 3\dot{3}$ and $\frac{1}{3}$ and 33%

Worked Example

This table shows the membership (in thousands) of two charity organisations.

	Royal Society for the Protection of Birds	World Wildlife Fund
1970	98	12
2000	1007	241

a For 1970 calculate the number who belonged to the World Wildlife Fund as a percentage of those who belonged to the Royal Society for the Protection of Birds.

b Show that the number who belonged to the World Wildlife Fund in 2000 is a greater proportion of the number who belonged to the Royal Society for the Protection of Birds than in 1970.

Answer

a $\frac{12}{98} \times 100\% = $ **12·24% (2 d.p.)**

Key ⬚12⬚ ⬚÷⬚ ⬚98⬚ ⬚×⬚ ⬚100⬚ ⬚=⬚

b In 1970 the proportion was 12·24%.

In 2000 the proportion was $\frac{241}{1007}$.

$\frac{241}{1007} \times 100\% = 23·93\%$ (2 d.p.)

Key ⬚241⬚ ⬚÷⬚ ⬚1007⬚ ⬚×⬚ ⬚100⬚ ⬚=⬚

23·93% > 12·24%

There was a greater proportion in 2000.

Exercise 1

1 Write the answers to each of these as a fraction, decimal and percentage.

 a James added enough water to 375 mℓ of a copper sulphate solution to make it up to 500 mℓ.
 What proportion of the 500 mℓ solution was the 375 mℓ solution?

 b A pendulum swings 50 mm high in its first swing.
 On its second it swings 38 mm.
 What proportion of the first swing height did it reach on its second swing?

2 In an experiment a mixture was made by adding
  540 mg iron filings
  320 mg sulfur
  340 mg sand.

> Think about what you are finding the fraction and percentage of.

 a About what fraction of the mixture is sulfur?
 b About what percentage of the mixture is iron filings?
 c In another mixture with a total mass of 874 g, 420 g are iron filings.
 Which mixture has the greater proportion of iron filings?

3 Draw a 4 by 6 rectangle.
 Shade four parts that are $\frac{1}{3}$, $\frac{1}{4}$, $\frac{1}{6}$ and $\frac{1}{8}$. The four parts must not overlap.
 What percentage of the rectangle is left unshaded?

Number

4 Which of < or > goes in the box?

a $\frac{1}{4}\ \square\ \frac{1}{3}$ b $\frac{2}{5}\ \square\ \frac{1}{2}$ c $\frac{3}{4}\ \square\ \frac{16}{20}$ d $\frac{2}{3}\ \square\ \frac{5}{8}$ e $\frac{5}{6}\ \square\ 0\cdot8$

f $80\%\ \square\ \frac{15}{20}$ g $3\frac{1}{2}\ \square\ 320\%$ h $66\%\ \square\ \frac{2}{3}$ i $62\frac{1}{2}\%\ \square\ \frac{17}{25}$ j $55\%\ \square\ \frac{5}{9}$

5 Put these in order from smallest to largest.
Use your calculator for part **d** only.

a $\frac{1}{3},\ \frac{3}{4},\ \frac{5}{12},\ \frac{5}{6}$ b $0\cdot25,\ \frac{2}{5},\ 35\%,\ \frac{1}{2}$

c $0\cdot66,\ \frac{2}{3},\ 65\%,\ \frac{7}{9}$ ▦ d $\frac{19}{24},\ 0\cdot79,\ 80\%,\ \frac{29}{36}$

6 Georgia bought her mp3 player for £183·65 and paid a deposit of £50.
Jack bought his mp3 player for £209·65 and paid a deposit of £60.
Who paid the greater proportion of the full price as deposit?

7 In their last three netball games, Amanda got 38 shots in out of 45 shots taken and Mylene got 23 out of 31 shots.
Who was the better goal shooter? Explain your answer.

8 This shows the sales during 'Discount week' at a hardware store.

a What percentage of the total number of items sold were garden hoses?

b What percentage of the total amount taken was for garden hoses?

c Which was cheaper, outdoor table style A or style B?
Explain how you know.

Item	Number sold	Amount taken
Hammer	86	£1320·10
Garden hose	150	£562·50
Outdoor table style A	34	£6766·00
Outdoor table style B	6	£1350·00
Total	**276**	**£9998·60**

9 There are 60 dogs at the RSPCA.
6 of these are black.
Exactly half of the dogs are female.
From this information, what percentage of female dogs are black?
Choose the correct answer.

A 5% B 6% C 10%
D 20% E 50% F not possible to tell

10 Brass is made from copper and zinc.
This table gives the amount of copper and zinc in three different grades of brass.
Work out the percentage of copper in each grade.
Which grade has the highest percentage of copper?

	Brass	
	Copper	Zinc
Grade 1	325 g	175 g
Grade 2	480 g	320 g
Grade 3	620 g	380 g

11 A cup of coffee costs £1·75.

[SATs Paper 2 Level 7]

The diagram shows how much money different people get when you buy a cup of coffee.
Use a copy of the table.

Retailers		%
Growers		%
Others		%

Complete the table to show what **percentage** of the cost of a cup of coffee goes to retailers, growers and others.
Show your working.

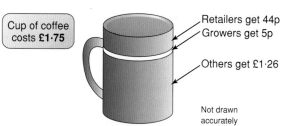

Cup of coffee costs £1·75

Retailers get 44p
Growers get 5p
Others get £1·26

Not drawn accurately

12 The table shows the average weekly earnings for men and women in 1956 and 1998.

[SATs Paper 2 Level 7]

	1956	**1998**
Men	£11·89	£420·30
Women	£ 6·16	£303·70

    a For **1956**, calculate the average weekly earnings for women as a percentage of the average weekly earnings for men.
Show your working and give your answer to 1 decimal place.
    b For **1998**, show that the average weekly earnings for women were a **greater proportion** of the average weekly earnings for men than they were in 1956.

13 $\frac{3}{4} > \frac{x}{5}$
What is the largest whole number x can be?

14 This diagram shows a circle and a square.
The circle touches the edges of the square.
What percentage of the diagram is purple?

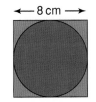

See page 332 for finding the area of a circle.

*15 The diagram models a rectangular rear windscreen of a car.
The windscreen wiper can rotate through 160°.

[SATs EP]

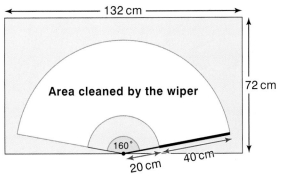

What percentage of the rear windscreen is cleaned by the wiper?
Show your working.

Review 1
a Marie is sitting a 3-hour examination. The examination started at 9:30 a.m. Marie left the exam room at 11:20 a.m. What percentage of the exam time did she use?
b One-fifth of a flagpole is painted red. One-third of the remainder is painted blue. The rest is painted white.
If the white portion is 7·2 m, what is the length of the flagpole?

Review 2 Put these in order, smallest to largest. Do not use your calculator.
0·58, $\frac{5}{8}$, 65%, $\frac{7}{12}$

Review 3 Four identical circles are placed in the square as shown so that the circles touch but do not overlap. They also touch the sides of the square.
If the circles each have diameter 6 cm, what percentage of the square is **not** covered by the circles?

Adding and subtracting fractions

Remember

To find **equivalent fractions** we multiply or divide both the numerator and denominator by the same number.

Examples $\dfrac{3}{4} \xrightarrow{\times 3} \dfrac{9}{12}$ $\dfrac{7\ \,28}{10\ \,40} = \dfrac{7}{10}$

Fractions can be **added and subtracted** easily when they have the same denominator.

Example $\dfrac{8}{15} + \dfrac{13}{15} = \dfrac{8 + 13}{15}$ add the numerators

$\qquad\qquad = \dfrac{21}{15}$

$\qquad\qquad = 1\dfrac{6}{15}$ write improper fractions as mixed numbers

$\qquad\qquad = \mathbf{1\dfrac{2}{5}}$ write fractions in their simplest form

To **add and subtract fractions with different denominators** we use equivalent fractions.

Example
$\dfrac{3}{8} + \dfrac{5}{24}$

1 Find LCM of the denominators.   LCM of 8 and 24 is 24.

2 Write equivalent fractions with this LCM. $\dfrac{3}{8} \xrightarrow{\times 3} \dfrac{9}{24}$

3 Add or subtract the fractions
(they now have the same denominator).
Give the answer in its simplest form.

$\dfrac{9}{24} + \dfrac{5}{24} = \dfrac{9 + 5}{24}$
$\qquad\qquad = \dfrac{14}{24}$
$\qquad\qquad = \dfrac{7}{12}$

You could use diagrams to help.

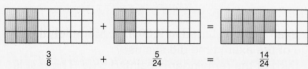

$\dfrac{3}{8} \qquad + \qquad \dfrac{5}{24} \qquad = \qquad \dfrac{14}{24}$

We can add and subtract fractions using the $\boxed{a^{b/c}}$ button on a calculator

Example $1\dfrac{7}{15} + \dfrac{19}{24}$ is found by keying $\boxed{1}\ \boxed{a^{b/c}}\ \boxed{7}\ \boxed{a^{b/c}}\ \boxed{15}\ \boxed{+}\ \boxed{19}\ \boxed{a^{b/c}}\ \boxed{24}\ \boxed{=}$

to get $\boxed{\;2\lrcorner 3 \lrcorner\; 120\;}$.

We read this as $2\dfrac{31}{120}$.

Exercise 2 **Except for question 4.**

T **1** Add $\frac{6}{10}$ and $\frac{6}{5}$. Now use an arrow ($\downarrow$) to show the result on the number line.

[SATs Paper 1 Level 6]

2 Calculate these.

a $\frac{3}{5} + \frac{4}{5}$ **b** $\frac{5}{8} + \frac{7}{8}$ **c** $1\frac{11}{12} - \frac{7}{12}$ **d** $2 - \frac{2}{3}$ **e** $3\frac{3}{8} + \frac{7}{8}$

3 Calculate these.

a $\frac{3}{8} + \frac{1}{4}$ **b** $\frac{5}{7} + \frac{3}{14}$ **c** $\frac{3}{4} - \frac{3}{8}$ **d** $\frac{11}{12} - \frac{1}{6}$ **e** $\frac{3}{5} + \frac{6}{10}$

f $\frac{7}{12} + \frac{5}{6}$ **g** $3\frac{9}{10} - 2\frac{4}{5}$ **h** $1\frac{3}{8} + 2\frac{2}{5}$ **i** $2\frac{7}{8} - 1\frac{2}{3}$ **j** $5\frac{1}{2} + \frac{4}{5}$

k $\frac{2}{5} + \frac{7}{8} + \frac{2}{3}$ **l** $\frac{9}{10} - \frac{2}{5} - \frac{1}{2}$ **m** $\frac{7}{8} + \frac{3}{4} - 1\frac{1}{2}$ **n** $1\frac{7}{24} + \frac{7}{12} - 1\frac{5}{6}$

o $\frac{5}{18} + \frac{5}{27}$ **p** $\frac{7}{18} + \frac{9}{24}$ **q** $\frac{7}{12} + \frac{15}{18} + \frac{19}{36}$

4 Use your calculator to find the answers to these.

a $\frac{8}{13} + \frac{1}{4}$ **b** $\frac{9}{15} - \frac{7}{25}$ **c** $\frac{1}{6} - \frac{3}{12}$ **d** $\frac{18}{27} + \frac{5}{19} - \frac{3}{8}$

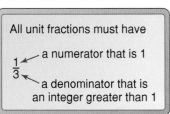

5 Jake's football team won $\frac{1}{3}$ of their games and drew $\frac{1}{2}$ of them.

a What fraction did they win or draw?

b What fraction did they lose?

6 Rani took a $\frac{1}{2}$ ℓ bottle of juice to school.

She drank $\frac{2}{5}$ ℓ.

a How much juice was left?

b Did she drink more or less than half the juice she took to school?

7 Find the next term in this sequence.

    $\frac{5}{8}$, $1\frac{1}{8}$, $1\frac{5}{8}$, $2\frac{1}{8}$, ...

8 $\frac{1}{3}$, $\frac{1}{8}$, $\frac{1}{5}$ are all examples of unit fractions.

[SATs Paper 1 Level 6]

> All unit fractions must have
>
> $\frac{1}{3}$ ⟵ a numerator that is 1
>
> ⟵ a denominator that is an integer greater than 1

The ancient Egyptians used only unit fractions.

For $\frac{3}{4}$, they wrote the sum $\frac{1}{2} + \frac{1}{4}$.

a For what fraction did they write the sum $\frac{1}{2} + \frac{1}{5}$?

Show your working

b They wrote $\frac{9}{20}$ as the sum of two unit fractions.

One of them was $\frac{1}{4}$.

What was the other?

Show your working.

∗c What is the biggest fraction you can make by adding two **different** unit fractions?

Show your working.

[Level 7]

79

9 These are fraction cards.

   a Which two cards make this true?

   b Which two cards make this true?

   c Which three cards make this true?

  *d Which three cards make the biggest answer for this?

      What is this answer?

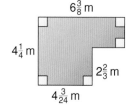

10 Find the perimeter of each of these shapes.

  a $8\frac{5}{8}$ cm   **b**

    $5\frac{3}{4}$ cm

      $6\frac{3}{8}$ m $4\frac{1}{4}$ m $2\frac{2}{3}$ m $4\frac{3}{24}$ m

11 Find the next term in each of these sequences.

  a $1\frac{1}{4}$, $1\frac{11}{12}$, $2\frac{7}{12}$, $3\frac{1}{4}$, ...

 *b $\frac{1}{2}$, 1, 2, $3\frac{1}{2}$, $5\frac{1}{2}$, 8, ...

* 12 The numbers $\frac{1}{2}$, p, q, $\frac{3}{4}$ form an ascending sequence with a constant difference.
What are the values of p and q?

13 a Shabeel wrote down this sequence.

      $1, \frac{1}{2}, \frac{1}{4}, \frac{1}{8}, \frac{1}{16}, \cdots$

   He began adding the terms of the sequence together.

      $1 + \frac{1}{2} + \frac{1}{4} + \frac{1}{8} + \frac{1}{16} + \cdots$

   Investigate what happens if he keeps adding more terms of the sequence.

  b Repeat **a** for this sequence.

      $1, \frac{1}{3}, \frac{1}{6}, \frac{1}{12}, \frac{1}{24}, \cdots$

* 14 What digits could * stand for to make these true?

  a $\frac{*}{5} + \frac{*}{*} = \frac{3}{*}$ **b** $\frac{*}{**} + \frac{*}{6} = \frac{9}{**}$

  Is there more than one way for each?

15 The nth term of a sequence is given by $T(n) = \frac{n}{n+1}$.

  a Write down the first six terms of the sequence.

  b What happens as n gets larger and larger?
    To help you investigate this, change the fractions in the sequence to decimals and draw
    a graph of the decimal values against term number.

  c Repeat **a** and **b** for $T(n) = \frac{n+1}{2n+1}$.

Review 1 Calculate these.

a $\frac{1}{6} + \frac{1}{9}$ b $\frac{3}{4} - \frac{5}{12}$ c $2\frac{2}{3} + \frac{5}{9}$ d $3\frac{1}{6} + \frac{2}{9}$

e $2\frac{3}{4} - 1\frac{1}{10}$ f $3\frac{5}{6} - 2\frac{1}{4}$ g $\frac{1}{2} + \frac{1}{6} + \frac{4}{15}$ h $1\frac{3}{8} + 2\frac{1}{2} - 3\frac{7}{12}$

Review 2 In a bag of coloured counters, $\frac{3}{16}$ are red, $\frac{1}{4}$ green and $\frac{5}{12}$ are yellow. The rest are blue.

a What fraction are blue?

b What is the smallest number of counters there could be?

Counters

Review 3 Use these unit fraction cards.

a Which two cards make this true?

i $\boxed{} + \boxed{} = \frac{5}{12}$ ii $\boxed{} - \boxed{} = \frac{1}{3}$

b Which three cards make this true?

$\boxed{} + \boxed{} - \boxed{} = \frac{5}{24}$

Review 4

a Find the next two terms in the sequence $\frac{5}{6}, 1\frac{1}{2}, 2\frac{1}{6}, 2\frac{5}{6}, ...$

b Find the values of a and b if $\frac{1}{4}, a, b, \frac{5}{8}$ forms a sequence with a constant difference.

Review 5 What number is exactly halfway between $1\frac{2}{3}$ and $2\frac{3}{5}$?

Multiplying and dividing fractions

Remember

To **multiply fractions** we
 – write whole numbers or mixed numbers as improper fractions
 – cancel if possible
 – multiply the numerators
 – multiply the denominators.

Examples $1\frac{3}{5} \times 1\frac{2}{3} \times 1\frac{3}{4} = \frac{{}^2 8}{{}^1 5} \times \frac{{}^1 5}{3} \times \frac{7}{4^1}$

$\qquad\qquad = \frac{2 \times 1 \times 7}{1 \times 3 \times 1}$

$\qquad\qquad = \frac{14}{3}$

$\qquad\qquad = \mathbf{4\frac{2}{3}}$

$\frac{1}{4}(3 - \frac{1}{3}) = \frac{1}{4} \times 2\frac{2}{3}$

$\qquad = \frac{1}{{}_1 4} \times \frac{8^2}{3}$

$\qquad = \mathbf{\frac{2}{3}}$

To divide by a fraction, multiply by the inverse.

Example $1\frac{1}{2} \div \frac{3}{8} = \frac{3}{2} \div \frac{3}{8}$

$\qquad\qquad = \frac{{}^1 3}{{}_1 2} \times \frac{8^4}{3^1}$ $\frac{8}{3}$ is the inverse of $\frac{3}{8}$

$\qquad\qquad = \frac{1 \times 4}{1 \times 1}$

$\qquad\qquad = \mathbf{4}$

We multiply by the inverse because $\frac{3}{8}$ goes into 1, $\frac{8}{3}$ times. $\frac{3}{8} \times \frac{8}{3} = \frac{1}{1} = 1$

Example $3\frac{3}{4} \div 4\frac{1}{8} = \frac{15}{4} \div \frac{33}{8}$

$\qquad\qquad = \frac{{}^5 15}{{}^1 4} \times \frac{8^2}{33^{11}}$

$\qquad\qquad = \mathbf{\frac{10}{11}}$

Number

1 a $2\frac{1}{4} \times \frac{2}{5}$ b $2\frac{2}{5} \times \frac{5}{8}$ c $(2\frac{3}{4})^2$ d $\frac{5}{8} \times \frac{4}{15} \times \frac{3}{12}$

   e $5 \div \frac{2}{5}$ f $\frac{3}{8} \div 1\frac{1}{4}$ g $\frac{4}{5} \times \frac{20}{25} \times \frac{15}{12}$ h $\frac{3}{5} \times \frac{20}{33} \div \frac{16}{22}$

   i $\frac{22}{7} \times 21 \times 21$ j $\frac{22}{7} \times 14 \times 14$ k $\frac{1}{3}(1 - \frac{1}{4})$ l $\frac{1}{5}(2 - \frac{1}{3})$

   m $(1\frac{1}{2})^3$ n $\dfrac{(1 - \frac{1}{3})}{(1 - \frac{2}{5})}$ o $\dfrac{(1 - \frac{2}{3})}{(1 - \frac{5}{8})}$ p $(2\frac{1}{4})^3 \div \frac{3}{4}$

2 a How many **sixths** are there in $3\frac{1}{3}$?   [SATs Paper 1 Level 6]

   b Work out $3\frac{1}{3} \div \frac{5}{6}$.
      Show your working.

3 Find the area of these picture frames.

   a $8\frac{3}{4}$ inches

      $4\frac{3}{7}$ inches

   b $12\frac{3}{8}$ inches

      $4\frac{1}{3}$ inches

4 The area of a coffee table is $1\frac{9}{40}$ m². It is $1\frac{2}{5}$ m long.

   a What is the width of the table?

   *b What is the maximum number of 'Gloss and glitter' magazines
       that will fit flat on the table if each magazine is 0·3 m by 0·15 m?
       Magazines must not overlap each other.

5 This diagram shows the courtyard outside Paulo's window.
   What is the area of the garden?

   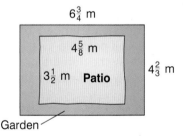

   $6\frac{3}{4}$ m

   $4\frac{5}{8}$ m

   $3\frac{1}{2}$ m **Patio**

   $4\frac{2}{3}$ m

   Garden

6 Which is larger, $\frac{2}{5} \times 1\frac{7}{8}$ or $4\frac{5}{6} \div 6\frac{2}{3}$?

7 What is the largest answer you can make using two of the
   fractions from the box and either × or ÷?

   $\frac{3}{5}$ $1\frac{5}{8}$ $3\frac{3}{4}$ $\frac{1}{8}$ $\frac{2}{3}$ $2\frac{7}{8}$

8 Look at this diagram.
   It is a square with sides of 1 metre.
   The largest triangle has area $\frac{1}{2}$ m².

   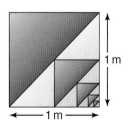

   1 m

   1 m

   a What are the areas of the next two largest triangles?

   *b What is the sum of $\frac{1}{2} + \frac{1}{8} + \frac{1}{32} + \frac{1}{128} + \cdots$ if you add an
       infinite number of terms? Use the diagram to help
       answer this question.

Review 1   Find these.

a $3\frac{3}{8} \times 1\frac{5}{9}$ b $3\frac{3}{4} \times 3\frac{1}{3}$ c $(3\frac{2}{3})^2$ d $2\frac{1}{2} \div \frac{1}{3}$

e $1\frac{4}{5} \div 2\frac{1}{10}$ f $7\frac{2}{5} \times \frac{3}{4} \div 3\frac{7}{10}$ g $\dfrac{\frac{2}{3} \times 2\frac{1}{4}}{\frac{5}{6}}$ h $1\frac{5}{6} - \frac{1}{2} \times 2\frac{1}{2}$

Review 2   A group of people are relaxing in a room. Three are asleep. Of the remainder, $\frac{1}{16}$ are playing table tennis, $\frac{3}{8}$ are watching T.V., and the rest are talking. If 18 are talking, how many people are in the room?

***Review 3**   James and Pete are training for the start of the football season.
James is walking at $5\frac{1}{4}$ km/h while Pete jogs at $7\frac{3}{8}$ km/h.
If they both train for $\frac{3}{4}$ hour, how far apart will they be if
a they travel in the same direction.
b they travel in opposite directions?

Adding and subtracting algebraic fractions

Discussion

One way of finding $\frac{2}{3} + \frac{4}{5}$ is like this.

$$\frac{2}{3} + \frac{4}{5} = \frac{\ }{15} + \frac{\ }{15}$$ Find a common denominator.
$$= \frac{10}{15} + \frac{12}{15}$$ Make equivalent fractions.
$$= \frac{22}{15}$$

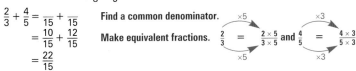

How could we use the same method to find $\frac{n}{m} + \frac{p}{q}$? **Discuss**.

$$\frac{n}{m} + \frac{p}{q} = \frac{\ }{\ } + \frac{\ }{\ }$$ ◄── Find a common denominator.

$$= \frac{\ }{\ } + \frac{\ }{\ }$$ ◄── Make equivalent fractions.

$$= \frac{\ }{\ }$$

Example $\dfrac{x}{2} + \dfrac{y}{3} = \dfrac{3x}{6} + \dfrac{2y}{6}$ **Make equivalent fractions with a common denominator of 6.**

$$= \frac{3x + 2y}{6}$$

Example $\dfrac{3}{a} + \dfrac{b}{c} = \dfrac{3c}{ac} + \dfrac{ab}{ac}$

$$= \frac{3c + ab}{ac}$$

$$\frac{3}{a} \xrightarrow{\times c} \frac{3c}{ac} \quad \text{and} \quad \frac{b}{c} \xrightarrow{\times a} \frac{ab}{ac}$$

This is linked to adding and subtracting fractions, page 78.

Exercise 4

1 Copy these and fill in the gaps.
a $\dfrac{a}{3} - \dfrac{b}{2} = \dfrac{\ }{\ } - \dfrac{\ }{\ }$

$$= \frac{\ }{\ }$$

b $\dfrac{2}{p} + \dfrac{4}{q} = \dfrac{\ }{pq} + \dfrac{\ }{pq}$

$$= \frac{\ }{pq}$$

c $\dfrac{3y}{4} - \dfrac{y}{3} = \dfrac{\ }{\ } - \dfrac{\ }{\ }$

$$= \frac{\ }{\ }$$

d $\dfrac{c}{d} + \dfrac{e}{f} = \dfrac{\ }{\ } + \dfrac{\ }{\ }$

$$= \frac{\ }{\ }$$

2 Simplify these.
a $\dfrac{y}{3} + \dfrac{2y}{3}$
b $\dfrac{a}{4} + \dfrac{a}{3}$
c $\dfrac{3}{x} + \dfrac{2}{x}$
d $\dfrac{5}{y} - \dfrac{3}{y}$
e $\dfrac{2}{x} - \dfrac{3}{y}$

f $\dfrac{7}{m} + \dfrac{3}{n}$
g $\dfrac{1}{p} + \dfrac{2}{q}$
h $\dfrac{p}{q} + \dfrac{r}{s}$
i $\dfrac{2x}{5} - \dfrac{x}{4}$
j $\dfrac{3x}{y} - \dfrac{2w}{z}$

3 Meryl wrote this.

For all values of n and m
$$\frac{1}{n} + \frac{1}{m} = \frac{2}{n+m}$$

Show that Meryl is wrong.

4 Find a fraction in the box that matches each of these.

a $\frac{a}{2} + \frac{a}{3}$ b $\frac{2}{a} + \frac{3}{b}$ c $\frac{a}{2} - \frac{a}{3}$ d $\frac{2}{a} - \frac{3}{b}$

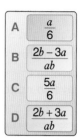

A $\frac{a}{6}$

B $\frac{2b-3a}{ab}$

C $\frac{5a}{6}$

D $\frac{2b+3a}{ab}$

5 Find two pairs of matching expressions.

$\frac{x}{3} - \frac{y}{4}$ $\frac{3}{x} + \frac{4}{x}$ $\frac{7}{x}$ $\frac{7}{x^2}$ $\frac{4x-3y}{12}$

T

6 Use a copy of this.
Add the expressions in two circles to find the expression in the square in between.

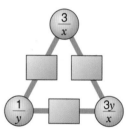

T

Review 1

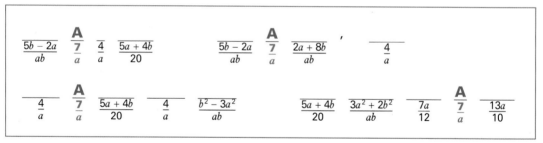

| $\frac{5b-2a}{ab}$ | **A**
$\frac{7}{a}$ | $\frac{4}{a}$ | $\frac{5a+4b}{20}$ | | $\frac{5b-2a}{ab}$ | **A**
$\frac{7}{a}$ | $\frac{2a+8b}{ab}$ | ' | | $\frac{4}{a}$ |

| $\frac{4}{a}$ | **A**
$\frac{7}{a}$ | $\frac{5a+4b}{20}$ | $\frac{4}{a}$ | $\frac{b^2-3a^2}{ab}$ | | $\frac{5a+4b}{20}$ | $\frac{3a^2+2b^2}{ab}$ | $\frac{7a}{12}$ | **A**
$\frac{7}{a}$ | $\frac{13a}{10}$ |

Use a copy of this box. Write the letter beside each above its answer in the box.

A $\frac{4}{a} + \frac{3}{a} = \frac{7}{a}$ G $\frac{a}{4} + \frac{a}{3}$ R $\frac{3a}{2} - \frac{a}{5}$ T $\frac{6}{a} - \frac{2}{a}$

C $\frac{5}{a} - \frac{2}{b}$ N $\frac{8}{a} + \frac{2}{b}$ S $\frac{a}{4} + \frac{b}{5}$ U $\frac{3a}{b} + \frac{2b}{a}$

E $\frac{b}{a} - \frac{3a}{b}$

Review 2 A pendulum is set swinging by pulling it out to one side. In the first swing the foot of the pendulum covers 1 m. In successive swings the distance covered is $\frac{1}{3}$ of the previous distance. After 3 swings the total distance covered is $1 + \frac{1}{3} + \frac{1}{9} = 1\frac{4}{9}$ m.

a How far does the foot of the pendulum cover in

i 4   ii 5 swings?

b Investigate what happens if the pendulum continues swinging in this way.

c *In theory* will the pendulum ever stop?

Recurring decimals

All fractions convert to either a terminating decimal or a **recurring decimal**.

Example $\frac{2}{5} = 0 \cdot 4$ ◄─── terminating decimal $\frac{1}{3} = 0 \cdot 33333333 \ldots$ ◄─── recurring decimal with repeating digits
$= 0 \cdot \dot{3}$

All **recurring decimals are exact fractions**.
Some you should recognise are

$0 \cdot 333333333 \cdots = \frac{1}{3} \left(\frac{3}{9}\right)$ $0 \cdot 666666666 \cdots = \frac{2}{3} \left(\frac{6}{9}\right)$ $0 \cdot 111111111 \cdots = \frac{1}{9}$

$0 \cdot 222222222 \cdots = \frac{2}{9}$ $0 \cdot 777777777 \cdots = \frac{7}{9}$ $0 \cdot 999999999 \cdots = \frac{9}{9} = 1$

We show which digits repeat using dots.

Examples $\frac{1}{7} = 0 \cdot 142857142857142857$
$= 0 \cdot \dot{1}4285\dot{7}$ ◄── Dots are placed above the first and last digit of the repeating sequence.

$\frac{8354}{9990} = 0 \cdot 8362362362 \ldots$
$= 0 \cdot 8\dot{3}6\dot{2}$ ◄── Only the 362 repeats so we put dots above the 3 and the 2.

Discussion

● Using a calculator, $\frac{2}{7}$ as a decimal is found by keying 2 $\div$ 7 $=$ to get

$$0.285714285$$

Is $\frac{2}{7}$ a recurring decimal?

Investigate the recurring decimals for $\frac{2}{7}, \frac{3}{7}, \frac{4}{7}, \ldots$
Investigate the decimals for $\frac{1}{2}, \frac{1}{3}, \frac{1}{4}, \frac{1}{5}, \ldots$
What factors must the denominator of a fraction have to be a terminating decimal?
Discuss.

What if the recurring decimal has some non-recurring and some recurring digits after the decimal point? What factor(s) must the denominator have? **Discuss**.

● We can find the fraction for $0 \cdot \dot{4}\dot{8}$ like this.

$$100 \times 0 \cdot \dot{4}\dot{8} = 48 \cdot 484848 \ldots$$
$$- \qquad 0 \cdot \dot{4}\dot{8} = \ \ 0 \cdot 484848 \ldots$$
$$99 \times 0 \cdot \dot{4}\dot{8} = 48$$
$$\therefore \qquad 0 \cdot \dot{4}\dot{8} = \frac{48}{99} \text{ or } \frac{16}{33}$$

Discuss how to find the fraction equivalent to $0 \cdot \dot{7}\dot{2}$.

A recurring decimal, such as $0 \cdot 34\dot{1}2\dot{6}$ can be written in the form $\frac{m}{n}$ as follows:

$$\times 100\ 000 \qquad 0 \cdot 34\dot{1}2\dot{6} = 34126 \cdot 126126126 \ldots$$
$$- \times \qquad 100 \qquad 0 \cdot 34\dot{1}2\dot{6} = \quad 34 \cdot 126126126 \ldots$$
$$\times \ 99\ 900 \qquad 0 \cdot 34\dot{1}2\dot{6} = 34\ 092 \cdot 000000000 \ldots \qquad \therefore \ 0 \cdot 34\dot{1}2\dot{6} = \frac{34\ 092}{99\ 900}$$
$$= \frac{947}{2775}$$

Discussion

To write $0.34\dot{1}2\dot{6}$ as a fraction, we multiplied firstly by 100 000 then by 100 and subtracted. Why did we choose to multiply by these numbers? **Discuss**.
What if we wanted to convert these to fractions? **Discuss**.

$0.6\dot{3}$ $0.\dot{2}6\dot{1}$ $0.03\dot{1}\dot{7}$

Exercise 5

1 Write these as fractions.
   a 0·333333333 ...   **b** 0·888888888 ...   **c** 0·555555555 ...   **d** 0·999999999 ...

2 Copy and complete these to find the fraction for each.

   a $0.\dot{3}\dot{6}$ $100 \times 0.\dot{3}\dot{6} = 36.363636 \ldots$
          $-\underline{\quad\quad} 0.\dot{3}\dot{6} = \underline{\quad} 0.363636 \ldots$
          $\underline{\quad} \times 0.\dot{3}\dot{6} = \underline{\quad}$
          $0.\dot{3}\dot{6} = \underline{\quad}$ or $\underline{\quad}$

   b $0.5\dot{8}2\dot{3}$ $10\,000 \times \underline{\quad\quad} = \underline{\quad\quad}$
          $-\quad 10 \times \underline{\quad\quad} = \underline{\quad\quad}$
          $\underline{\quad} \times 0.5\dot{8}2\dot{3} =$
          $= $

   c $0.305\dot{2}\dot{1}$ $100\,000 \times 0.305\dot{2}\dot{1} = \underline{\quad\quad}$
          $\underline{\quad\quad} \times 0.305\dot{2}\dot{1} = \underline{\quad\quad}$

3 Convert these recurring decimals to fractions.
   a $0.4\dot{6}$ **b** $0.23\dot{5}$ **c** $0.105\dot{7}$ **d** $0.\dot{4}\dot{5}$ **e** $0.3\dot{1}\dot{8}$
   f $0.204\dot{3}\dot{6}$ **g** $0.\dot{2}6\dot{1}$ **h** $0.2\dot{1}3\dot{4}$ **i** $0.03\dot{2}45\dot{1}$

Review 1 Copy and complete this to find $0.40\dot{7}\dot{2}$ as a fraction in its simplest form.
   $10\,000 \times 0.40\dot{7}\dot{2} = 4072.7272 \ldots$
   $\underline{\quad\quad} \times 0.40\dot{7}\dot{2} = \underline{\quad\quad}$
   $0.40\dot{7}\dot{2} = \underline{\quad\quad}$

Review 2 Another way of converting recurring decimals to fractions is as follows.

Example
$0.12\dot{3}\dot{6}$

1 Write the decimal in recurring form, e.g. $0.12\dot{3}\dot{6}$.

2 To find the numerator of the fraction: write down all the figures after the decimal point and subtract the non-recurring part.

1236 − 12 = 1224

3 To find the denominator: write down as many 9s as there are figures in the recurring part and the same number of zeros as there are digits in the non-recurring part.

9900

The fraction is $\frac{1224}{9900}$.

Then simplify to $\frac{34}{275}$.

Use this method and the method of **Review 1** to find these as fractions in their simplest form.
   a $0.3\dot{6}$ **b** $0.20\dot{3}\dot{7}$

Summary of key points

 We can **convert between fractions, decimals and percentages**.

We can compare proportions by converting them all to either fractions, decimals or percentages.

To compare fractions we write them as decimals or with a common denominator.

Examples $\frac{5}{8} > \frac{4}{7}$ because $0 \cdot 625 > 0 \cdot 571$ (3 d.p.)

$\frac{3}{4} < \frac{4}{5}$ because $\frac{15}{20} < \frac{16}{20}$

 To **add and subtract fractions with different denominators** we use equivalent fractions. We find the LCM of the denominators then write both fractions with this as the denominator.

Examples $2\frac{3}{4} + 1\frac{1}{3} = \frac{11}{4} + \frac{4}{3}$

$= \frac{33}{12} + \frac{16}{12}$

$= \frac{49}{12}$

$= \mathbf{4\frac{1}{12}}$

> When adding, we can add the whole numbers first.

$4\frac{1}{5} - 2\frac{1}{2} = \frac{21}{5} - \frac{5}{2}$

$= \frac{42}{10} - \frac{25}{10}$

$= \frac{17}{10}$

$= \mathbf{1\frac{7}{10}}$

We can add and subtract fractions using the $\boxed{a^{b/c}}$ key on a calculator.

See page 78 for an example.

 When **multiplying fractions** we

1 write whole numbers or mixed numbers as improper fractions

2 cancel if possible

3 multiply the numerators

   multiply the denominators.

> You can cancel any numerator with any denominator.

When we **divide fractions** we use the inverse rule.

To divide by a fraction, multiply by the inverse.

See page 81 for examples.

 We **add and subtract algebraic fractions** using the same methods we use for number fractions.

Examples $\frac{p}{4} + \frac{q}{5} = \frac{5p}{20} + \frac{4q}{20}$ make equivalent fractions with a common denominator of 20

$= \frac{5p + 4q}{20}$

$\frac{4}{x} - \frac{x}{y} = \frac{4y}{xy} - \frac{x^2}{xy}$

$= \frac{4y - x^2}{xy}$

$\frac{4}{x} \xrightarrow{\times y} \frac{4y}{xy}$ $\frac{x}{y} \xrightarrow{\times x} \frac{x^2}{xy}$

> **E** All fractions convert to **terminating or recurring decimals**.
>
> All recurring decimals are exact fractions.
>
> We can write a recurring decimal as a fraction using this method.
>
> *Example* To write $0 \cdot 45\dot{1}\dot{2}$ as a fraction:
>
> $$10\ 000 \times 0 \cdot 45\dot{1}\dot{2} = 4512 \cdot 1212 \ldots$$
>
> multiply so the first digit of the second group of repeating digits is in the tenths place
>
> $$-\quad 100 \times 0 \cdot 45\dot{1}\dot{2} = 45 \cdot 1212 \ldots$$
>
> multiply so the first digit of the first repeating digit is in the tenths place then subtract
>
> $$9900 \times 0 \cdot 45\dot{1}\dot{2} = 4467$$
>
> $$0 \cdot 45\dot{1}\dot{2} = \frac{4467}{9900}$$
>
> $$= \frac{1489}{3300}$$

Test yourself   **Except questions 2, 4 and 9.**

1 Put these in order from smallest to largest. **A**

   $\frac{12}{20}$,   130%, $\frac{35}{25}$,   0·9

2 A factory canteen sells sandwiches, pies, chips and sausage rolls. **A**
   In one week the sales were recorded as follows.

   a What percentage of the takings came
      from chips?

   b What percentage of the food purchased
      was pastries (pies or sausage rolls)?

   c Which item was cheapest to buy?

   d If the price of chips was increased by
      $12\frac{1}{2}$%, what would the percentage
      increase in total takings be?

Item	Number sold	Amount taken
Sandwiches	72	£ 79·20
Pies	85	£106·25
Chips	104	£ 98·80
Sausage rolls	60	£ 69·00
Total	**321**	**£353·25**

3 a In a magazine there are three adverts on the same page.   [SATs Paper 1 Level 6]

> Advert 1 uses $\frac{1}{4}$ of the page
>
> Advert 2 uses $\frac{1}{8}$ of the page
>
> Advert 3 uses $\frac{1}{16}$ of the page

   In **total**, what **fraction** of the page do the three adverts use?
   Show your working.

   b

> Cost of advert.
>
> £10 for each $\frac{1}{32}$ of a page.

   An advert uses $\frac{3}{16}$ of a page. How much does the advert cost?

4 This table shows the average rainfall in mm for a city in January and June in 2002 and 2004.

	2002	2004
January	116	142
June	54	62

  a For 2002 find the rainfall in June as a percentage of the rainfall in January. Give your answer to 1 decimal place.

  b Show whether the rainfall in June as a percentage of the rainfall in January was greater in 2004 or in 2002.

5 Calculate these.

  a $\frac{5}{12} + \frac{2}{3}$ **b** $\frac{5}{12} - \frac{3}{8}$ **c** $\frac{7}{12} + \frac{1}{6} - \frac{5}{8}$ **d** $2\frac{2}{3} + 1\frac{4}{5}$ **e** $3\frac{5}{8} - 1\frac{4}{5}$

6 Find the next term in each of these sequences.

  a $1\frac{7}{8}, 2\frac{1}{2}, 3\frac{1}{8}, 3\frac{6}{8}, 4\frac{3}{8}, ...$ *$\ast$**b** $\frac{2}{3}, 1, 1\frac{1}{3}, 2\frac{2}{3}, 4, ...$

7 In a science class, Amelia's group used $\frac{1}{6}$ of a 60 g jar of sodium carbonate. Simon used $\frac{3}{5}$ of what was left. How many grams were left in the bottle after this?

8 a $8 \times 3\frac{3}{4}$ **b** $2\frac{2}{5} \times 3\frac{1}{8}$ **c** $\frac{2}{5} \times \frac{3}{4} \times \frac{1}{2}$ **d** $1\frac{1}{3} \times 3\frac{2}{5}$ **e** $(1\frac{1}{2})^2$

9 Find these using your calculator.

  a $(3\frac{5}{9} + 1\frac{4}{7}) \times \frac{3}{4}$ **b** $2\frac{3}{8} \times (1\frac{3}{5} - 1\frac{7}{20})$

10 Jenny's bedroom measures $3\frac{1}{2}$ m by $3\frac{1}{4}$ m.

  a What is the area of the floor?

  b A door in the room is $\frac{4}{5}$ m wide. What length of skirting board will be needed to go around the room?

11 Work these out.

  a $8 \div \frac{2}{5}$ **b** $\frac{3}{10} \div 5$ **c** $\frac{5}{6} \div \frac{2}{3}$ **d** $3\frac{1}{4} \div \frac{5}{6}$ **e** $3\frac{1}{3} \div \frac{2}{5}$ **f** $\dfrac{1 - \frac{1}{4}}{1 - \frac{2}{5}}$

12 Simplify these.

  a $\frac{4}{m} + \frac{5}{m}$ **b** $\frac{2x}{3} + \frac{x}{5}$ **c** $\frac{a}{x} - \frac{b}{y}$ **d** $\frac{2p}{m} + \frac{3q}{n}$

13 Convert these to fractions in their simplest forms.

  a $0.7\dot{1}$ **b** $0.05\dot{7}$ **c** $0.3\dot{2}\dot{5}$ *$\ast$**d** $0.64\dot{8}2\dot{3}$

5 Percentage and Proportional Changes

You need to know

✓ percentage changes   page 6

✓ ratio and proportion   page 6

···· Key vocabulary ··

direct proportion,   proportional to,   proportionality,   unitary method

Golden rectangles

Measure the length and width of this rectangle.

Find the ratio   length : width   or $\frac{\text{length}}{\text{width}}$.

It should be very close to 1·6 : 1.

This is called the **golden ratio** (1·618 : 1).

1·618 is an approximate value.

Rectangles with this ratio are said to be 'appealing to the eye'.

Buildings such as The Parthenon in Greece are built to this ratio.

Construct a golden rectangle as follows.

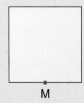

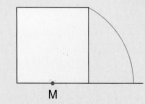

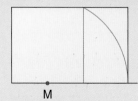

Draw a square (any size). Mark the mid-point of the base.

Extend the base. Place your compass point on M and draw an arc as shown.

Draw the rectangle so that it just encloses the arc.

Percentage change

Remember

We calculate a **percentage increase or decrease** using a single calculation.

Example Dallas bought an antique rocking horse for £5300.
She sold it for a profit of $12\frac{1}{2}\%$.

100%	$12\frac{1}{2}\%$

$$112\tfrac{1}{2}\% = 100\% + 12\tfrac{1}{2}\%$$

Selling Price = $112\frac{1}{2}\%$ of £5300

$\qquad\qquad\quad = 1{\cdot}125 \times £5300 \qquad 112\tfrac{1}{2}\% = 1{\cdot}125$

$\qquad\qquad\quad = $ **£5962·50**

We calculate **an increase or decrease as a percentage** by
- finding the fractional increase
- converting this to a percentage.

% increase $= \dfrac{\text{actual increase}}{\text{original amount}} \times$ **100%**

% decrease $= \dfrac{\text{actual decrease}}{\text{original amount}} \times$ **100%**

Worked Example

In an experiment, 39 g of powder decreased to 28 g.
What percentage decrease is this?

Answer

Actual decrease in mass = 39 g − 28 g

$\qquad\qquad\qquad\qquad\quad = 11$ g

% decrease $= \dfrac{\text{actual decrease}}{\text{original amount}} \times 100\%$

$\qquad\qquad\; = \dfrac{11}{39} \times 100\%$

$\qquad\qquad\; = $ **28% (nearest percentage)**

We **calculate the original amount**, given the final amount and the percentage or fractional
increase or decrease, using inverse operations, the unitary method or using algebra.

Worked Example

Sinead bought a dishwasher in this sale and paid £445·25.
What was the original price?

35% off
all appliances

Answer

If 35% has been taken off the price, Sinead paid 65% of the original price.

Using inverse operations	**Finding 1% first**	**Using algebra**
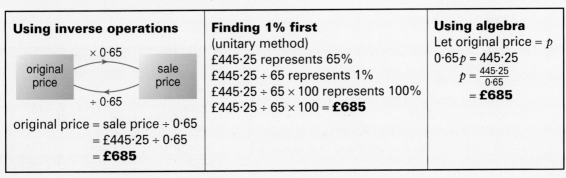	(unitary method) £445·25 represents 65%	Let original price = p
	£445·25 ÷ 65 represents 1%	$0{\cdot}65p = 445{\cdot}25$
original price = sale price ÷ 0·65	£445·25 ÷ 65 × 100 represents 100%	$p = \dfrac{445{\cdot}25}{0{\cdot}65}$
$\qquad\qquad\quad$ = £445·25 ÷ 0·65	£445·25 ÷ 65 × 100 = **£685**	$\qquad$ = **£685**
$\qquad\qquad\quad$ = **£685**		

Number

Exercise 1

1 Chandia bought a digital camera in a 20% off sale.
  The original price of the camera was £1650.
  a How much did Chandia pay for the camera?
  b Chandia sold the camera 3 months later and made a 15% loss.
    How much did she sell it for?

2 a How much, including VAT, did the stereo cost before the sale?
  b How much did the stereo cost during the sale?

3 The population of London in 1901 was 6 506 889.
  Over the next 100 years, the population increased by 10·22%.
  What was the population of London in 2001?

4 This table shows the membership numbers
  in 1960 and 2000 for four tennis clubs.
  a In 1960, for every member Champs had,
    how many did Central have?
  b Patrick thinks that from 1960 to 2000
    Country membership increased by 100%.
    Is Patrick right?
    Explain your answer.

Club	Membership 1960	Membership 2000
Reds	222	642
Central	1558	3402
Country	13	26
Champs	166	276

5 A report on the number of teenagers in an area said:
      There are 139 000 teenagers.
      Almost 55% of them are female.
  a The percentage was rounded to the nearest whole number, 55.
    What is the smallest value the percentage could have been, to one decimal place?
    A 54·1%   B 54·2%   C 54·3%   D 54·4%   E 54·5%
    F 54·6%   G 54·7%   H 54·8%   I 54·9%
  b What is the smallest number of female teenagers that there might be in the area?
  c An appendix to the report gave some exact figures for two
    years.
    Calculate the percentage increase in the number of teenagers
    in the area.

Number of teenagers	
1995	136 540
2000	139 123

  d This table gives the percentage of teenagers in 2000 and 2004
    who were female.
    Decide which of these statements is true.
    A In 2004 there were more female teenagers than in 2000.
    B In 2004 there were fewer female teenagers than in 2000.
    C There is not enough information to tell whether there were more or fewer female
      teenagers.
    Explain your answer.

2000	54·7%
2004	54·6%

6 This table shows what was sold at the school fair.
  a What percentage of items sold were
    second-hand?
  b What percentage of the takings were from the
    second-hand stall?
  c Was the average price of items sold at the
    craft stall or at the second-hand stall cheaper?

Stall	Number of items sold	Takings
Cake	84	£231·50
Clothing	186	£927·80
Craft	51	£464·50
Second-hand	364	£1242·25
Total	685	£2866·05

7 Joel used some scales to weigh powder for a science experiment.
He was told the powder might weigh 20% more or 20% less than the reading shown.
The powder weighed 65 g on the scales.
What is the greatest and smallest mass the powder might be?

8 How much cheaper is it to pay cash?

9 a One calculation below gives the answer to the question   [SATs Paper 2 Level 7]
   What is 70 increased by 9%?
   a Write down the correct one.

   (70 × 0·9)   (70 × 1·9)   (70 × 0·09)   (70 × 1·09)

   b Choose one of the other calculations.
   Write a question **about percentages** that this calculation represents.
   c Now do the same for one of the remaining two calculations.
   d Copy and fill in the missing decimal number.
   To decrease by 14%, multiply by _____

10 Mel bought a plant.
The height of the plant increased 75% to 14 cm in the first month.
How high was the plant when Mel bought it?

11 a Paul wants to buy a pair of trousers that were £89·70 before
   the sale.
   He estimated he would save about £14 in the sale.
   How might he have estimated this?
   b Paul's brother bought trousers for £85 in the sale.
   How much were they before the sale?

12 A dog at the RSPCA weighs 18·4 kg.
This is an increase of 80% on its mass when it first arrived.
How much did the dog weigh when it first arrived at the RSPCA?

13 A shop had a sale. All prices were reduced by 15%.
A pair of shoes cost £46·75 in the sale.
What price were the shoes before the sale?

14 a Both sides of the rectangle were increased by 22%.
   What is the area of the new rectangle?
   b The area of the square is increased by 31%.
   What is the length of each side of the new square?

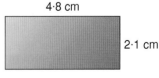

4·8 cm

Area = 30 cm²

2·1 cm

15 A company spent £860 on advertising.
As a direct result, the company's profits rose 15% to £6500.
Was the advertising worthwhile from a financial point of view?
Justify your answer.

16 Tandia read that

The population of a town had increased 10% to 8950.
She worked out that before the increase the population was $\frac{8950}{1.1} = 8136.\dot{3}\dot{6}$
In fact the population before the increase was 8136.
Why is Tandia's answer different?

∗17 VAT of $17\frac{1}{2}$% was added to an electricity bill.
What percentage of the total bill is VAT?

∗18 A two-month-old puppy weighs 3·6 kg.
It has increased in mass by 450% since birth.
How much did the puppy weigh at birth?

∗19 How much water must be added to 24 litres of a 80% concentrate solution to make a 70% concentrate solution?

Review 1   Toystores bought children's cars for £85. They sold them to make a profit of 12%.
a   What was the selling price of the cars?
b   In the post-Christmas sale all toys were reduced by 15%. What did the car cost in the sale?

Review 2
a   The school roll increased by $12\frac{1}{2}$% from 1998 to 2003. If the roll was 1272 in 1998, what was it in 2003?
b   In 2000 the population of Littletown was 21 725. In 2003 the population fell to 19 869. What percentage decrease is this? Round your answer sensibly.

Review 3   The price of a new car is £18 000 plus VAT of $17\frac{1}{2}$%.
a   What is the cash price including VAT?
If I purchase the car on hire purchase the deposit is 18% of the cash price and the monthly repayments are £740 a month for 3 years.
b   What is the total cost of buying the car on hire purchase?
c   What percentage more do you pay by hire purchase than if you pay cash?

Review 4   Jeremy sold his VCR for £247·95. He made a loss of 13%.
How much did he pay for his VCR?

Review 5   In the tropics plants grow very quickly. A plant was planted before the wet season. At the end of the wet season it had grown by 225% to 58·5 cm. What was the original height of the plant?

Puzzle

1   The difference between decreasing a number by 7% and increasing it by 8% is 75. Find the number.

2   A sum of money is to be divided between three cousins. The first gets 20% of the money, the second gets 60% of what is left and the third gets £24.
How much do the other two cousins get?

Repeated percentage change

Worked Example
A shop dropped its prices for a sale by 30%.
Once the sale had finished, 10% was added back to the sale price.
By what one number should the shop multiply the pre-sale price to get the post-sale price?

Answer
Sale price = 70% of pre-sale price
$\qquad\quad = 0.7 \times$ pre-sale price
post-sale price = 110% of sale price
$\qquad\qquad\quad = 110\%$ of $(0.7 \times$ pre-sale price)
$\qquad\qquad\quad = 1.1 \times 0.7 \times$ pre-sale price
$\qquad\qquad\quad = \mathbf{0.77 \times}$ **pre-sale price**
The shop should multiply the pre-sale price by **0.77** to get the post-sale price.

If the multipliers needed are $a\%$ and $b\%$ then the total multiplier is $ab\%$.

We use **index notation** when repeating the same percentage change, such as when calculating compound interest.

Worked Example
Kim put £165 into her bank account with an annual compound interest rate of 5%.
How much will she have at the end of 5 years?

Answer
At end of first year she will have $\qquad$ £165 × 1.05.
At end of second year she will have $\quad$ £165 × 1.05 × 1.05.
At end of third year she will have $\qquad$ £165 × 1.05 × 1.05 × 1.05.
At end of fourth year she will have $\quad$ £165 × 1.05 × 1.05 × 1.05 × 1.05.
At end of fifth year she will have $\qquad$ £165 × 1.05 × 1.05 × 1.05 × 1.05 × 1.05.

We calculate this as $165 \times (1.05)^5$.
Key 165 × 1.05 x^y 5 = to get **£210.59** (nearest penny).

Discussion

Make a general statement about what the total multiplier would be if we increased an amount by $a\%$, decreased it by $b\%$ and then increased it by $c\%$.

Exercise 2

1 What one number would you multiply by to
   a increase by 8% then 12%
   b increase by 10% then decrease by 10%
   c decrease by 8% then decrease by 10%?

2 *Benny's* bought clothing from a manufacturer then added 40% mark up.
In a sale, *Benny's* reduced all clothing by 40%.
  a What would a coat that *Benny's* bought from the manufacturer for £159 cost in the sale?
  b Did *Benny's* make a profit on items sold in the sale?
  Explain.
  c What single number should *Benny's* multiply the price they paid by to get the price in the sale?

3 A baby octopus increases its mass by about 8% each day.
It began life with a mass of 220 g.
  a How much will it weigh after a day?
  b How much will it weigh after 3 days?
  c What single number (including an index) could you multiply 220 by to find its mass after a week?
  *d If it were to increase its mass by between 7% and 9% each day, what would the upper and lower bounds be for its mass after a week?

4 Andrea deposited £1100 in her bank account at an annual compound interest rate of 6%.
How much will she have in total after 3 years?

5 A shop reduces its prices by 15% for a sale.
On the last day of the sale, the prices are reduced by a further 20%.
Show that the prices on the last day are reduced by 32% in total.

SALE
15% off
LAST DAY
FURTHER 20% OFF

6 a Work out the total amount, to the nearest penny, in these bank accounts given the initial deposit (P), the annual compound interest rate (R) and the number of years (n),
  i $P = £5000$, $R = 4\%$, $n = 5$ years   **ii** $P = £6843$, $R = 6\%$, $n = 10$ years
  iii $P = £8650$, $R = 5\frac{1}{2}\%$, $n = 4$ years   **iv** $P = £3845$, $R = 6\frac{1}{2}\%$, $n = 6$ years
  v $P = £836 \cdot 55$, $R = 5\frac{1}{4}\%$, $n = 8$ years   **vi** $P = £72\,516$, $R = 5 \cdot 7\%$, $n = 7$ years
  *b Which of these formulae would give the amount, A, after n years at $R\%$ compounding interest if P is invested?
  A $A = \frac{PnR}{100}$ **B** $A = P(1 + \frac{R}{100})^n$ **C** $A = (\frac{PR}{100})^n$

7 In its first year, a company made a profit of £6 357 840. In **each** of the next three years the profit increased by 10%.
  a What profit did the company make in the third year?
  b What profit did the company make in the fourth year?
  c What total profit did the company make in its first four years of operation?

8 The value of Merilyn's apartment went up and down in value each time she had it valued.
The first time it had increased 10%, the second time it had increased 10%, the third time it had decreased 10% and the fourth time it had decreased 10%.
Find the percentage change over the four valuations.

9 Rupert bought a horse, Harmony Way, and then sold it to Karina a year later for 5% profit.
Karina sold the horse two years later for a 10% profit.
Which of these is true?
  A Harmony Way increased in value by 15% over three years.
  B Harmony Way increased in value by more than 15% over three years.
  C Harmony Way increased in value by less than 15% over three years.

10 In **1995**, the Alpha Company employed 4000 people. [SATs Paper 1 Level 8]
For **each** of the next **2 years**, the number of people employed increased by **10%**.

1995	employed 4000 people
1996	employed 10% more people
1997	employed 10% more people

a Tony said:

'Each year, the Alpha company employed another 400 people.'
Tony was wrong. Explain why.

b Write the calculation below which shows how many people worked for the company in **1997**.

$4000 \times 0{\cdot}1 \times 2$ $4000 \times 0{\cdot}1^2$ $(4000 \times 0{\cdot}1)^2$

$4000 \times 1{\cdot}1 \times 2$ $4000 \times 1{\cdot}1^2$ $(4000 \times 1{\cdot}1)^2$

c Look at these figures for the Beta Company:

1995	employed n people
1996	employed 20% **fewer** people
1997	employed 10% **more** people

Write an expression using n to show how many people the company employed in **1997**.
Show your working and write your expression as simply as possible.

11 After a 10% freight charge and then a 10% fee for insurance had been added, the price of a couch was £363. What was the original price?

***12** The mass of packets of Trent's icing sugar was increased by 25% for a sales promotion. William used $\frac{1}{5}$ of a packet to ice a cake. Victoria weighed the remainder and found there was 500 g left. How many grams were in packets of Trent's icing sugar before the promotion?
A 1000 g   **B** 695 g   **C** 500 g   **D** 417 g

Review 1   Francie bought platters for £22 each. She sold them to make a profit of 35%.
a At what price did she sell the platters?
b In the sale she reduced the price by 30%. What did the platters cost in the sale?
c Did Francie make a profit on the platters sold in the sale? Explain.
d What single number should Francie multiply the price she bought the platters for by to obtain the sale price?

Review 2   In April, an airline increases all of its fares by 20%. In September of the same year it reduces all its fares by $12\frac{1}{2}\%$.
What is the net percentage increase in fares from April to September?

Review 3   Find the total amount, to the nearest penny, if
a £6000 is invested for 8 years at a compound interest rate of $5\frac{1}{2}\%$
b £7500 is invested for 5 years at a compound interest rate of $6\frac{1}{4}\%$.

Review 4   Bob's building business is growing. Each month he employs 8% more workers.
a If he starts the year with 175 employees, how many workers will he have after 3 months?

He continues to employ at this rate.
b How many employees will he have at the end of the year?
c Which of these calculations could Bob have done to work out how many employees he will have at the end of the year?
A $175 \times 1{\cdot}08$ **B** $175 \times 0{\cdot}08 \times 12$ **C** $175 \times 1{\cdot}08^{12}$ **D** $175 + 1{\cdot}08^{12}$

Review 5   Sharon worked in a dress shop. She priced all dresses by adding $17\frac{1}{2}$% VAT plus a profit. To do this she multiplied the purchase price by 1·316.
  a   What is the price of a dress costing the shop £75?
* **b**   What percentage profit is the shop owner making?

 Puzzle

  1   The original price is £15. After two mark-ups the final price is £20·70. If one of the mark-ups is 15%, what is the other?

  2   What percentage discount do the staff at Hats Galore get if they can buy at cost, hats that have been marked-up 40%?

  3   As sales increased, the Baby Factory reduced its prices. What percentage change in the amount of income would occur if sales increased by 40% but prices dropped by 5%?

Solving ratio and proportion problems

Remember
Ratio compares part to part.

We can **compare ratios** by putting them both into the form $1 : m$ or $m : 1$.

Example   The ratio of goals scored to shots at goal for two netball players is given.

	goals scored : shots at goal
Julie	8 : 12
Darlene	15 : 24

We can compare these by putting them both in the form $1 : m$.

We only write ratios with decimals when we are comparing them, otherwise we always write them with whole numbers.

Julie scored 1 goal for every 1·5 shots.
Darlene scored 1 goal for every 1·6 shots.
Julie is a better goal shooter.

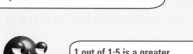

1 out of 1·5 is a greater success rate than 1 out of 1·6.

When two variables are **directly proportional**, if one doubles, the other doubles, if one increases by a factor of 4, so does the other and so on.

Example   If in 100 g of cereal, there is 3 g of fat, then in 500 g of cereal (5 times as much) there will be 15 g (5 times as much) of fat.
In x g of cereal there will be $\frac{x}{100} \times 3$ g of fat.
$\frac{x}{100}$ is called the constant **multiplier**.

Worked Example

The ratio of red paint to yellow paint in orange paint is 5 : 3.

How much red paint is needed to mix with 400 mℓ of yellow paint?

Answer

$$\text{red : yellow} = 5 : 3$$
$$\text{So red paint} = \tfrac{5}{3} \text{ of yellow paint}$$
$$(\text{similarly yellow paint} = \tfrac{3}{5} \text{ of red paint})$$
$$\text{red paint needed} = \tfrac{5}{3} \text{ of yellow paint needed}$$
$$= \tfrac{5}{3} \times 400$$
$$= \mathbf{666\tfrac{2}{3} \ m\ell \text{ or } 666 \cdot 7 \ m\ell \ (1 \ d.p.)}$$

Unitary method

red paint needed for 1 mℓ of yellow paint $= \tfrac{5}{3}$

red paint needed for 400 mℓ of yellow paint $= \tfrac{5}{3} \times 400$
$$= \mathbf{666 \cdot 7 \ m\ell \ (1 \ d.p.)}$$

Constant multiplier

amount required : amount given $= 400 : 3$

constant multiplier $= \tfrac{400}{3}$

red paint required for 400 mℓ yellow paint $= \tfrac{400}{3} \times 5$
$$= \mathbf{666 \cdot 7 \ m\ell \ (1 \ d.p.)}$$

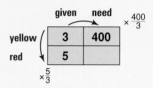

Discussion

● This rectangle is enlarged by a scale factor of 2.
  What is the ratio of corresponding lengths on the enlarged
  rectangle and the original rectangle? **Discuss**.

4 cm

2 cm

● Triangle A is enlarged to give triangle B.

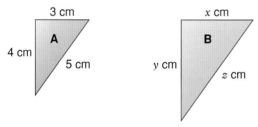

Corresponding lengths on this triangle and its enlargement are in the ratio 2 : 3.
What are the values of x, y and z? **Discuss**.

● The green cuboid is enlarged to the purple cuboid.

Corresponding lengths on the enlargement and the original cube are in the ratio 4 : 1.
What is the scale factor for the enlargement? **Discuss**.
What are the values of l, m and n? **Discuss**.
What is the ratio of the areas of the end faces? **Discuss**.
What is the ratio of the volumes of the cuboids? **Discuss**.

Link to
enlargement.

Number

Exercise 3

1 Two parts of this square design are shaded black.
Two parts are shaded grey.

[SATs Paper 1 Level 6]

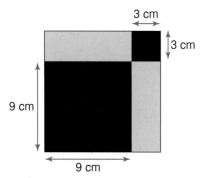

3 cm

3 cm

9 cm

9 cm

Show that the ratio of black to grey is 5 : 3.

2 Write these as ratios in their simplest form.

  a $1\frac{1}{2} : 2$ **b** $3\cdot5 : 3$ **c** $\frac{1}{4} : 1\frac{1}{4}$ **d** $4\cdot5 : 2\cdot5$

  e $1\frac{3}{4} : 1\frac{1}{2}$ **f** 3 mm : 1 cm   **g** 15 min : 1 hour   **h** 350 mℓ : 1 litre

3 Use a method of your choice to do these. Round your answers sensibly.

  a 5 miles is approximately equal to 8 km.
  About how many kilometres are equal to 138 miles?

  b 10 pieces of fish cost £8·60.   **c** 6 filled baps cost £9.
  How much do 18 pieces cost?   How much do 15 filled baps cost?

  d 7 posters cost £15·75.   **e** 12 apples cost £5·64.
  How much would 12 cost?   How much will 97 apples cost?

4 The angles in a triangle are in the ratio 1 : 2 : 3.
Find the sizes of the three angles.

5 Julie makes a potting mix for her plants from loam, peat and sand
in the ratio 7 : 3 : 2 respectively.
She uses 5 kg of peat to make the compost.

  a How much loam and sand does she use? Round sensibly.

  b How much of each does she use to make 30 kg of compost?

6 Two parts grout paste to 3 parts sand makes grouting mix for tiling.

  a How much of each is needed to make 1500 cm^3 of grouting mix?

  b Paul has 100 cm^3 of grout paste and 200 cm^3 of sand.
  What is the maximum amount of grouting mix he can make?

7 This shows the prices of some sweets.

How much would Mel pay for

  a 800 g of white stars   **b** 800 g of black cats

  c 500 g of red hearts and 1·2 kg of black cats.

  d 0·4 kg of white stars and 0·3 kg of red hearts?

8 For £5 you would get about US $9·10.
How many US dollars would you get for these amounts?
a £8   **b** £12   **c** £160   **d** £2750

9 The ratio of flour to sugar in a pastry mix is 3 : 2.
How much sugar is needed to mix with these amounts of flour?
a 500 g   **b** 1·2 kg   *c 2$\frac{7}{8}$ kg

10 Write the ratio 84 : 127 in the form 1 : m.

11 Alexander made two bowls of punch each using different amounts of punch concentrate
and lemonade.
The ratio of punch concentrate to lemonade in the first bowl was 3 : 8 and in the second
bowl was 5 : 13.
Which bowl had a greater proportion of punch concentrate? Explain.

12 Logan was growing a plant for a science experiment.
On 1 August, he measured the height as 11·5 cm.
On 1 September, the height had increased by **30%**.
What was the ratio of the height of the plant on 1 August to the height on 1 September?
Write the ratio in its simplest form without decimals or fractions.

13 A packing company has two sizes of container.
One is an enlargement of the other.
Corresponding lengths are in the ratio 1 : 3.
a Find the values of x and y.
b Find the ratio of the areas of the dark
shaded rectangles on A and B.
c Find the ratio of the volumes of the cuboids.

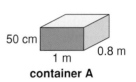
container A

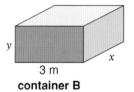
container B

14 Toby timed the advertisements and programmes on TV
on Friday and Saturday. He wrote each as a ratio.
By changing each ratio to the form 1 : m, decide which
day had the greater proportion of advertisements.
Explain.

Friday
ads : programmes
97 : 432

Saturday
ads : programmes
47 : 211

15 A large ocean liner usually travels at about 38 miles per hour.
It uses about 4·5 litres of fuel for every 15 feet travelled.
Calculate, to the nearest litre, how much fuel it would use in
an hour, travelling at its usual speed.

1 mile = 5280 feet

16 Rosalind and Sirah both guessed the height of a tall tree.
Rosalind guessed it was 82 m high.
Sirah guessed it was 97 m high.
The tree cast a shadow of 12·1 m.
At the same time they measured the shadow cast by a 2 m high fence.
It was 26·7 cm long.
Whose guess was closest?

17 The pentagonal bases of these two
prisms are congruent.
Prism A is 8 cm high and has a volume
of 400 cm³.
What is the volume of prism B?

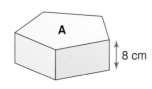
A
8 cm

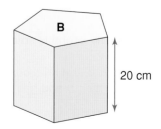
B
20 cm

Number

*18 Look at the table: [SATs Paper 2 Level 8]

	Earth	Mercury
Mass (kg)	5.98×10^{24}	3.59×10^{23}
Atmospheric pressure (N/m²)		2×10^{-8}

What is the **ratio** of the mass of Earth to the mass of Mercury?
Write your answer in the form $x : 1$.

*19 Olivia's group played a game of 'Guess my length'.
This table shows the results.
They divided each guess by the actual length to get an
accuracy ratio.

Name	Guess	Actual length
Olivia	6 cm	7·2 cm
Grace	40 cm	29 cm
Ryan	2 m	3 m
Keith	20 cm	25 cm

Example   Olivia's accuracy ratio = 0·83 : 1 (2 d.p.)

a   Find the accuracy ratio for each of the others.
b   Which guess was most accurate? Explain your answer.
c   Which guess was least accurate?

*20 A 300 mℓ bowl is filled with cream and milk in the ratio 4 : 1.
Another 300 mℓ bowl is filled with cream and milk in the ratio 3 : 1.
If both bowls are poured into one larger bowl, what is the ratio of cream to milk?

Review 1
a   In 5 hours Keri earns £52·50. How much will she earn in 18 hours?
b   A car travels 120 km using 35 ℓ of fuel. How far will the car travel on 63 ℓ of fuel?

Review 2
a   The angles of a triangle are in the ratio 2 : 3 : 4.
Find the sizes of the three angles.
b   A punch needs 375 mℓ orange juice, 100 mℓ lime juice and 1·5 ℓ ginger ale. Write the ratio
orange juice : lime juice : ginger ale in its simplest form.
c   At a party the ratio of smokers to non-smokers was 3 : 7. If there were 12 smokers at the
party, how many were there altogether?

Review 3   Concrete is made from cement and gravel in the ratio 1 : 6.
a   How much cement is needed if John has 8 bags of gravel?
b   Jack needs 10 m³ of concrete. How much of each ingredient is needed?

Review 4   For £4 you would get about $9·50 Australian.
a   How many Australian dollars would you get for £18?
b   How many pounds would you get for $18 Australian?

Review 5   Aunt Betty's recipe makes 24 cookies. It needs 2 cups flour, 120 g butter and
$\frac{1}{2}$ cup icing sugar.
If she wants to make 40 cookies, how much of each ingredient will she need?

Review 6   One fruit concentrate is sold in 250 mℓ packs. The instructions are 'add water to
make 1 litre of fruit drink'.

Another brand is sold in 400 mℓ packs. The instructions are 'add 1 litre of water to make fruit
drink'.
a   Write the ratio *concentrate : water* in the form 1 : n for each brand.
b   Which brand has the greater proportion of water?

Review 7 Adam wants to estimate the height of a flagpole. He paces out the length of its shadow and finds it is $8\frac{1}{2}$ paces. He paces his friend's shadow to be 3 paces. His friend is 174 cm tall. What is the approximate height of the flagpole?

Review 8

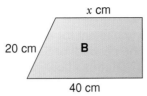

Trapezium B is an enlargement of trapezium A.
a Find the values of x and y.
*b If the area of A is 91 cm², what is the area of B?

Proportional relationships

There is 25 g of protein in 100 g of cheese spread.
This table shows the amount of protein in different amounts of the spread.

Spread	100 g	200 g	300 g	400 g
Protein	25 g	50 g	75 g	100 g

The ratio   *mass of protein : mass of spread*   is *always* 25 : 100 or 1 : 4.
It is constant.

Mass of protein is directly proportional to mass of spread.
We write   mass of protein ∝ mass of spread.   ∝ means 'is proportional to'.

When variables are directly proportional the graph is a straight line through the origin.

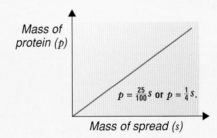

Discussion

● This table shows the distance travelled by a car.
For $t = 4$, the value of the ratio $\frac{s}{t}$ is $\frac{8}{4} = 2$.
What is the value of the ratio $\frac{s}{t}$ for other values of t?
Is t directly proportional to s? **Discuss**.

t (min)	1	2	3	4	5
s (km)	2	4	6	8	10

Can the relationship between s and t be written as $s = kt$ where k is the constant of proportionality? If so, what is the value of k?
What does the distance/time graph look like?
Discuss.

- To make scones, $\frac{1}{2}$ cup of milk is needed for every 2 cups of flour.
  Discuss how to fill in this table.

Cups of flour (f)	2	4	6	8	10
Cups of milk (m)	$\frac{1}{2}$				

Is *cups of flour : cups of milk* constant for each set of values on the table?
How many cups of milk would be needed for 5 cups of flour?
Is cups of milk $\propto$ cups of flour? **Discuss** how you can tell.
What does the graph of m plotted against f look like?
Discuss.
What would the equation of the line be? **Discuss**.

Some variables are **inversely proportional** to each other.
As one increases by a constant amount the other decreases by a constant amount.

Example   Joe chooses an amount to save each week.
The time taken to save £1000 is **inversely proportional** to the amount Joe chooses to save.
If the amount he chooses doubles then the time taken halves.

If y is inversely proportional to x we say $y \propto \frac{1}{x}$.
$y = k \times \frac{1}{x}$ $y = \frac{k}{x}$.
This means the value of xy is constant.

Discussion

The following table gives the time taken to travel 200 km at various average speeds.

Average speed, v (km/h)	20	40	50	80	100
Time, t (hours)	10	5	4	2·5	2

For $t = 10$, $vt = 200$.
What is vt for other values of t?
Is v inversely proportional to t? **Discuss**.

Some variables are proportional to the **square** of another variable.

Example   The energy of a moving object is proportional to the square of its speed.
If the speed doubles, the energy increases by a factor of 2^2 or 4.
If the speed increases by a factor of 3, the energy increases by a factor of 3^2.

If y is proportional to the square of x we say $y \propto x^2$.
The ratio $y : x^2$ is constant.

$y = kx^2$.

This means $\frac{y}{x^2}$ is a constant because $\frac{y}{x^2} = k$, where k is a constant.

Discussion

● Copy and fill in this table for the perimeter, P, of squares of side length l.

l	2	4	5	8	10	15	20	30
p	8							

Is the ratio $P:l$ constant?
Is P directly proportional to l or inversely proportional to l? How can you tell?
What does the graph of P versus l look like? **Discuss**.

● This table shows the time taken to earn £100 at different rates of pay.

Hourly rate, r (in £)	2	4	5	8	10
Hours, h	50	25	20	12·5	10

What happens to the number of hours taken if the hourly rate doubles?
Is h directly proportional to r?
Work out $r \times h$ for each set of values in the table.
What do you notice? **Discuss**.
Is h inversely proportional to r? How can you tell?

● A ball is shot into the air at various speeds, v, and the height it reaches, s, is measured.
This table gives the results.
Is the ratio $s:v$ constant?

v (m/s)	5	10	15	20	25	30
s (m)	1·25	5	11·25	20	31·25	45

Work out the values of v^2 from the table.
Is the ratio $s:v^2$ constant? **Discuss**.
What can you say about the relationship between s and v?

● **Discuss** how to work out from a table of values, if the variables are directly proportional, inversely proportional or if one variable is proportional to the square of the other.

Worked Example
Rob wanted to know if there was a relationship between the volume of gas inside a sealed container and the pressure.
He did an experiment and recorded these results.

Pressure, p (N/m^2)	50	100	200	400
Volume, v (m^3)	16	8	4	1

Is pressure directly proportional or indirectly proportional to volume? How can you tell?

Answer
As pressure increases, volume decreases so it is possible that pressure is inversely proportional to volume.
To test this, find the value of vp for each set of values given in the table.
These are 800, 800, 800, 800
so $v \propto \dfrac{1}{p}$.

We can test if one variable is directly proportional to another using a spreadsheet.

Example To test if weight is proportional to mass, input values of weight and mass into a spreadsheet and find if the ratio $\dfrac{\text{weight}}{\text{mass}}$ is constant.

	A	B	C	D	E	F	G	H	I
1	Mass	2	5	9	15	46	79	104	123
2	Weight	19.6	49	88.2	138	544.8	774.2	1019.2	1205.4
3	Weight/Mass	=B2/B1	=C2/C1	=D2/D1	=E2/E1	=F2/F1	=G2/G1	=H2/H1	=I2/I1

We can plot the graph of mass versus weight using a **graph plotter**.

> ⭐ **Practical**
>
> **You will need** a spreadsheet package.
> Ask your teacher for the **Proportional Relationships** ICT worksheet.

Exercise 4

1 For each of these write down the relationship between the two variables using $\propto$.
  a Pressure (P) is proportional to Force (F)
  b The surface area (SA) of a sphere is proportional to the square of its radius (r)
  c The wavelength of sound (λ) is inversely proportional to the frequency (f).
  ***d** The intensity of light (I) is inversely proportional to the square of the distance to the light source (d)

2 For each of the following pairs of variables, which is true?
  A $y \propto x$ **B** $y \propto \frac{1}{x}$ **C** $y \propto x^2$

> Check if the ratio $\frac{y}{x}$ or $\frac{y}{x^2}$ is constant or if xy is constant.

You could use a spreadsheet to check.

  a

x	0·1	0·7	0·9	1·3	2·1	4·3
y	0·3	2·1	2·7	3·9	6·3	12·9

  b

x	5	10	15	20	25	30
y	36	18	12	9	7·2	6

  c

x	3	4	5	6	7	8
y	18	32	50	72	98	128

  d

x	4	8	12	16	20	24
y	48	24	16	12	9·6	8

3 Daniel drew some parallelograms between a set of parallel lines.
This meant they all had the same height.
He filled in this table of values for the base and area of each parallelogram.

Base, b (cm)	3	2·4	5·6	0·6	1·3
Area, A (cm^2)	24	19·2	44·8	4·8	10·4

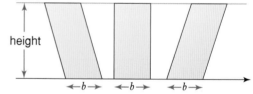

  a Write the relationship between A and b as $A \propto$ _____.
    Justify your answer.
  b Find $\frac{b}{A}$ for each set of values in the table.
  c $A = kb$. What is the value of k, the constant of proportionality?

4 The formula for the circumference of a circle is $C = \pi d$.
  a What happens to C if d is doubled?
    What happens to C if d is halved?
  b Write the relationship between C and d as $C \propto$ _____.
  c What would the graph of C versus d look like?

5 The area (A) and radius (r) of some circles are given.
  a Calculate $\frac{A}{r}$ for each set of values in the table.
    The value of A has been rounded to 1 d.p.
    Is the ratio $\frac{A}{r}$ close to being a constant value?
    Is A directly proportional to r?

r cm	1·2	3	4·6	8·2	9·6	12
A cm^2	4·5	28·3	66·5	211·2	289·5	452·4

  b Calculate Ar for each set of values. Is A inversely proportional to r?
  c Calculate r^2 for each value of r given in the table.
  d Calculate $\frac{A}{r^2}$ for each set of values. Round your answers to 1 d.p. Is it a constant value?
    Is A directly proportional to r^2?
  ***e** Using your answer to **d** write a formula relating A and r? What do you notice about k?

6 Simone has £60. She wants to take some friends to a concert. The tickets range from £5 for the cheapest seats to £20 for the best seats. The table gives the number of people Simone can buy tickets for at each different ticket price.

Price, P (in £)	20	15	12	10	5
Number, n	3	4	5	6	12

   a Write the relationship between n and P using a $\propto$ sign.

 * **b** Write a formula for the relationship between P and n.

*7 This table gives the frequency and wavelength of some radio waves.

Wavelength, λ (metres)	212·0	384·0	485·5	355·8
Frequency, f (cycles per sec)	$1·42 \times 10^6$	$7·84 \times 10^5$	$6·2 \times 10^5$	$8·46 \times 10^5$

Link to Science

Which of these is true? Justify your answer.

   A $\lambda \propto f$ **B** $\lambda \propto \dfrac{1}{f}$ **C** $\lambda \propto f^2$

Note The wavelengths and frequencies have been rounded and so when looking for a constant you will only get an approximate constant ratio.

*8 Find the relationship between current and power in an electrical circuit from this table of values.

Power, P (watts)	108	243	588	867	1452
Current, I (amps)	3	4·5	7	8·5	11

   a Write the relationship as $P \propto$ _____.

   b Write an equation for the relationship $P =$ _____

*9 **a** Tandia wrote down the height of a plant after 1, 2 and 3 weeks.
     This table gives her results.
     She said 'The plant will be 2·6 m tall after one year'.
     Is Tandia correct? Explain.

Week	1	2	3
Height (cm)	5	10	15

   b Give two real-life examples where the variables appear to be directly proportional for a few values but in fact are not.

Review 1   Decide whether $y \propto x$, $y \propto x^2$, $y \propto \dfrac{1}{x}$ for each of the following tables.

a

x	7	9	15	22	30
y	12·25	20·25	56·25	121	225

b

x	0·2	0·3	0·5	1·2	2·0
y	30	20	12	5	3

c

x	0·2	0·5	0·8	1·2	1·8
y	0·5	1·25	2	3	4·5

Review 2   The radius and volume of a number of cylinders of the same height are measured. The results are given in this table. The volume has been rounded to 1 d.p.

Radius, r (cm)	1·2	1·8	2·5	3·2	4·8
Volume, V (cm³)	27·1	61·1	117·8	193	434·3

   a Calculate $\dfrac{V}{r}$ for each set of values in the table.
     Is V directly proportional to r?

   b Calculate Vr for each set of values.
     Is V inversely proportional to r?

   c Calculate r^2 for each value of r in the table then calculate $\dfrac{V}{r^2}$.
     Is V directly proportional to r^2?

*Review 3   Some molecules are made out of 2 atoms. The moment of inertia, I, and the distance between the nuclei of the atoms, r, is given for four such molecules.

Distance, r (10^{-8} cm)	0·90	1·28	1·42	1·62
Moment of inertia, I (10^{-4} g cm^{-2})	1·34	2·66	3·31	4·31

You will learn about moment of inertia in advanced physics.

Find the relationship $I \propto$ _____.

---

*Discussion

● For each statement, **discuss** whether the quantities are likely to be directly proportional or inversely proportional or not proportional.

The mass of a sheet of paper and its area.
The mass of a sheet of paper and its length, if the width is always the same.
The number of chocolates in a box and the size of the box.
The cost of loose chocolates and the mass of chocolates.
The rate of pay and the earnings for a 40 hour week.
The age of a kitten and its mass.
The cost of a phone call and the time taken for this call.
The area of an isosceles triangle and the length of a side.
The size of an exterior angle and the number of sides of regular polygons.
The base length and height of right-angled triangles which have the same area.

● Is the perimeter of a square proportional to the length of a side?
Is the area of a square proportional to the length of a side?
Is the area of a square proportional to the perimeter? **Discuss**.

● **Discuss** what happens to the volume of a cube if the length of an edge is doubled.
What if the length of an edge is trebled?
What if the length of an edge is halved?

Each edge on a cube is five times the length of an edge of the cube shown. What can you say about the volumes of these two cubes? **Discuss**.

Summary of key points

When there is a **percentage increase or decrease** we can calculate the new amount in a single step.

Example   Trudi bought a bike for £210.
She later sold it for a loss of 10%.
Selling price = 90% of £210
$\qquad\qquad\quad = 0·9 \times 210$
$\qquad\qquad\quad = £189$

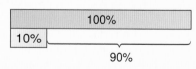

To **calculate an increase or decrease as a percentage** find the fractional increase or decrease first.

Example   The local school roll increased from 250 to 270 in one year.
The percentage increase $= \dfrac{\text{actual increase}}{\text{original number}} \times 100\%$
$\qquad\qquad\qquad\qquad = \dfrac{20}{250} \times 100\%$
$\qquad\qquad\qquad\qquad = 8\%$

When there is a percentage change we can **calculate the original amount** if we are given the final amount. We use **inverse operations**, the **unitary method** or **algebra**.

Example   After a 25% increase a box of chocolates was £3·40.

We can find the original price in one of these ways.

Using inverse operations	**Finding 1% first**	**Using algebra**
× 1·25 Original price of chocolates → Final price £3·40 ÷ 1·25	(unitary method) £3·40 represents 125% £3·40 ÷ 125 represents 1% £3·40 ÷ 125 × 100 represents 100% £3·40 ÷ 125 × 100 = **£2·72**	Let the original price = c. $1·25 × c = £3·40$ $c = \frac{£3·40}{1·25}$ = **£2·72**
original price = final price ÷ 1·25 = £3·40 ÷ 1·25 = **£2·72**		

B   When there is **more than one percentage change** we can work out a single number to multiply by.

Example   A house was valued at £250 000. Its value increased by 10% one year, then by 15% the following year.

new value = 1·1 × 1·15 × 250 000   1·1 × 1·15 = 1·265

= 1·265 × 250 000

= £316 250

When the **same percentage change is repeated**, index notation can be used.

Example   Sam deposited £500 in an account with an annual compound interest rate of 5%.

After 1 year she will have   £500 × 1·05.

After 2 years she will have   £500 × 1·05 × 1·05 = £500 × $1·05^2$.

After 6 years she will have   £500 × $1·05^6$.

After n years she will have   £500 × $1·05^n$.

C   **Proportional reasoning** can be used to solve problems.

We can use the **unitary method** or the **constant multiplier method**.

Example   If 8 muffins cost £6·75 we can find the cost of 12 muffins in two ways.

Unitary method	**Constant multiplier method**
Cost of 1 muffin = $\frac{6·75}{8}$ Cost of 12 muffins = $12 × \frac{6·75}{8}$ = **£10·13** (to nearest p)	muffins required : muffins given = 12 : 8 $\frac{12}{8}$ is the constant multiplier. Cost of 12 muffins = $\frac{12}{8} × 6·75$ = **£10·13** (to nearest p)

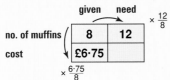

D We can solve **ratio and proportion problems**.

Ratio compares part : part.

Proportion compares part : whole.

 E Variables may be

1 directly proportional to each other.

As one increases by a constant amount, so does the other.

Example If y is directly proportional to x, we write $y \propto x$. $\quad y = kx$. This means $\frac{y}{x}$ is a constant.

The graph will be a straight line through the origin with a positive slope.

2 inversely proportional to each other.

As one increases by a constant amount the other decreases by a constant amount.

Example The time taken by a boat travelling at constant speed is inversely proportional to the speed.

If speed doubles, time halves.

We write time $\propto \frac{1}{\text{speed}}$. $\quad$ **time** $= \frac{k}{\text{speed}}$

This means speed $\times$ time will be constant.

3 directly proportional where one variable is proportional to the square of another.

Example The area of a circle is proportional to the square of its radius.

$A \propto r^2$

$A = kr^2$

This means $\frac{A}{r^2}$ is a constant.

Test yourself

1 A company made a profit of £750 245 in 1990.
Over the next 10 years the profit increased by 11·3%.
What was the profit in 2000?

2 The population of a rural town was 2035 in 1990.
By 2000 it had decreased to 1905.
What was the percentage decrease in population?

3 One of the calculations in the box gives the answer to 'what is 60 decreased by 7%?'
   a Write down the correct answer.
   b For each of the other calculations in the box, write down a possible question about percentages that it represents.
   c What goes in the gap?
      To increase an amount by 6·5%, multiply by _____.

60 × 1·07
60 × 0·3
60 × 0·93
60 × 1·7

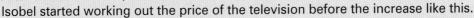

4 After a 20% increase, a television sold for £228.
Isobel started working out the price of the television before the increase like this.

  Price before increase × 1·2 = £228
  So price before increase = £228 _____ = _____

Finish Isobel's working to find the price before the increase.

5 This carton of juice contains 20% more juice than an ordinary carton.
How much juice does an ordinary carton contain?

20% more juice!
Just Juice
600 ml

6 Mr Lee weighed some tomatoes.
They weighed 3·5 kg.
He knew the tomatoes might weigh 15% more or 15% less than the reading shown.
What is the biggest and smallest mass the tomatoes might have been?

7 Samuel bought a bike. A year later he sold it to Tony, making a 5% profit.
Another year later Tony sold it to Dave at a 15% loss. What single number could you
multiply the price Samuel paid for the bike by to work out what Dave paid for it?

8 Stanley deposited £850 in a bank account at an annual compound interest rate
of 6·5%.
 a Which of these gives the amount Stanley will have after 3 years?
  $850 \times 1 \cdot 065 \times 3$ $850 \times 1 \cdot 065^3$ $850 \times 0 \cdot 065^3$
  $850 \times 0 \cdot 065 \times 3$ $(850 \times 1 \cdot 065)^3$ $(850 \times 0 \cdot 065)^3$
 b How much will Stanley have in total after 7 years?
 c How much interest had he made altogether after 7 years?

9 This recipe makes 24 muffins:

Apple Muffins

450 g flour
6 tsp baking powder
1½ cups milk
2 eggs
240 g sugar
2 chopped apples
90 g butter

Rewrite the recipe to make 16 muffins. Round sensibly.

10 Juice is mixed in the ratio shown.
 a Tom has 75 ml of orange concentrate. What
  is the most juice he can make?
 b How much of each ingredient will Paige use
  if she makes 6·5 ℓ of juice?

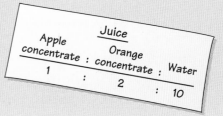

Juice		
Apple concentrate :	Orange concentrate :	Water
1 :	2 :	10

11 Sophie looked at the labels on two snack bars.
In one bar the ratio of fat to carbohydrate is 4 : 83.
In the other it is 5 : 104.
Which bar has the smaller proportion of fat?

12 The formula for the volume of a cone is
$\frac{1}{3}$ × area of base × height.
The volume of cone A is 33 cm^3. It is 9 cm tall.
Cone B is 12 cm tall and the area of its base is
the same as the area of the base of cone A.
What is the volume of cone B?

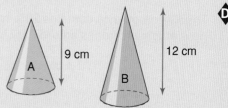

13 A 500 mℓ bottle is filled with oil and vinegar in the ratio 1 : 3. A 400 mℓ bottle is
filled with oil and vinegar in the ratio 2 : 3.
Susie decides to combine the two bottles of dressing.
What will be the ratio of oil to vinegar then?

14 Write down the relationship between each of the two variables using ∝.
  a Pressure (P) is proportional to temperature (T).
  b Solubility product, (K_{sp}), of silver chloride is proportional to the square of its
     solubility (s).
  c Resistance (R) is inversely proportional to current (I).

15 The formula for the area of a trapezium is $A = \frac{1}{2}(a+b)h$.
  a What happens to the area if h is made 3 times bigger
     but a and b stay the same?
  b What happens to the area if h is halved but a and b stay the same?
  c Write the relationship between A and h as $A \propto$ _____.
     Assume a and b stay the same.

16 Saskia wrote down the surface area (SA) of some cubes of side l cm.
  a Calculate $\frac{SA}{l}$ for each set of values in the table.
  b Is the ratio $\frac{SA}{l}$ constant?
  c Calculate $SA \times l$ for each set of values.
     Is (SA) inversely proportional to l?
  d Calculate l^2 for each value of l.
  e Calculate $\frac{SA}{l^2}$ for each set of values.
     Is (SA) directly proportional to l^2?

SA (cm²)	0·24	6	37·5	54	96	105·84
l (cm)	0·2	1	2·5	3	4	4·2

Algebra Support

In algebra **letters stand for numbers**.

$$2x + 5, \quad 6q - 1, \quad \frac{5r}{10} \quad \text{and} \quad 3a + 4b \quad \text{are all linear expressions}.$$

$$2x + 5 = 12, \quad 6q - 1 = 29, \quad \frac{5r}{10} = 25 \quad \text{and} \quad 3a + 4b = {}^-6 \quad \text{are all linear equations}.$$

$F = ma$ is a **formula**. It gives the relationship between unknowns that stand for something specific. In $F = ma$, F is force, m is mass and a is acceleration.

The relationship between the input, x and the output, y, of a function machine is called a **function**.
$y = 3x - 2$ is a function. The graph of a linear function is always a straight line.

An **inequality** has $>$, $<$, $\geqslant$ or $\leqslant$.
$y > 5$ and $6 < x \leqslant 8$ are inequalities.

Practice Questions 7, 10

Equations

$x + 7 = 16$ is a **linear equation**. x has a particular value.
We can **solve linear equations** using **inverse operations** or by **transforming both sides**.

Inverse operations

Example
$$3p + 2 = 18$$
$$3p = 18 - 2 \qquad \text{The inverse of adding 2 is subtracting 2.}$$
$$3p = 16$$
$$p = \frac{16}{3} \qquad \text{The inverse of multiplying by 3 is dividing by 3.}$$
$$p = 5\tfrac{1}{3} \text{ or } 5\cdot\dot{3}$$

Transforming both sides

Example
$$\frac{5x + 3}{4} + 7 = {}^-2$$
$$\frac{5x + 3}{4} + 7 - \mathbf{7} = {}^-2 - \mathbf{7} \qquad \text{subtract 7 from both sides}$$
$$\frac{5x + 3}{4} = {}^-9$$
$$\frac{5x + 3}{4} \times \mathbf{4} = {}^-9 \times \mathbf{4} \qquad \text{multiply both sides by 4}$$
$$5x + 3 = {}^-36$$
$$5x + 3 - \mathbf{3} = {}^-36 - \mathbf{3} \qquad \text{subtract 3 from both sides}$$
$$5x = {}^-39$$
$$\frac{5x}{\mathbf{5}} = \frac{{}^-39}{\mathbf{5}} \qquad \text{divide both sides by 5}$$
$$x = {}^-\mathbf{7\cdot8}$$

To solve an equation with an **unknown on both sides**, first transform both sides to get the unknowns on one side.

Example
$$5y - 3 = 2y + 9$$
$$5y - 3 - \mathbf{2y} = 2y - \mathbf{2y} + 9 \qquad \text{subtract } 2y \text{ from both sides}$$
$$3y - 3 = 9$$
$$3y - 3 + \mathbf{3} = 9 + \mathbf{3} \qquad \text{add 3 to both sides}$$
$$3y = 12$$
$$\frac{3y}{\mathbf{3}} = \frac{12}{\mathbf{3}} \qquad \text{divide both sides by 3}$$
$$y = \mathbf{4}$$

We subtract the smaller amount from both sides.

$x^2 - 10 = 159$ is a **non-linear equation**.
It can be solved to get an exact answer.

$$x^2 - 10 = 159$$
$$x^2 = 159 + 10$$
$$x^2 = 169$$
$$\sqrt{x^2} = \sqrt[\pm]{169}$$
$$x = +\textbf{13 or } {}^-\textbf{13}$$

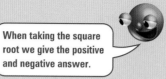

When taking the square root we give the positive and negative answer.

We can solve harder non-linear equations using **trial and improvement** or using a **calculator**, **spreadsheet** or **graph plotting software**.

Practice Questions 8, 14, 41, 45, 48, 49, 50

Expressions

Algebraic operations follow the same rules as arithmetic operations.
We can **simplify algebraic expressions** by collecting like terms, by using the index laws or by cancelling.
Before **collecting like terms** we must **expand any brackets**.

Example $3(p - 4) - 2(p + 5) = (3 \times p) + (3 \times {}^-4) + ({}^-2 \times p) + ({}^-2 \times 5)$
$$= 3p - 12 - 2p - 10$$
$$= p - 22$$

Sometimes we use **index laws** when simplifying expressions.

$$x^a \times x^b = x^{a+b} \qquad\qquad x^a \div x^b = x^{a-b} \qquad\qquad (x^a)^b = x^{ab}$$

Examples $4a^2 \times 2a^5 = 4 \times 2 \times a^{2+5}$ $\dfrac{{}^3 6m^4}{{}_1 2m} = 3m^{4-1}$ $\dfrac{{}^3 36x^3}{{}_2 24x} = \dfrac{3x^{3-1}}{2}$
$$= 8a^7 \qquad\qquad\qquad = 3m^3 \qquad\qquad\qquad = \dfrac{3x^2}{2}$$

Factorising an expression is the inverse of multiplying out a bracket.

Examples $5a + 10 = 5(a + 2)$ $3y^2 + y^3 = y^2(3 + y)$

	a	2	
5	$5a$	10	5 is the HCF of $5a$ and 10.

	3	y	
y^2	$3y^2$	y^3	y^2 is the HCF of $3y^2$ and y^3.

We take out the highest factor possible.

We can find the value of an expression by **substituting values** for the unknown. We use the **order of operations** rules.

Examples   If $m = \textbf{4}$ then
$$3m - 2 = 3 \times \textbf{4} - 2 \qquad\qquad 5(m + 2) - 7 = 5(\textbf{4} + 2) - 7$$
$$= 12 - 2 \qquad\qquad\qquad\qquad\quad = 5 \times 6 - 7$$
$$= \textbf{10} \qquad\qquad\qquad\qquad\qquad = 30 - 7$$
$$= \textbf{23}$$

Practice Questions 2, 9, 11, 24, 30, 31, 39, 40, 42

Formulae

We can substitute into **formulae** to find one unknown if we know the value of the other variables.

Example $v = \frac{s}{t}$ v = speed in km/h, s = distance in km, t = time in hours

If s = 120 km and $t = 1\frac{1}{2}$ hours,

$v = \frac{120}{1 \cdot 5}$

  = 80 km/h

We can **rearrange formulae** so the subject is different.

Example $V = IR$ V = voltage, I = current, R = resistance can be rearranged to make I the subject.

We can use a flow chart.

Start with subject required $\rightarrow I \rightarrow$ [multiply by R] $\rightarrow IR$

$\frac{V}{R} \leftarrow$ [divide by R] $\leftarrow V \leftarrow$ Return with current subject

$I = \frac{V}{R}$

Or we can transform both sides.

$V = IR$

$\frac{V}{R} = \frac{IR}{R}$

$\frac{V}{R} = I$

$I = \frac{V}{R}$

Practice Questions 28, 36

Sequences

We can write a **sequence** if we are given the **first term and the rule** for finding the next term. This is called a **term-to-term rule**.

Example   **1st term** $^-2$,   **rule** divide by $^-2$ gives

$^-2, 1, \frac{^-1}{2}, \frac{1}{4}, \frac{^-1}{8}, \frac{1}{16}, \dots$

We can write a sequence if we know the **rule for the nth term**.

Example   If the rule for the nth term is $T(n) = 120 - 4n$, the sequence is 116, 112, 108, 104, 100, ...

When a sequence goes up in equal steps it is called a **linear sequence** or arithmetic sequence. The starting number is a and the constant difference is d.

To be able to continue a sequence we must know either the term-to-term rule or the rule for the nth term. To find **the rule for the nth term of a linear sequence** we can use a difference table.

Example   For the sequence   3, 7, 11, 15, 19, ...

Term number	1	2	3	4	5
Term	3	7	11	15	19
Difference		4	4	4	4

constant difference between consecutive terms

The rule will be $T(n) = \mathbf{4}n + ?$

$T(\mathbf{1}) = 3$ $4 \times \mathbf{1} + ? = 3$ $4 - 1 = 3$

$T(\mathbf{2}) = 7$ $4 \times \mathbf{2} + ? = 7$ $8 - 1 = 7$

The rule must be $T(n) = 4n - \mathbf{1}$.

The constant difference gives the number multiplying n.

Practice Questions 1, 5, 15, 16, 17, 18, 21, 22, 25, 27, 37, 46, 47

Algebra

Functions

This is a **function machine**.

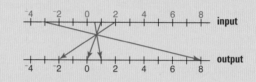

$x \rightarrow$ | add 2 | $\rightarrow$ | multiply by 3 | $\rightarrow y$

$x + 2$ $3(x + 2)$

The rule for this function machine is written as $y = 3(x + 2)$ or $x \rightarrow 3(x + 2)$.

If we are given the input we can find the output.

Example   1, 2, 0·5, ⁻3 → | add 1 | → | multiply by 3 | → 6, 9, 4·5, ⁻6

$$(1 + 1) \times 3 = 6$$
$$(2 + 1) \times 3 = 9$$
$$(0·5 + 1) \times 3 = 4·5$$
$$(⁻3 + 1) \times 3 = ⁻6$$

We can show the input and output of a function in a **table** or on a **mapping diagram**.

Example $x \rightarrow ⁻2(x - 1)$

Input	Output
1	0
2	⁻2
0·5	1
⁻3	8

If we are given the input and output we can **find the rule** for a function machine.

Example   3, 6, 9, 12 → | ? | → | ? | → 5, 11, 17, 23

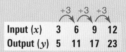

	+3	+3	+3
Input (x)	3	6	9	12
Output (y)	5	11	17	23

Each time the input increases by 3 the output increases by 6.
Try 'multiplying by 2' for the first operation.
From the first input and output, we find that the second operation is 'subtract 1'.
The rule is $y = \textbf{2}x - \textbf{1}$.

Check this is correct for the other input and output values.

Practice Questions   3, 4, 6, 12, 20, 23, 26, 32

Graphs

(2, 3) is a **coordinate pair**.
Each of the coordinate pairs (⁻1, ⁻3), (0, ⁻1), (1, 1), (2, 3) satisfies the rule $y = 2x - 1$.

$y = mx + c$ represents a **straight-line graph**.
m is the gradient and c is the y-intercept.
If m is positive the graph slopes ╱.
If m is negative the graph slopes ╲. (╲)

Sometimes the equation of a straight-line graph is not written in the form $y = mx + c$.
If we want to find m and c we must rearrange it into that form.

Example $2y - 6x = 5$

$$2y = 5 + 6x \quad \text{add 6x to both sides}$$
$$y = \frac{5}{2} + \frac{6x}{2} \quad \text{divide both sides by 2}$$
$$y = 3x + \frac{5}{2}$$

Lines **parallel to the x-axis** have equation $y = a$ where a is any number.
Lines **parallel to the y-axis** have equation $x = b$ where b is any number.

Practice Questions 19, 29, 34, 35, 38

Real-life graphs

To draw a **real-life graph** we must

- construct a table of values using a formula or relationship
- choose suitable scales for the axes
- plot the points accurately
- draw a line though the points if it is sensible to do so
- give the graph a title and label the axes.

Variables which are in 'direct proportion' always give a straight-line graph.

Example   This graph shows the charge against the length of lesson in hours for a music teacher. Charge and length of lesson are in direct proportion.

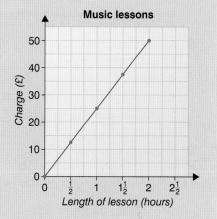

Practice Questions 13, 33, 43, 44

Practice Questions

1 Write down the first six terms of these sequences.
   a **1st term** 3   **rule** add 0·5
   b **1st term** 2500   **rule** divide by 5
   c **1st terms** 2, 3   **rule** add the two previous terms together
   d **1st term** 2   **rule** add consecutive even numbers 2, 4, 6, ...

2 Write each of these expressions in its simplest form.
   a $3x^2 + 2x + 4x^2 + x$ **b** $8x^2 + 9 + 2x^2 - 7$ **c** $5x^2 - 3 + 2x^2 + 8$
   d $7x^2 - 3 - 5x^2 + 6$ **e** $2(n + 2) + (n - 3)$ **f** $4(m - 2) + 3(m + 4)$
   g $4(r + 2) - 3(r - 4)$ **h** $5(f - 2) - (3 - f)$

3 Find the output for these.
   a   3 → [add 5] → ?
          $2\frac{3}{4}$ → [add 5] → ?
          0·36 → [add 5] → ?

   b   4 → [subtract 1] → [multiply by 3] → ?
          $1\frac{1}{2}$ → [subtract 1] → [multiply by 3] → ?
          1·75 → [subtract 1] → [multiply by 3] → ?

4
   6, 1, 7, 4 → [?] → [?] → 24, 4, 28, 16

   Jamie was asked to fill in the missing operations.
   He wrote: 6, 1, 7, 4 → [divide by 3] → [multiply by 12] → 24, 4, 28, 16

   Simon said the two operations could be replaced with one.
   What is this operation?

5 Write down the terms generated by this flow chart.

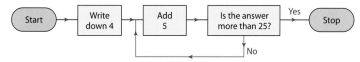

6 A function machine changes n to the number $3(n-1)$.
What number will it change these to?
a 2   **b** 10   **c** 0   **d** $\frac{1}{2}$ **e** $^-3$

7 In each of the following there is one equation, one expression, one formula and one function.
Which is which?
a $c-d=8$, $y=3x+5$, $4x-2$, $E=\frac{1}{2}mv^2$ E is energy, m is mass, v is speed

b $P=\frac{F}{A}$, P is pressure, F is force, A is area $\frac{3x}{5}$, $y=\frac{1}{3}x-2$, $3x-2=7$

8 Solve these equations.
Show your working.
a $8y-1=15$ **b** $2p+5=10$ **c** $5m+2=13$ **d** $20=2x+8$
e $4=\frac{12}{n}$ **f** $8=\frac{56}{m}$ **g** $\frac{n+7}{12}=4$ **h** $\frac{3a+6}{5}=2$
i $2\cdot4w+5\cdot9=14\cdot3$ ⊞**j** $6\cdot3m-2\cdot7=12\cdot42$ ⊞**k** $4\cdot6b+2\cdot7=^-11\cdot56$

9 Write an expression for the perimeter of each shape in its simplest form.
a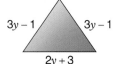
$3y-1$ $3y-1$
$2y+3$

b
$3a+4$
$2a+1$

c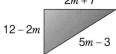
$2m+7$
$12-2m$
$5m-3$

10 Decide which inequality best describes each statement.
  a In an exam, every pupil got less than 85%.
     A $m>85\%$ **B** $m<85\%$ **C** $m\geqslant85\%$ **D** $m\leqslant85\%$
  b Jared always takes between 20 and 25 minutes to walk to school.
     A $20<t<25$ **B** $20\leqslant t<25$ **C** $20<t\leqslant25$ **D** $20\leqslant t\leqslant25$

11 It is Imogen's birthday and she is n years old.
This table gives the ages of Imogen's brothers and sisters.
  a What would be written for a in the table?
  b What would be written for b in the table?
  c In two year's time Imogen will be $n+2$ years old.
     Write a simplified expression for the age of each of her brothers and sisters in two years time.

	Expression	Words
Teagan	$n+2$	two years older than Imogen
Greta	$n-3$	a
Matthew	$2n$	b

12 a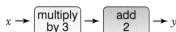
$x \rightarrow$ [multiply by 3] $\rightarrow$ [add 2] $\rightarrow y$
Copy and fill in this table.

x	1	2	3	4	5
y					

Write a function for this function machine.
  b The order of the operations is changed.

$x \rightarrow$ [add 2] $\rightarrow$ [multiply by 3] $\rightarrow y$
Copy and fill in this table.

x	1	2	3	4	5
y					

Write a function for this function machine.
  c What happens when the order of the operations is changed?

13 This is a distance–time graph for Jack's journey from Oxford to Hull.

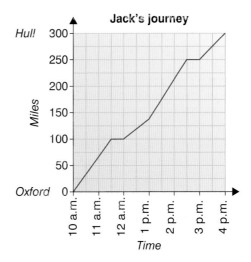

Jack's journey

Hull

Oxford

Miles

Time

    a Jack stopped after 100 miles. At what time did he stop and for how long?

    b What time did he arrive at Hull?

    c For about an hour of the journey he had to travel quite slowly in heavy traffic. Which hour do you think this was?

    d How far was he from Hull when he stopped the second time?

14 a The number in each rectangle is found by adding the two numbers above it.
Write and solve an equation to find the number, n in the red rectangle.

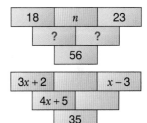

    b Use a copy of this.
Fill in the missing expressions.
Write and solve an equation to find the value of x.

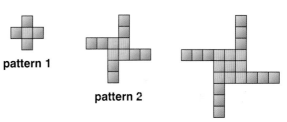

15 Joey makes a sequence with some purple and blue tiles.

    a How many purple tiles will there be in pattern 8?

pattern 1

    b How many purple tiles will there be in pattern 15?

pattern 2

    c Predict what numbers go in the gaps for the number of blue tiles.

pattern 3

Pattern number	1	2	3	4	5	...	20
Number of blue tiles	4	8	12	__	__	...	__

    d What sequence does the number of blue tiles make?
What is the term-to-term rule for it?

    e Write an expression for the number of blue tiles b, in pattern n.
Justify your expression by referring to the diagrams.
Check that your expression gives you the numbers you predicted in **c**.

16 These are the rules for the nth terms of some sequences.
Write the sequences.

    a $T(n) = 3n$ **b** $T(n) = 4n - 1$ **c** $T(n) = 48 - 4n$
    d $T(n) = 0.2n - 3$ **e** $T(n) = n^2 + 5$ **f** $T(n) = 2n^2 - 1$

17 Describe the sequences in questions **16a–e** using a term-to-term rule,
e.g. **first term _____ rule _____** .

18 a Gwyneth was asked to continue the sequence 3, 5, 8, ...
How might she do this?
Give two possible answers.

    b What do you need to know to be able to write down how a sequence continues with certainty?

Algebra

19 Write true or false for these.
  a The graph of $y = 2x - 4$ will have a steeper slope than the graph of $y = x - 3$.
  b The graph of $y = x - 2$ crosses the y-axis at $(0, {}^-2)$.
  c The graph of $y = {}^-5x + 1$ has a gradient which slopes $\diagup$.
  d The graph of $x = 4$ is a vertical line.
  e The graph of $y = {}^-3$ is a vertical line.
  f The graph of $y = \frac{x}{2} + 3$ has a steeper gradient than the graph of $y = {}^-2x + 5$.

20 These are some function machines.
  Write their rules in the form $y = $ ____ or $x \rightarrow$ ____ .
  a $x \rightarrow$ [multiply by 4] $\rightarrow$ [add 2] $\rightarrow y$
  b $x \rightarrow$ [subtract 1] $\rightarrow$ [multiply by 3] $\rightarrow y$
  c $x \rightarrow$ [divide by 2] $\rightarrow$ [add 6] $\rightarrow y$
  d $x \rightarrow$ [subtract 3] $\rightarrow$ [divide by 3] $\rightarrow y$

21 Predict the next three terms of these sequences.
  a 43, 39, 35, 31, 27, ...
  b 81, 27, 9, 3, 1, ...
  c 46, 34, 22, 10, ${}^-2$, ...
  d 1, 3, 4, 7, 11, ...

22 ___, ___, ___, 12
  What might the missing terms be?
  Explain your rule.

T

23 Use a copy of this. $x \rightarrow$ [subtract 1] $\rightarrow$ [multiply by 3] $\rightarrow y$
  a Fill in the table for the function machine given.
  b Fill in your mapping diagram for the function machine. Use inputs of 0, 1, 2, 3, 4, 5.

Input	Output
3	
1	
5	
0	

  ${}^-3\ {}^-2\ {}^-1\ 0\ 1\ 2\ 3\ 4\ 5\ 6\ 7\ 8\ 9\ 10\ 11\ 12$

  ${}^-3\ {}^-2\ {}^-1\ 0\ 1\ 2\ 3\ 4\ 5\ 6\ 7\ 8\ 9\ 10\ 11\ 12$

  c Write the rule for the function machine as $y = $ ____ and $x \rightarrow$ ____ .

24 Simplify these expressions.
  a $2 \times 6x$
  b ${}^-3 \times 2y$
  c $3a \times 4a$
  d $5d \times d$
  e $9q \times {}^-3q$
  f $3a \times 4a^2$
  g $2n \times 6n^2$
  h $\frac{y^4}{y^2}$
  i $\frac{m^3}{m^5}$
  j $\frac{15x^4}{5x^2}$
  k $\frac{35y^2}{15}$
  *l $\frac{12p^3}{18p}$
  *m $\frac{14x^2}{21x^5}$

25 Mark said '$T(n) = 4n - 3$ generates ascending numbers with a difference of 4 which are all 1 more than a multiple of 4.'
  It starts at 1.
  Describe these sequences in a similar way.
  a $T(n) = 4n + 2$
  b $T(n) = 50 - 5n$
  c $T(n) = 4n + 12$

26 Find the input values.
  a ___, ___, ___ $\rightarrow$ [subtract 5] $\rightarrow$ [divide by 2] $\rightarrow$ 18, 5·2, 14·6
  b ___, ___, ___ $\rightarrow$ [multiply by 3] $\rightarrow$ [add 2] $\rightarrow$ 23, 38·6, ${}^-4\cdot3$

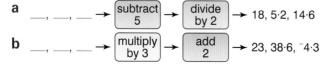

27 Find a sequence with the rule 'to find the next term add ☐' that has every fourth number an integer.
Write down the first term and a value that could go in the box.

28 'Can it' make cylindrical cans of all different sizes.
The formula for the volume of a cylinder is $V = \pi r^2 h$, where V is the volume, r is radius and h is height.
Use $\pi = 3.14$.
a Find V if $r = 5$ cm and $h = 12$ cm.
b Find h if $V = 36$ cm^3 and $r = 6$ cm.
c Find r if $V = 45$ cm^3 and $h = 15$ cm.

29 Find the gradients of these lines.
a

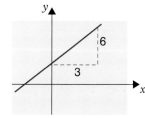
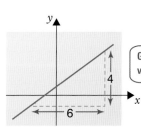
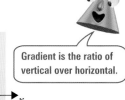

Gradient is the ratio of vertical over horizontal.

30 When $x = 3.5$ find the value of these.
a $4x$ **b** $2x - 5$ **c** $3(x - 10)$ **d** $5(x - 10)^2$ **e** $\frac{x^2}{2}$

31 Factorise these.
a $3n + 3$ **b** $8x + 12$ **c** $15y - 5$ **d** $12y + 20$ **e** $14x - 22$
f $15 - 18x$ **g** $y^2 + y^3$ **h** $18n^2 + 24n$ **i** $10m^2 - 2m^3$

32 Choose the correct answer.
a If $3x + 5 = y$ then   **A** $x = \frac{y+5}{3}$ **B** $x = \frac{y}{3} - 5$ **C** $x = \frac{y-5}{3}$.
b If $2(a + 3) = y$ then   **A** $a = \frac{y-3}{2}$ **B** $a = \frac{y}{2} - 3$ **C** $a = \frac{y}{2} - 6$.

33 Use a copy of this grid.
Plot a distance–time graph for two car journeys.
1 Millie left home at 9 a.m. She travelled to her grandparents at a constant speed for 40 km. She arrived there at 9:45 a.m.
She stayed for 30 minutes then travelled home, arriving at 11:30 a.m.
2 Eva left home at 9:30 a.m. She travelled 35 km in 45 minutes. She then slowed down and travelled the next 5 km in 30 minutes.
She stopped for an hour.

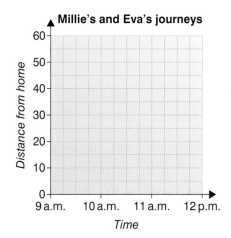

34 Write each of the following linear equations in the form $y = mx + c$.
a $y - x = 3$ **b** $y - x = 6$ **c** $y + x = {}^-5$ **d** $2y = 3x + 4$
e $3y = x - 5$ **f** $4y = {}^-x - 4$ **g** $x + 2y = 6$ **h** $x - 2y = 5$

Algebra

35 Match these sketches with an equation from the box.

a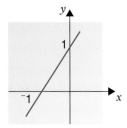
b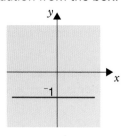
c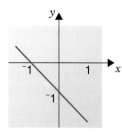

d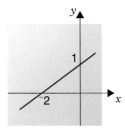
e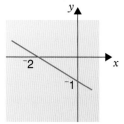

A $y = ^-1$
B $y = ^-x - 1$
C $y = \frac{1}{2}x + 1$
D $y = x + 1$
E $y = ^-\frac{1}{2}x - 1$

36 Rearrange these formulae to give the subject stated.

a $F = ma$, $a = $ _____
b $A = \frac{1}{2}bh$, $h = $ _____
c $V = lbh$, $l = $ _____

d $V = u + at$, $u = $ _____
e $V = u + at$, $t = $ _____

37 What are a and d in these arithmetic sequences?

a 4, 1, $^-2$, $^-5$, ...
b 0·4, 0·8, 1·2, 1·6, ...
c $1\frac{3}{8}$, $1\frac{1}{8}$, $\frac{7}{8}$, $\frac{5}{8}$, $\frac{3}{8}$, ...
d 0·75, 0·5, 0·25, 0, $^-0·25$, ...

See page 115 if you have forgotten what a and d stand for.

38 a Draw axes with both x- and y-values from $^-6$ to 6.
  On these axes draw and clearly label these lines.
    $y = 2x - 1$ $y = ^-x + 2$

b Will $y = 2x - 1$ go through the point (7, 13)? Explain how you can tell.

c Will the point $(7\frac{1}{2}, 9\frac{1}{2})$ lie on the line $y = ^-x + 2$? Explain how you can tell.

d The point ($^-6$, ___) lies on the line $y = ^-x + 2$. What is the missing y-coordinate?

39 Write these different expressions for the area of the blue section.
  Show that all three are equivalent.

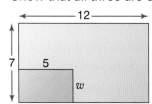

40 Evaluate these expressions if   **a** $n = 0·4$ **b** $n = ^-3$.

  i $2n^2 - 3$ **ii** $3n^3 + 2n$ **iii** $\frac{4n - 3}{2}$ **iv** $\frac{2n^2 + 1}{3}$

41 Write and solve an equation for each of these.

a The length of a rectangular field is three times its width.
  Its perimeter is 96 cm.
  Find its area.

b In a triangle ABC, $\angle$A is $\frac{1}{3}$ of $\angle$B, $\angle$C is 5 times $\angle$A.
  Find all three angles.

42 Eliot, Jenni and Jim each have a bag of sweets.
They do not know how many are in each bag.
They know that
- Jenni has two more sweets than Eliot
- Jim has four times as many sweets as Eliot.

  a If Eliot has x sweets, write an expression using x for
  the number of sweets in Jenni's and Jim's bags.
  b If Jenni has y sweets, write an expression using y for
  the number of sweets in Eliot's and Jim's bags.
  c If Jim has w sweets, which of these expressions
  gives the number of sweets in Jenni's bag?
  A $4w + 2$ **B** $4w - 2$ **C** $\frac{w}{4} + 2$ **D** $\frac{w}{4} - 2$ **E** $\frac{w+2}{4}$ **F** $\frac{w+2}{4}$

43 Write down a possible explanation for the shape of these graphs.

a

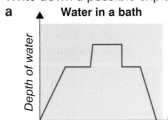

b

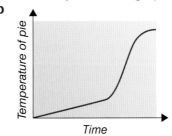

44 Water flowed steadily at the same rate into each of these containers.

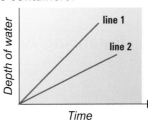

A depth of water against time graph was drawn.
Which line on the graph represents each container?
Explain your answer.

45 Solve these equations.
  a $4(n + 1) = 12$ **b** $3(m - 2) = 9$ **c** $3(n - 4) + 6 = 30$
  d $6(4a - 3) = 14 \cdot 4$ **e** $6x + 8 + 3x - 4 = 22$ **f** $6p + 9 - 3p + 2 = 26$
  g $6(m - 2) - 3(m + 2) = 12 \cdot 6$ **h** $4(x - 2) = 16 - 2(x + 3)$ **i** $3(a - 5) - 5(a + 2) = {}^-35$

46 Find the rule for these sequences.
  Write the rule as $T(n) = \underline{\hspace{1.5cm}}$
  a 3, 6, 9, 12, ...   **b** 6, 11, 16, 21, ...   **c** 12, 10, 8, 6, ...   **d** 40, 36, 32, 28, ...

47 Which term of question **46b** is equal to 101?

48 Find the exact solutions for $y^2 + 20 = 164$.
There are two solutions.

49 Use a calculator to find the solutions to these to 2 d.p.
  a $x^3 = 42$ **b** $5x^3 = 56$

50 Use trial and improvement to find the solution's to these to 2 d.p.
  a $x^3 - 11x = 72$ **b** $m^3 - m = 98$

6 Algebra and Equations

You need to know

 equations   page 113

Key vocabulary

identically equal to (≡),   identity,   inequality,   linear equation,   region,   simultaneous equations,   solution

 Balancing Act

If

and

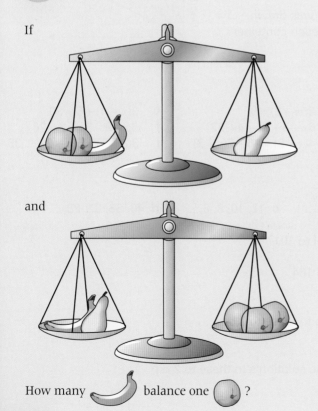

How many 🍌 balance one 🍎 ?

Understanding algebra

Discussion

Discuss the similarities and differences between equations, formulae and functions.

See page 113 for examples of these.

An **identity** is true for **all** values of the unknown.

Example $3(x + 2) \equiv 3x + 6$

= means 'is equal to'.

$\equiv$ means 'is identically equal to'.

It doesn't matter what value we substitute for x, $3(x + 2)$ will **always** have the same value as $3x + 6$.

Example   Show that $5(x - 1) \equiv 5x - 5$ is an identity.
Choose $x = 3$.
$5(\mathbf{3} - 1) = 5(2)$ and $5x - 5 = 5 \times \mathbf{3} - 5$
$\qquad\qquad = 10$ $= 10$

Choose $x = {}^{-}2$.
$5({}^{-}\mathbf{2} - 1) = 5({}^{-}3)$ and $5x - 5 = 5 \times {}^{-}\mathbf{2} - 5$
$\qquad\qquad = {}^{-}15$ $= {}^{-}10 - 5$
$\qquad\qquad\qquad\qquad\qquad\qquad\qquad = {}^{-}15$

Check some more yourself.

The left-hand side of an identity equals the right-hand side for *any* chosen values of x.

Exercise 1

T

1 Use a copy of this.
Match these statements with the best name.
The first one is done.

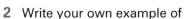

| Equation |
| Function |
| Formula |
| Identity |

● $y = 7x + 8$
● $v = u + at$ where v is final velocity in m/s, u is initial velocity in m/s, a is acceleration in m/s^2 and t is time in seconds
● $2x + 3x \equiv 5x$
● $3x + 8 = 4x - 7$
● $V = IR$ where V is voltage in volts, I is current in amps and R is resistance in ohms
● $3(x + 4) \equiv 3x + 12$
● $y = \frac{x}{4} - 2$
● $\frac{2x + 4}{7} = x - 3$

2 Write your own example of
a a formula   **b** an equation   **c** a function.

3 Show that $3x + 7x + 2x - 5 \equiv 12x - 5$ is true for
a $x = 2$ **b** $x = 10$ **c** $x = {}^{-}3$.
What do we call $3x + 7x + 2x - 5 \equiv 12x - 5$?

Algebra

4 Show that $7(x + 4) \equiv 7x + 28$ is an identity.

Review 1 Write equation, formula, function or identity for each of these.

a $4x - 2 = x + 5$

b $y = 2x - 1$

c $I = \frac{PRT}{100}$ where I = simple interest, P = principal, R = rate of interest and T = time in years

d $8y - 3y - 2 \equiv 5y - 2$

e $y = \frac{3x - 1}{4}$

f $3x - 1 = 5(x + 1)$

g $V = \pi r^2 h$ where V = volume of cylinder, r = radius of base and h = height of cylinder

h $7(p + 3) \equiv 7p + 21$

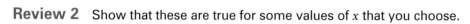

Review 2 Show that these are true for some values of x that you choose.
You could try $x = 0, 4, \,^-3$.

a $2x + 6 \equiv 2(x + 3)$ **b** $4(x - 3) \equiv 3x - 9$ **c** $3(2x + 1) \equiv 6x + 3$

Solving linear and non-linear equations

We can solve **linear equations** using inverse operations or by transforming both sides.

Discussion

• **Discuss** the most efficient way to solve these equations.

$$\frac{3(2y - 8)}{7} = 21 \qquad 3(n - 4) = 4(n + 2) + 12 \qquad \frac{2}{5} = \frac{3}{p}$$

• Kayla solved the equation $\frac{2}{3}(x - 1) = \frac{5}{6}(2x + 3)$ by transforming both sides.

$$\frac{2}{3}(x - 1) = \frac{5}{6}(2x + 3)$$

$$\frac{2}{3}(x - 1) \times 6 = \frac{5}{6}(2x + 3) \times 6 \qquad \text{Multiply both sides by 6.} \quad \boxed{\text{6 is the LCM of 3 and 6.}}$$

$$\left(\frac{2}{3} \times 6\right)(x - 1) = \frac{5}{6} \times 6(2x + 3)$$

$$4(x - 1) = 5(2x + 3)$$

$$4x - 4 = 10x + 15 \qquad \text{Expand the brackets.}$$

$$-4 = 6x + 15 \qquad \text{Subtract } 4x \text{ from both sides.}$$

$$-19 = 6x \qquad \text{Subtract 15 from both sides.}$$

$$x = -3.17 \ (2 \text{ d.p.}) \qquad \text{Divide both sides by 6 and write } x \text{ on the left-hand side.}$$

Discuss Kayla's method.
Could she have used inverse operations to solve the equation?

Laney started to solve the equation $\frac{y - 3}{4} = \frac{y - 5}{5}$ like this.

$$\frac{y - 3}{4} = \frac{y - 5}{5}$$

$$\frac{y - 3}{4} \times 20 = \frac{y - 5}{5} \times 20 \qquad \text{Multiply both sides by 20.}$$

Discuss how she could continue, to find y.

Worked Example

Solve these: **a** $5t + 6 = 7t - 3$ **b** $\frac{3}{4}(b - 4) = \frac{b + 3}{3}$

Answer

a

$$5t + 6 = 7t - 3$$
$$5t - \mathbf{5t} + 6 = 7t - \mathbf{5t} - 3 \qquad \text{Subtract } 5t \text{ from both sides.}$$
$$6 = 2t - 3$$
$$6 + \mathbf{3} = 2t - 3 + \mathbf{3} \qquad \text{Add 3 to both sides.}$$
$$9 = 2t$$
$$\frac{9}{2} = \frac{2t}{2} \qquad \text{Divide both sides by 2.}$$
$$4\frac{1}{2} = t$$
$$t = \mathbf{4\frac{1}{2}}$$

> Subtract $5t$ rather than $7t$ from both sides because it is the least amount of t.

> You don't need to write down these side notes every time.

b

$$\frac{3}{4}(b - 4) = \frac{b + 3}{3}$$
$$\frac{3}{4}(b - 4) \times \mathbf{12} = \frac{b + 3}{3} \times \mathbf{12} \qquad \text{Multiply both sides by 12.}$$
$$\frac{3}{1\,4} \times 12^{3}(b - 4) = \frac{b + 3}{1\,3} \times 12^{4}$$
$$9(b - 4) = 4(b + 3)$$
$$9b - 36 = 4b + 12 \qquad \text{Expand the brackets.}$$
$$5b = 48 \qquad \text{Subtract } 4b \text{ from and add 36 to both sides.}$$
$$b = \mathbf{9{\cdot}6} \qquad \text{Divide both sides by 5.}$$

Remember

A **non-linear equation** has terms with indices greater than 1, such as x^2, x^3, ...

Worked Example

Solve these.   **a** $x^2 - 100 = 224$ **b** $\frac{16}{y + 5} = y + 5$

Answer

a $x^2 - 100 = 224$
$$x^2 = 324 \qquad \text{Add 100 to both sides.}$$
$$x = \sqrt[\pm]{324} \qquad \text{Take the positive } and \text{ negative square root of both sides.}$$
$$x = \mathbf{18 \text{ or } {}^-18}$$

b $\frac{16}{y + 5} = y + 5$
$$16 = (y + 5)(y + 5) \qquad \text{Multiply both sides by } (y + 5).$$
$$\sqrt[\pm]{16} = y + 5 \qquad \text{Take the square root of both sides.}$$
$$y + 5 = +4 \text{ or } y + 5 = {}^-4$$
$$y = {}^-\mathbf{1} \text{ or } y = {}^-\mathbf{9}$$

Discussion

Sometimes we use trial and improvement to solve a non-linear equation.

Discuss how to solve these using trial and improvement and a calculator or a spreadsheet.

$$x^3 + 5x = 79{\cdot}4 \qquad 3x^3 - x + 4 = 82{\cdot}3$$

Algebra

Exercise 2

1 Solve these equations.
   a $3n + 4 = n + 13$
   b $7 + 5m = 8m + 1$
   c $8y + 17 = 4y + 19$
   d $9b + 3 = 5b + 13$
   e $12 + x = 5x - 2$
   f $8 + 3p = 9p + 4$
   g $25 - 2y = 6y + 5$
   h $14 + 10k = 15k + 5$
   i $5 - 2b = 3b + 25$
   j $\frac{5y}{2y + 3} = 5$
   k $\frac{4m + 2}{3m - 1} = 2$
   l $\frac{7p}{2p - 3} = 8$

2 Find the values of t and r. **[SATs Paper 2 Level 6]**

   $\frac{2}{3} = \frac{t}{6}$

   $\frac{2}{3} = \frac{5}{r}$

3 Look at the equations. **[SATs Paper 2 Level 6]**

   $3a + 6b = 24$

   $2c - d = 3$

   a Use the equations to work out the value of the expressions below.
   The first one is done for you.

   $8c - 4d = \underline{12}$

   $a + 2b = \underline{}$

   $d - 2c = \underline{}$

   b Use one or both of the equations to write an expression that has a value of 21.
   $\underline{} = 21$

4 Write and solve an equation to find x.

   a 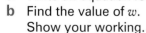 $2x + 9$ $3x + 4$

   b  $5x + 12$ $8x - 9$

5 Write and solve an equation for these.
   a In PQR, $\angle$Q is half of $\angle$P and $\angle$R is three-quarters of $\angle$P.
   Find the size of all the angles in the triangle.
   b Sanjay and Niki have the same number of CDs.
   They both store some of their CDs in cases that hold n CDs.
   Sanjay has two full cases of CDs plus eighteen loose CDs.
   Niki has three full cases of CDs plus six loose CDs.
   Write and solve an equation to find how many CDs each case holds.

   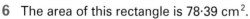

6 The area of this rectangle is 78·39 cm².
   a Write an equation for the area of the rectangle.
   b Find the value of w.
   Show your working.
   You may find this table helpful.

   $w + 5$
   w

w	$w + 5$	$w(w + 5)$	
6	11	66	too small
7	12	84	too large

7 Solve these equations to get an exact answer.

   a $m^2 - 4 = 12$ **b** $p^2 - 200 = 376$ **c** $6 = \frac{150}{x^2}$ **d** $\frac{4}{m+3} = m + 3$

8 Solve these equations using trial and improvement. Give the answers to 1 d.p.

   a $c^3 - c = 75 \cdot 2$ **b** $3b^3 + b = 427$ **c** $18 \cdot 4 = a(a^2 + 7)$

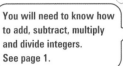

9 Solve these equations.

   a $5d - 8 = {}^-28$ **b** $4x - 8 = {}^-12$ **c** $5m + 6 = {}^-9$

   d ${}^-17 = 8a + 5$ **e** $\frac{x+2}{2} = {}^-4$ **f** $\frac{{}^-4x+2}{2} - 3 = {}^-10$

> You will need to know how to add, subtract, multiply and divide integers. See page 1.

10 Kade solved the equation $12 - 4x = 4$ like this.

$$12 - 4x = 4$$
$$12 = 4 + 4x \qquad \text{add } 4x \text{ to both sides to make } 4x \text{ positive}$$
$$12 - 4 = 4x \qquad \text{subtract 4 from both sides}$$
$$8 = 4x$$
$$\frac{8}{4} = x \qquad \text{divide both sides by 4}$$
$$2 = x$$

Use Kade's method or another method to solve these.

   a $8 - 2x = 3$ **b** $20 - 4x = 8$ **c** $12 - 9x = 24$ **d** ${}^-3x - 8 = {}^-11$

   e $4(4 - x) + 7 = 8$ **f** $\frac{{}^-3}{n} - 4 = \frac{7}{2n}$

11 Solve these equations. Show your working.   [SATs Paper 1 Level 7]

   a $4 - 2y = 10 - 6y$ **b** $5y + 20 = 3(y - 4)$ **c** $\frac{9y}{2y+1} = 9$ **d** $\frac{9}{y+2} = y + 2$

12 a Pupils started to solve the equation $6x + 8 = 4x + 11$ in different ways.   [SATs Paper 1 Level 7]

   For each statement below, write true or false.

   i $6x + 8 = 4x + 11$ **ii** $6x + 8 = 4x + 11$

        so $14x = 15x$ so $6x + 4x = 11 + 8$

   iii $6x + 8 = 4x + 11$ **iv** $6x + 8 = 4x + 11$

        so $6x = 4x + 3$ so $2x + 8 = 11$

   v $6x + 8 = 4x + 11$ **vi** $6x + 8 = 4x + 11$

        so $2x = 3$ so ${}^-3 = {}^-2x$

   b A different pupil used trial and improvement to solve the equation $6x + 8 = 4x + 11$.

       Explain why trial and improvement is not a good method to use.

13 Solve these equations.

   a $3(x + 4) = x + 16$ **b** $2(m + 3) = 4(m - 1)$

   c $3p - (p + 1) = p - 2$ **d** $a + 3(a - 1) = 2a$

   * **e** $3(2b - 1) = 5(4b - 1) - 4(3b - 2)$ **f** $2(y - 0 \cdot 3) - 3(y - 1 \cdot 3) = 4(3y + 3 \cdot 1)$

14 Solve these.

   a $\frac{12}{n+1} = \frac{21}{n+4}$ **b** $\frac{5}{n+3} = \frac{4}{n+5}$ **c** $\frac{y-2}{3} = \frac{y+4}{4}$ **d** $\frac{7-m}{3} = \frac{m+2}{2}$

   e $\frac{4a-2}{3} = \frac{5a+1}{5}$ **f** $\frac{3}{5}(m + 2) = \frac{3}{4}(2m - 2)$ **g** $\frac{5}{8}(2p + 4) = \frac{1}{2}(3p - 4)$

* **15** Jeanna used this diagram to find two consecutive even numbers with a difference of 100 between their squares.
She let one number be y and the other be $y + 2$.
Write and solve an equation to find the two numbers.

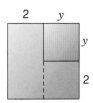

Algebra

∗16 The product of three consecutive odd numbers is 68 757.
Use the solution to $x^3 = 68\ 757$ to help you find the numbers.
Explain how you did this.

∗17 This cuboid has a square base of side x cm.
It has height 30 cm.
The total surface area of the cuboid is 600 cm².
Write an equation for the surface area of the cuboid.
Solve it to find x, correct to one decimal place.
Use trial and improvement using a calculator or spreadsheet.

∗18 The two rectangles below have the **same area**.   **[SATs Paper 1 Level 8]**

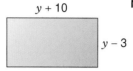

Not drawn accurately

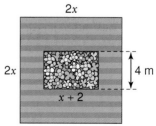

You need to know how to expand two expressions of the form $(x + a)(y + b)$. See page 161.

Use an algebraic method to find the value of y.
You **must** show your working.

Review 1   Solve these equations.
a $3x + 5 = x + 8$ **b** $6a - 4 = 2a + 3$ **c** $3y + 8 = 10 - 2y$ **d** $\frac{2m - 6}{3} = 8$
e $\frac{3x}{x + 1} = 2$ **f** $\frac{5}{n} = \frac{2}{3}$

Review 2   Write and solve an equation to find x.
a

$(4x + 12)° \quad (x - 2)°$

b

$8x - 15$

rectangle

$5x + 12$

Review 3   Solve these equations to get an exact answer.
a $t^2 + 16 = 25$ **b** $8(x + 2)^2 = 392$

Review 4   Solve these equations.
a $3(b + 2) = 5(2 - b)$ **b** $\frac{2x - 1}{3} = \frac{x + 3}{5}$ **c** $4(a - 2·3) = a + 5·5$ **d** $\frac{3}{4}(k - 3) = \frac{2}{5}(4 + k)$

Review 5
a A first class ticket costs two and a half times as much as a third class ticket. If the total cost of 6 first class and 10 third class tickets is £187·50, write an equation to show this information. Solve it to find how much a third class ticket costs.
∗b Mrs Smith has a square garden of side $2x$ metres. She has a rectangular flower bed in the garden of length $(x + 2)$ m and width 4 m. The rest is lawn.
  i Explain why the area of the lawn is $4x^2 - 4(x + 2)$.
  ii Use trial and improvement to find the value of x if the lawn has area 29·44 m².

$2x$

$2x$

4 m

$x + 2$

Practical

A **You will need** a spreadsheet package.
Ask your teacher for the **Trial and Improvement** ICT worksheets.

B Use a graphical calculator to solve this equation.
$$x^3 + x = 30$$

Puzzle

1 A number plus its square equals 30.
What might the number be?
Is there more than one answer?

2 **What if** a number plus its cube equals 30?

Simultaneous equations

Discussion

Discuss how to use 'trial and improvement' to solve the following problems. Solve them.

Problem 1   A total of £8616 was received from ticket sales for
the Clown Show.
How many tickets were sold?

Clown Show
Tickets £12

Problem 2   800 tickets were sold for the Buskers Festival,
for a total of £8540.
How many tickets were sold at each price?
Is there more than one answer to this problem?

Buskers Festival
Adults   £12
Students   £8

Problem 3   800 tickets were sold for the Circus, for a total of £8092.
How many tickets were sold at each price?
Can you find more than one answer for this problem?
What if you were also told that only 50 tickets were
available for door sales and all of these were sold?

For **problem 2**, the following two equations are true. $12a + 8s = 8540$
$$a + s = 800$$

What does a stand for? What does s stand for?

Discuss possible equations for *problem 1* and *problem 3*.

Discuss the following for each of the above problems.
How many unknowns are there? How many equations are needed to find these unknowns?

Simultaneous equations are equations which need to be solved together to find the value of the unknowns.

Example

$7x + 4y = 41$
$5x - 4y = {}^-5$

We can use trial and improvement, adding and subtracting equations, a substitution method or a graphical method to solve simultaneous equations.

Elimination method – Adding and subtracting equations

Discussion

- $7x + 4y = 41$
  $5x - 4y = {}^-5$ are simultaneous equations.

  We can illustrate them using scales.

  If we add the left-hand sides and add the right-hand sides, would the scales still balance? **Discuss**.

  What is the left-hand side total?
  What is the right-hand side total?
  How can you use this to find x? **Discuss**.
  What is the value of x?
  How can you use this to find y? **Discuss**.

- Look at the equations $\begin{aligned} 4x + 5y &= 22 \\ 3x - 2y &= 5 \end{aligned}$. What happens if you add the equations?

  What happens if you subtract? **Discuss**.

- Try adding the pairs of equations below.
  Can you solve the resulting equation to find the value of one of the variables?
  What if you subtract one equation from the other? **Discuss**.

$2x + y = 4$	$4a + 2b = 1$	$3p + 2q = 12$	$3x + y = 4$
$x - y = 5$	$8a - 2b = {}^-4$	$p + 2q = 8$	$3x + 4y = {}^-2$
$3a + 2b = 9$	$3x - 2y = 8$	$5p + 2q = 3$	
$2a - b = {}^-1$	$4x - 5y = 6$	$2p + 3q = 1$	

By **'adding or subtracting' the equations**, sometimes one of the unknowns is eliminated. If it is, we can solve the resulting equation to find the other unknowns.

Worked Example

Solve these. $7x + 4y = 41$
                   $5x - 4y = {}^-5$

This is why it is called the 'elimination' method.

Answer

$$7x + 4y = 41$$
$$+ \ 5x - 4y = {}^-5$$
$$\overline{12x \qquad = 36} \qquad \text{adding the equations}$$
$$x = \mathbf{3} \qquad \text{dividing both sides by 12}$$

We can substitute $x = 3$ into one of the equations to find y.

$$7x + 4y = 41$$
$$21 + 4y = 41$$
$$4y = 20 \qquad \text{subtracting 21 from both sides}$$
$$y = \mathbf{5} \qquad \text{dividing both sides by 4}$$

The solution is $x = \mathbf{3}$, $\qquad y = \mathbf{5}$.

Always check the solution by substituting the values of the unknowns into the equation you didn't use to find the second unknown.

Example For the previous worked example:

If $x = 3$, $\quad y = 5$ then LHS of $5x - 4y = 5 \times 3 - 4 \times 5 = {}^-5$. ✔

So LHS = RHS.

LHS means left-hand side.
RHS means right-hand side.

Exercise 3

1 Solve these simultaneous equations.

a $\ 3x + 2y = 14$	**b** $\ a + b = 3$	**c** $\ 2x - y = 8$	**d** $\ 5p + 4q = 8$
$\ \ \ 5x - 2y = 18$	$\ \ \ 3a - b = 1$	$\ \ \ {}^-2x + 3y = {}^-12$	$\ \ \ 3p + 4q = 4$
e $\ 2l + 5m = 18$	**f** $\ 5x - 2y = 17$	**g** $\ p + 3q = 4$	**h** $\ 10l - m = {}^-2$
$\ \ \ 2l + 7m = 26$	$\ \ \ 3x - 2y = 9$	$\ \ \ p + q = 3$	$\ \ \ 10l - 4m = 7$
i $\ 5a - 4b = {}^-6$	**j** $\ 3x + 2y = 4$	**k** $\ a - 6b = {}^-18$	
$\ \ \ 2a - 4b = 0$	$\ \ \ 4x + 2y = 7$	$\ \ \ 3a - 6b = {}^-24$	

Review

a $\ 5x - 3y = 19$	**b** $\ 3p + q = 5$	**c** $\ 6a + 5b = {}^-1$
$\ \ \ 4x + 3y = {}^-1$	$\ \ \ 2p + q = 2$	$\ \ \ 6a + 2b = 5$

Sometimes we have to **multiply both sides of one or both of the equations** to enable us to eliminate one of the unknowns.

Worked Example

Solve these simultaneous equations.

$$3x - 2y = {}^-19$$
$$2x + 4y = 6$$

Answer

Both equations need the same number in front of x or y so that when we add or subtract the equations either x or y is 'eliminated'.

Multiply $3x - 2y = {}^-19$ by 2:

$$(3x - 2y = {}^-19) \times 2 = 6x - 4y = {}^-38 \qquad \text{multiplying each term by 2}$$

Add the two equations we have now:

$$6x - 4y = {}^-38$$
$$\underline{2x + 4y = 6}$$
$$8x \qquad = {}^-32 \qquad \text{adding the equations}$$
$$x \qquad = \mathbf{{}^-4} \qquad \text{dividing both sides by 8}$$

Algebra

When $x = {}^-4, \quad 2x + 4y = 6$
$$2 \times {}^-4 + 4y = 6$$
$${}^-8 + 4y = 6$$
$$4y = 14$$
$$y = \textbf{3·5}$$

The solution is $x = {}^-\textbf{4,} \qquad y = \textbf{3·5}$.

Check If $x = {}^-4, y = 3·5$ then LHS of $3x - 2y = 3 \times {}^-4 - 2 \times 3·5 = {}^-19$. ✓
So LHS = RHS

Note We checked the solutions using the equation we *didn't* use to find y.

Worked Example

Solve these simultaneous equations. $\quad 3x + 4y = 6$
$\qquad\qquad\qquad\qquad\qquad\qquad\quad 2x - 3y = {}^-13$

Answer

We must multiply both equations by something so that we get the same number of either xs or ys.

$\qquad (3x + 4y = 6) \times 3 \quad = \quad 9x + 12y = 18 \qquad$ multiplying both sides by 3
$\qquad (2x - 3y = {}^-13) \times 4 = \quad 8x - 12y = {}^-52 \qquad$ multiplying both sides by 4
$\qquad\qquad\qquad\qquad\qquad\quad 17x \qquad\qquad = {}^-34 \qquad$ adding the equations
$\qquad\qquad\qquad\qquad\qquad\qquad x = {}^-\textbf{2} \qquad$ dividing both sides by 17

When $x = {}^-2$, $9x + 12y = 18$ becomes ${}^-18 + 12y = 18$
$\qquad\qquad\qquad\qquad\qquad\qquad\qquad\qquad 12y = 36 \qquad$ adding 18 to each side
$\qquad\qquad\qquad\qquad\qquad\qquad\qquad\qquad x = \textbf{3} \qquad$ dividing both sides by 12

The solution is $x = {}^-\textbf{2}, y = \textbf{3}$.

Check If $x = {}^-2, \qquad y = 3$ then LHS of $2x - 3y = 2y - 2 - 3 \times 3 = {}^-13$. ✓
So LHS = RHS

Exercise 4

1 Solve these simultaneous equations.

   a $3p + 4q = 29$ **b** $2x - 6y = {}^-11$ **c** $3a - 8b = 1$ **d** $4m + n = 4$
       $5p - 2q = 5$ $3x - 2y = {}^-13$ $5a - 4b = {}^-3$ $6m - 2n = 13$

   e $2a + 3d = 7$ **f** $4p - 5q = 17$ **g** $3a + 5b = 7$ **h** $2x - 3y = 5$
       $3a - 4d = 2$ $3p - 4q = 13$ $5a - 2b = {}^-9$ $3x + 4y = {}^-18$

   i $\quad 5l + 7m = 0·5$
       $10l - 6m = {}^-29$

2 Solve these simultaneous equations.

   a $3x + 2y = 0$ **b** $y = x + 2$ **c** $2x + y = 0$ **d** $4x + 3y = 11$
       $3y - x = 11$ $y + x = {}^-4$ $x = y + 3$ $3x + 2y = 9$

   e $x - y = {}^-2$ **f** $2y = 5 - x$ **g** $3p - 4q = 21$ **h** $1 - l + 5m = 0$
       $y = 2x - 1$ $3x + 5y - 11 = 0$ $5q = 2p - 17·5$ $m - 2l = 2·5$

*3 *The solutions are $x = 2, y = {}^-1$. Write down some pairs of simultaneous equations with these solutions.*

Review Solve these simultaneous equations.

   a $3a + 10m = 22$ **b** $4x - 3y = 23$ **c** $5a + 3b = {}^-6$ **d** $2x - 3y = 4$
       $2a - 5m = 3$ $8x - 2y = 26$ $4a + 5b = 3$ $y + x = 7$

   e $l = 3m - 1$ **f** $2p = 23 + 5q$ **g** $3y - 2b = 2·5$
       $m + 2l = 12$ $3q - 4p = {}^-18$ $2b + 4y = {}^-13$

Substitution method

Discussion

Look at these simultaneous equations.
$$2x + y = 1$$
$$3x + 4y = 6$$

We can make y the subject of the first equation.

$y = 1 - 2x$ subtracting 2*x* from both sides

We can now substitute for y in the second equation.

$3x + 4(1 - 2x) = 6$

Can you solve this equation to find x? **Discuss.**

What if the equations were $\quad \begin{array}{l} 3a + 2b = 12 \\ a + 2b = 8 \end{array} \quad$ or $\quad \begin{array}{l} 5x + 2y = 3 \\ 2x + 3y = {}^-1 \end{array}$?

Worked Example

Solve these simultaneous equations. $\quad \begin{array}{l} y = 1 - 2x \\ 3x + 4y = 6 \end{array}$

Answer

$y = 1 - 2x \qquad \dots (1)$
$3x + 4y = 6 \qquad \dots (2)$

Substitute the expression for y from equation (1) into equation (2) to get

$\quad 3x + 4(1 - 2x) = 6$
$\qquad 3x + 4 - 8x = 6 \qquad$ expanding brackets
$\qquad\quad {}^-5x + 4 = 6$
$\qquad\qquad {}^-5x = 2 \qquad$ subtracting 4 from both sides
$\qquad\qquad\quad x = {}^-0\cdot4 \qquad$ dividing both sides by $^-5$

When $x = {}^-0\cdot4$, $y = 1 - 2 \times ({}^-0\cdot4) \qquad$ substituting x into $y = 1 - 2x$
$\qquad\qquad\qquad = 1\cdot8$

The solution is $x = {}^-\mathbf{0\cdot4}, \qquad y = \mathbf{1\cdot8}$.

Check If $x = {}^-0\cdot4$, $y = 1\cdot8$ then LHS of $3x + 4y = 6 = 3 \times ({}^-0\cdot4) + 4 \times 1\cdot8 = {}^-1\cdot2 + 7\cdot2 = 6$. ✓
LHS = RHS

Worked Example

Use the substitution method to solve these simultaneous equations. $\quad \begin{array}{l} 5a - 2m = 14 \\ 3a + m = 4 \end{array}$

Answer

$5a - 2m = 14 \qquad \dots (1)$
$3a + m = 4 \qquad\quad \dots (2)$

From (2), $m = 4 - 3a$.

Substituting in (1) we get $\quad 5a - 2(4 - 3a) = 14 \qquad$ substituting for m
$\qquad\qquad\qquad\qquad\quad 5a - 8 + 6a = 14 \qquad$ expanding the brackets
$\qquad\qquad\qquad\qquad\qquad\quad 11a - 8 = 14$
$\qquad\qquad\qquad\qquad\qquad\qquad 11a = 22$
$\qquad\qquad\qquad\qquad\qquad\qquad\quad a = 2$

When $a = 2$, $m = 4 - 3 \times 2 \qquad$ substituting a into $m = 4 - 3a$
$\qquad\qquad\qquad = {}^-2$

The solution is $a = \mathbf{2}, m = {}^-\mathbf{2}$.

Check If $a = 2$, $m = 2$ then LHS $= 3a + m = 3 \times 2 + ({}^-2)$
$\qquad\qquad\qquad\qquad\qquad\qquad = 6 + ({}^-2)$
$\qquad\qquad\qquad\qquad\qquad\qquad = 4$
$\qquad\qquad\qquad\qquad\qquad\qquad = $ RHS. ✓

Algebra

1 Use the substitution method to solve these simultaneous equations.

  a $x = y - 2$
    $2x + 3y = 21$

  b $m = l + 4$
    $3l + 2m = 3$

  c $x = 2y + 1$
    $3x - 4y = 7$

  d $a = 4b - 2$
    $3a + 2b = 15$

  e $p = 3 - 4q$
    $5q = 3 - 2p$

  f $l - 3a = 11$
    $5l + 2a = 4$

  g $3x + y = 7$
    $2x - 5y = 16$

  h $2b - 7a = 11$
    $6b + a = {}^-11$

2 Which of these simultaneous equations can be easily solved using the substitution method?
  Explain why.

  a $p = q - 7$
    $2p + 3q = 6$

  b $b = 2a + 1$
    $2a - 3b = 3$

  c $x - 2y = {}^-10$
    $3y + 4x = 4$

  d $2a - 3b = 4$
    $5a + 3b = {}^-11$

  e $5p + 4q = 23$
    $3p - 5q = {}^-1$

  f $m = l - 1$
    $3m + 4l = {}^-20 \cdot 5$

  g $p - 6q = 5$
    $2p - 4q = 14$

  h $x = 3y - 2 \cdot 5$
    $y = 6x - 2$

  i $5y + x = {}^-7$
    $3x - 10y = {}^-1$

  j $a - b - 7 = 0$
    $3a - 2b = 18$

  k $2x - 3y = 0$
    $5y - 4x = 1$

3 Solve the equations in question **2** using either the substitution method or by adding and subtracting the equations.

Review 1 Use the substitution method to solve these simultaneous equations.

a $a = b + 4$
  $2a + 3b = 3$

b $x = 4y - 3$
  $2y - 3x = 4$

c $4p + q = 5$
  $2p - 3q = 20$

Review 2 Use a method of your choice to solve these.

a $3a + 2b = 3$
  $6a + 5b = 3$

b $l = 4 - m$
  $5l - 6m = 9$

c $y = 5x - 4$
  $3y - 10x = {}^-9$

d $2p + 3q + 4 = 6$
  $3p + 2q - 1 = {}^-8$

Solving simultaneous equations graphically

Discussion

$x + y = 1$

x	$^-2$	0	1
y	3	1	0

$2x - y = {}^-4$

x	$^-2$	0	1
y	0	4	6

The lines $x + y = 1$ and $2x - y = {}^-4$ can be drawn by drawing a straight line through the points given on the table.
What are the coordinates of the point which lies on *both* lines?

Solve the simultaneous equations $\begin{array}{c} x + y = 1 \\ 2x - y = {}^-4 \end{array}$ using the adding

or subtracting equations method or the substitution method.
Compare the values of x and y you found with the coordinates of the point where these lines meet. What do you notice? **Discuss**.

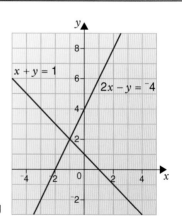

Practical

You will need a graph plotter or graphical calculator.

A Draw graphs for some of the simultaneous equations you solved in **Exercises 4 and 5**. Draw each pair on the same set of axes.
Could all of these simultaneous equations be solved by drawing graphs?

B Draw the graphs of $x + y = 7$ and $3x + 3y = 21$.
What do you notice?
Is it possible to find solutions to these equations algebraically?
How many solutions will these two equations have?

C Draw the graphs of $y = 3x - 2$ and $y = 3x + 1$.
What do you notice?
Is it possible to find solutions to these equations algebraically?

D Draw the graphs of these three equations.

$$y = 2x - 7 \qquad y = {}^-x + 2 \qquad y = 3x - 10$$

Do these equations have a common solution? How can you tell?

Exercise 6

1 Use the graph to write down the solutions of the following simultaneous equations.

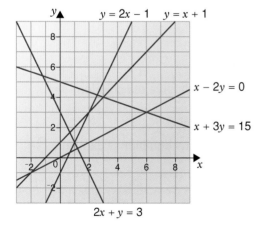

a $y = 2x - 1$
$y = x + 1$

b $x - 2y = 0$
$x + 3y = 15$

c $2x + y = 3$
$y = 2x - 1$

d $y = x + 1$
$x - 2y = 0$

Algebra

For questions 2 to 8 use the following graph.

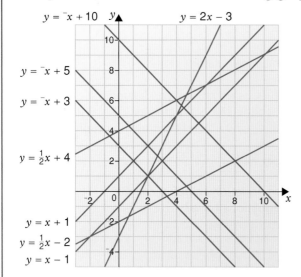

$y = {}^-x + 10$ $y = 2x - 3$

$y = {}^-x + 5$

$y = {}^-x + 3$

$y = \frac{1}{2}x + 4$

$y = x + 1$

$y = \frac{1}{2}x - 2$

$y = x - 1$

2 **a** Write $x + y = 10$ as $y = $ ___ $+ 10$.

   b Write down the coordinates of the point where the lines $x + y = 10$ and $y = \frac{1}{2}x + 4$ meet.

   c Use your answer to **b** to solve the simultaneous equations $x + y = 10$ and $y = \frac{1}{2}x + 4$.

3 **a** Rearrange $y = 3 - x$ as $y = $ ___ $+ 3$.

   b Rearrange $y + 3 = 2x$ as $y = 2x - $ ___.

   c Use the graph to solve the simultaneous equations $y = 3 - x$ and $y + 3 = 2x$.

4 Use the graph to solve the simultaneous equations $x + y = 5$ and $y - x = 1$.

5 Use the graph to decide which of the following pairs of simultaneous equations have the solution $x = 8$, $y = 2$. Explain your answer.

Pair 1 $y = 4 + 0 \cdot 5x$ **Pair 2** $y = 0 \cdot 5x - 2$
          $y = x + 1$ $y = 10 - x$

6 Use the graph to write down simultaneous equations which have the solution $x = 4 \cdot 5, y = 5 \cdot 5$.

*7 Is it possible to find a solution to these pairs of simultaneous equations? Explain.

   a $y = x + 1$ **b** $y = {}^-x + 10$

       $y = x - 1$ $y = {}^-x + 5$

*8 Is it possible to find a common solution for these three equations? Explain.

       $y = 2x - 3$ $y = x - 1$ $y = {}^-x + 3$

9 Draw lines to solve these simultaneous equations.
   On your sets of axes, have both x and y from $^-4$ to 8.

   a $x - y = 3$ **b** $2x - y = {}^-1$ **c** $y - 2x = 3$ **d** $x + y = 2$

       $2x + y = 0$ $x - y = 1$ $y - x = 5$ $y = 2x - 1$

   e $2x + y = 4$

       $y = 2x + 2$

For Reviews 1 to 4 use the following graph

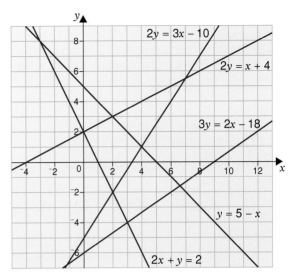

Review 1   Write down the solution of these simultaneous equations.

$2x + y = 2$
$3x - 2y = 10$

Review 2
a   Rearrange $2y - 3x = {}^-10$ as $2y = $ _____.
b   Rearrange $2y - x = 4$ as $2y = $ _____.
c   Write down the coordinates of the point where the lines $2y - 3x = {}^-10$ and $2y - x = 4$ meet.
d   Use your answer to **c** to solve the simultaneous equations $2y - 3x = {}^-10$ and $2y - x = 4$.

Review 3   Use the graph to decide which of these pairs of simultaneous equations has the solution $x = 0, y = 2$.

Pair 1 $2y - 3x = {}^-10$
             $y = 5 - x$

Pair 2 $2x + y = 4$
             $x + 2y = 4$

Pair 3 $2x + y = 2$
             $x - 2y = {}^-4$

Review 4

∗a   Is it possible to find a solution to the pair of simultaneous equations $\begin{array}{l} x + y = 5 \\ x + y = {}^-1 \end{array}$?
     Explain.
b   Is it possible to find a common solution for the three equations $\begin{array}{l} x + y = {}^-1 \\ 2y = x + 4 \\ 2x + y = 2 \end{array}$?
    Explain your answer.

Review 5   Solve these simultaneous equations graphically.
a $x + y = {}^-1$
    $2x + y = 1$
b $y = 2x - 1$
    $x - y = {}^-2$

Using simultaneous equations to solve problems

Problems which have two unknowns may be **solved using simultaneous equations** if two equations can be written down.

1 Assign a letter to each of the unknowns.
2 Write down two equations for the unknowns. This means we must interpret the words to make a mathematical sentence.
3 Solve these two simultaneous equations.
4 Check the solutions.

Algebra

Worked Example

Tickets for a concert cost £12 for an adult and £8 for a student. 800 tickets were sold altogether for a total of £8540. How many tickets were sold to adults and how many to students?

Answer

1 Let a be the number of tickets sold to adults and s be the number sold to students.

2   $a + s = 800$
   $12a + 8s = 8540$

3 The adding and subtracting method could be used as follows:

$a + s = 800$ $12a + 12s = 9600$ multiplying both sides by 12
$12a + 8s = 8540$ becomes $12a + 8s = 8540$
                                 $4s = 1060$ subtracting the equations
                                  $s = 265$ dividing both sides by 4

If $s = 265$, $a + s = 800$ becomes $a + 265 = 800$
                                  $a = 535$ subtracting 265 from both sides

4 If $s = 265$, $a = 535$ then $12a + 8s = 12 \times 535 + 8 \times 265 = 8540$, which is correct. ✓

Exercise 7

1 The difference between two numbers p and q is 21.

   $p - q = 21$

   a The two numbers p and q add to 95. Write an equation for this.
   b Solve the two simultaneous equations to find the two numbers.

2 a Lance owned 500 Reece Holdings shares and 200 Fletcher shares. Altogether Lance's shares were worth £2300.
   $500r + 200f = 2300$ is an equation for Lance's shares.
   What does r stand for?
   What does f stand for?

   b Anne owned 50 Reece Holdings shares and 100 Fletcher shares. In total, Anne's shares were worth £550.
   Write another equation, using r and f, for Anne's shares.

   c Solve the two simultaneous equations from **a** and **b** to find how much each Reece Holdings share and each Fletcher share was worth.

3 In the Barnway School hall, some rows seat 25 people and some seat 15 people. There are 26 rows altogether. When the hall is full it seats 550 people.
   a Let x be the number of rows which seat 25 and y the number of rows which seat 15. Write down two equations for x and y.
   b Solve these equations to find the number of rows which seat 25 people and the number of rows which seat 15 people.

4 A bakery was asked to bake a number of pies in a number of days.
   If the bakery bakes 160 pies each day, then they will have baked 50 pies too few.
   If the bakery bakes 180 pies each day, then they will have baked 50 pies too many.
   How many pies was the bakery asked for?
   How many days will it take to bake them?

5 The ages of an elderly couple add to 154. If the wife is six years older than her husband, how old is the wife?

6 Yesterday, Chun's bank balance was twice as large as Katie's. After Chun had deposited an extra £15 today, her bank balance was £100 more than Katie's. How much did Chun have in her bank account yesterday?

7 Find two numbers such that twice their sum is 66 and their difference is 3.

8 Each year a school has a concert of readings and songs.
Last year the concert had 3 readings and 9 songs.
It lasted 120 minutes.
The year before the concert had 5 readings and 5 songs.
It lasted 90 minutes.
This year the school plans to have 5 readings and 7 songs.
Use simultaneous equations to estimate how long the concert will last.
Call the time estimated for a reading x minutes, and the time estimated for a song y minutes.
You **must** show your working.

[SATs Paper 2 Level 8]

9 a In a plant sale, small trees were £3 and large trees £5. On one day, 18 of these were sold for a total of £60.
   One equation for this information is $3s + 5l = 60$.
   What do s and l stand for?
 b Copy and complete this table for $3s + 5l = 60$.

s	0	10	20
l			

 c Draw a set of axes like those shown.
   On your axes, draw the line $3s + 5l = 60$.
 d Write another equation which involves s and l.
 e Draw another line on your graph to find how many small trees and how many large trees were sold.

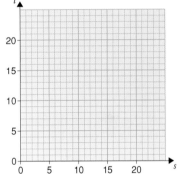

10 In the Ziggy night-club, members pay £50 per year and £5 per night. Non-members pay £10 per night.
A formula which gives the cost for a member who goes to this night-club n times in one year is $C = 50 + 5n$.
 a Use a copy of these axes. On your axes, draw the graph of $C = 50 + 5n$.
 b Copy and complete the formula $C = $ _____ for a non-member who goes to this night-club n times in one year.
   On your graph, draw another line for this formula.
 c Damian often goes to the Ziggy night-club. Use your graph to find the number of times he could go before it becomes cheaper for him to be a member.

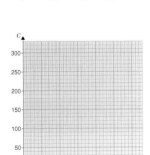

*11 Make up a problem for this pair of simultaneous equations.
$$s + d = 47$$
$$s + 2d = 78$$

*12 Find a pair of numbers satisfying $9x - 2y = 68$ such that one number is four times the other.
Is there more than one answer?

Algebra

Review 1 206 people have reserved seats to Brighton on the Newton coaches. They have paid a total of £1856. The adult's fare is £11 and the child's fare is £6.
a One equation for this information is $11a + 6c = 1856$. What does a stand for? What does c stand for?
b Write another equation for a and c.
c Solve the two equations to find how many adults and how many children had reserved seats.

Review 2 The Great Outdoors Boating Company has 20 large and 5 small canoes for hire. The New Canoe Company has 10 large and 15 small canoes for hire. The Great Outdoors Boating Company canoes can carry a maximum of 185 people while the New Canoe Company canoes can carry a maximum of 155 people.
How many people can each of the large canoes carry?

Review 3
a The Eastlake Fun Park entrance fee is £5. Each ride costs £1·50. A formula for this information is $C = 5 + 1·5n$. What does n stand for? What does C stand for?
Copy and complete this table for $C = 5 + 1·5n$.

n	0	4	8
C			

b Draw a set of axes like those shown.
Draw the line $C = 5 + 1·5n$.

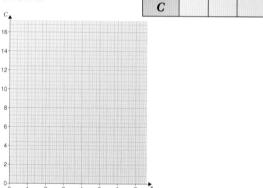

c At the Westerfield Fun Park, the entrance fee is £2 and each ride costs £2.
Copy and complete this formula for the total cost for n rides.

$$C = \underline{\quad} + 2n.$$

Draw another line on your graph for this.
d Glen went to the Eastlake Fun Park on Saturday and to the Westerfield Fun Park on Sunday. Both days he had the same number of rides for the same total cost.
Explain how you can use your graph to find the number of rides Glen had each day. How many rides was this? Including the entrance fee, what did these cost at Eastlake Fun Park?

Solving inequalities

Discussion

- If $n + 4 = 6$, n can have only one value. What is this one value?
  Which integer values could n have if $n + 4 > 6$?
  Is 2·5 also a solution for $n + 4 > 6$? Is 2·1? Is $2\frac{1}{9}$?
  Can you list *all* the solutions for $n + 4 > 6$? **Discuss**.

- Which of the following show *all* the solutions for $n + 4 > 6$? **Discuss**.
  What do you think ● and ○ mean?
  What do the arrows mean? **Discuss**.

Worked Example

Show these on the number line. List the integer solutions.

a $x > {}^-2$ b $a \leqslant 3$ c $1 \leqslant n < 4$

Answer

a  ○ means the value ⁻2 is not included.

The integer solutions are ⁻1, 0, 1, 2, ...

b ← —┼—┼—┼—┼—●—┼→
   ⁻1 0 1 2 3 4
● means the value 3 is included.

The integer solutions are ..., ⁻1, 0, 1, 2, 3.

c ┼—●—┼—┼—○—┼
   0 1 2 3 4 5
● means the value 1 is included; ○ means the value 4 is not included.

The integer solutions are 1, 2, 3.

Worked Example

d satisfies each of these inequalities.

$21\cdot36 \leqslant d \leqslant 21\cdot78$ and $21\cdot93 > d > 21\cdot46$

Mark the solution set on a number line.

Answer

Because the inequalities overlap we need to use two lines.
The solution set that satisfies *both* inequalities is shown.

21·36   21·46   21·78   21·93

21·46   21·78

Exercise 8

1 Write down the inequalities displayed below. Use n for the variable.

a ——————●——————→
        ⁻2

b ←——————●——————
            4

c ←——————○——————
            1

d ——————●——————→
        2·5

e ——————●——————○——
        ⁻3   4

f ○——————————○
   ⁻3   5

g ——○——————————●——
     1   4

h ——○——————————○——
     2   6·8

i ——○——————————○——
    ⁻1½   2

j ——●——————————○——
    ⁻3   2·5

k ——○——————————○——
    ⁻4   0

2 a n is a prime number.
         Write down all the values of n that make $n < 30$ true.

   b m is a fraction.
         Write down a value of m that makes $\frac{1}{3} < m < \frac{1}{2}$ true.

3 Show the following inequalities on a number line. List the integer solutions.

   a $x \geqslant 4$ b $x < 2$ c $a > {}^-2$ d $p \leqslant 0$
   e $1 \leqslant n \leqslant 4$ f ${}^-2 < n < 2$ g ${}^-1 \leqslant a < 2$ h ${}^-2 < x \leqslant 3$

*4 q satisfies each of these inequalities.
   Mark the solution set on a number line.

   a $18\cdot12 \leqslant q \leqslant 19\cdot47$ b $53\cdot61 < q \leqslant 54\cdot28$
      and $19\cdot86 > q > 18\cdot37$ and $55\cdot27 \geqslant q > 53\cdot92$

Review 1 Show each of these on a number line. List the integer solutions.

a ${}^-4 \leqslant x < 3$ b $p \leqslant 2$ c $n > {}^-3$ d ${}^-6 < a \leqslant 0$

Algebra

***Review 2** x satisfies each of these inequalities.
Mark the solution set on a number line.

$21{\cdot}23 < x < 41{\cdot}62$

$47{\cdot}5 > x \geqslant 27{\cdot}9$

Discussion

$2 < 8$ is true.

$2 + \mathbf{3} < 8 + \mathbf{3}$

When we add 3 to *both* sides, is the inequality still true? **Discuss.**
What if we added 5, subtracted 4, added 7, subtracted 9, ... to both sides? **Discuss.**

$2 \times \mathbf{3} < 8 \times \mathbf{3}$

When we multiply both sides of the inequality by 3, is it still true? **Discuss.**
What if we multiply both sides by 4, divide both sides by 2, multiply both sides by 1·5, divide both sides by 0·5, ...? **Discuss.**

What if we multiply both sides by ⁻1? What happens?
What if we divide both sides by ⁻1?
Discuss.

If we start with an **inequality** that is true, it will stay true if we

add or subtract the same number to both sides
multiply or divide both sides by the same **positive** number.

If we multiply or divide both sides by the same **negative** number, the inequality sign must be reversed.

Examples

$^{-}4 > ^{-}8$
$^{-}4 + 6 > ^{-}8 + 6$ adding a positive number
$2 > ^{-}2$

$^{-}3 < 4$
$^{-}3 \times 6 < 4 \times 6$ multiplying by a positive number
$^{-}18 < 24$

$8 > ^{-}4$
$8 \times ^{-}1 < ^{-}4 \times ^{-}1$ reverse the inequality when multiplying or dividing by a negative number
$^{-}8 < 4$

We **solve inequalities** in the same way as we solve equations.
If you multiply or divide both sides by a negative number – reverse the direction of the inequality sign.

Worked Example
Solve $2n - 1 \leqslant 15$. Show the answer on a number line.

Answer
$2n - 1 \leqslant 15$
$2n \leqslant 16$ adding 1 to both sides
$n \leqslant \mathbf{8}$ dividing both sides by 2

Note The inequality sign stays the same throughout because we did not multiply or divide by a negative number.

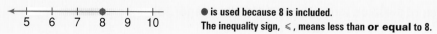

● is used because 8 is included.
The inequality sign, $\leqslant$, means less than **or equal to** 8.

144

Worked Example
Solve $2(3 - 2x) < {}^-3$.　　　Show the answer on a number line.

Answer
$2(3 - 2x) < {}^-3$
　　$6 - 4x < {}^-3$　　　expanding brackets
　　　${}^-4x < {}^-9$　　　subtracting 6 from both sides
　　　　$x > \frac{{}^-9}{{}^-4}$　　　dividing both sides by ${}^-4$ (so change direction of inequality sign)
　　　　$x > \mathbf{2 \cdot 25}$　

$2 \cdot 25$

Worked Example
y satisfies both of these inequalities.　　　　　　　$6 - 3y \leqslant 15$
　　　　　　　　　　　　　　　and　$5y + 3 \leqslant 13$

Mark the solution set for y on a number line.

Answer
$6 - 3y \leqslant 15$
　${}^-3y \leqslant 9$　　　subtracting 6 from both sides
　　$y \geqslant {}^-3$　　　dividing both sides by ${}^-3$ (so change the direction of the inequality sign)

and　$5y + 3 \leqslant 13$
　　　$5y \leqslant 10$　　　subtracting 3 from both sides
　　　$y \leqslant 2$　　　dividing both sides by 5

so　${}^-3 \leqslant y \leqslant 2$
$\begin{array}{ccccccccccccc} & | & \bullet & | & | & | & | & | & \bullet & | & | & | & | \\ {}^-4 & {}^-3 & {}^-2 & {}^-1 & 0 & 1 & 2 & 3 & 4 & 5 & 6 & 7 \end{array}$

Exercise 9

1　Solve these inequalities.
　　a　$x + 7 > 12$　　**b**　$n - 3 \leqslant {}^-1$　　**c**　$a + 8 \geqslant 2$　　**d**　$4n < 12$　　**e**　$n - 2 < 3$
　　f　$2a - 1 < 5$　　**g**　$4x + 5 \geqslant {}^-3$　　**h**　$\frac{a}{7} - 2 < 1$　　**i**　$\frac{n+3}{2} < {}^-2$　　**j**　$\frac{1+2n}{3} \geqslant 4$

2　Solve these inequalities. Show the solution set on a number line.
　　a　$2n > 5$　　**b**　$2n \geqslant {}^-5$　　**c**　${}^-2n > 5$　　**d**　${}^-2n \geqslant {}^-5$　　**e**　$3 - 2x \leqslant 4$
　　f　$2 - 5x > {}^-8$　　**g**　$\frac{{}^-5a}{3} < 4$　　**h**　$\frac{{}^-2a}{5} \geqslant {}^-1$　　**i**　$2 - \frac{5n}{4} > 1$　　**j**　$\frac{3 - 2a}{2} \leqslant {}^-4$

3　Find the smallest integer value for which **a**, **d**, **e** and **h** are true and the largest integer value for which **b**, **c**, **f** and **g** are true.
　　a　$2(3 + 2n) > 10$　　**b**　$2(2n + 1) \leqslant {}^-5$　　**c**　$2(2 - 3n) \geqslant 7$　　**d**　$4(1 - n) \leqslant 15$
　　e　$6n - n \geqslant {}^-20$　　**f**　$2n + 5 > 5n - 2$　　**g**　$2n + 1 < 3$　　**h**　$n - 6n < 4$

4　Solve these inequalities. Show your working.　　　　　　**[SATs Paper 1 Level 8]**
　　a　$\frac{2(2y + 7)}{3} < 2$　　**b**　$\frac{4(7 - 2y)}{12} > 1$

5　Solve these inequalities.
　　a　${}^-1 < x + 3 < 4$　　**b**　$2 \leqslant a - 3 \leqslant 7$　　**c**　${}^-5 \leqslant 5n < 25$　　**d**　${}^-5 < \frac{n}{4} < 2$

6　What can you say about n if:　**a**　$3n + 4 \leqslant {}^-11$　　**b**　${}^-3 < n - 2 < 8$

7　Write down the values of n, where n is an integer, such that
　　a　$3(2 + n) > 10$　　**b**　$2 - 3n \leqslant {}^-5$　　**c**　$3n - 9 < 7 - 4n$.

*8　Find all the possible integer values of m that satisfy $2m < 19$ *and* $3m > 5$.

Algebra

*9 Solve these inequalities, marking the solution set on a number line.

a $5p + 3 > 12$
    and $p < 3$

b $2(y - 2) \leq 1$
    and $y \geq {}^-1$

c $4 - 3m \leq 16$
    and $2m + 4 \leq 8$

*10 A bus can carry a maximum of 46 passengers.
    A school wants to take 5 adults and as many groups of
    4 children as possible on the bus.
    a   Which of these inequalities is true for the bus?

    A $4n + 5 > 46$ B $4n + 5 \leq 46$
    C $4n - 5 < 46$ D $4n - 5 \geq 46$

    b   Solve the inequality to find the maximum number of
        groups of 4 children the bus can carry.

Review 1   Solve these inequalities. Show each solution set on a number line.

a $2a + 5 \geq 9$ b $3 + 5a < {}^-7$ c $2(3 + 2x) > {}^-1$
d $\frac{n+5}{2} \leq 6$ e $2 + n < 5n - 4$ f $\frac{-2x}{3} \geq 5$

Review 2   Write down the values of n, where n is an integer, such that

a ${}^-8 < 2n \leq 6$ b $5 - 2n < 12$.

Review 3   Find all possible integer values of n that satisfy $2n > {}^-5$ and $5n < 21$.

*Review 4**   Solve these inequalities, showing each solution set on a number line.

a $2a - 4 < 3$
    and $a \geq {}^-2$

b $3(b - 1) > {}^-5$
    and $5 - 4b > 3$

Linear inequalities in two variables

$2x + 3 \geq 1$ is a **linear inequality in one variable**, x. A linear inequality in one variable can be
shown on a number line.

$2x + y \geq 1$ is a **linear inequality in two variables**, x and y. A linear inequality in two variables
can be shown on a set of axes.

Discussion

● The red shaded region can be described by a set of three
  inequalities.
  Which three are they? **Discuss.**

$x > y$	$x > 0$	$y > 0$	$x < 2$	$x > 2$
$x \geq 2$	$x \leq 2$	$y \geq 2$	$y \leq 2$	$y > x + 1$
$y < x + 1$	$y \geq x + 1$	$y \leq x + 1$		

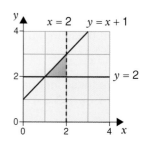

Note   A solid line on the graph indicates that the values **on** the
           line are included in the region.
           A dashed line indicates that the values on the line are
           not included in the region.

● The line $y = {}^-1$ is shown on this graph.
In which region of this graph are the following
inequalities true: $y > {}^-1$, $y \geqslant {}^-1$, $y < {}^-1$, $y \leqslant {}^-1$?
Discuss. Test points to check.

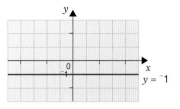

The line $x = 2$ is shown on this graph.
In which region of this graph are the following
inequalities true: $x > 2$, $x \geqslant 2$, $x < 2$, $x \leqslant 2$? **Discuss**.

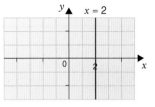

In which region of this graph are both $x \geqslant 2$ and $y \leqslant {}^-1$
true? **Discuss**.

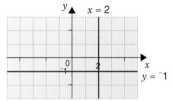

● $x = 2$, $y = {}^-1$
Is the inequality $2x + y > 1$ true for these values of
x and y? Where, in relation to the line $2x + y = 1$, is the
point $(2, {}^-1)$? **Discuss**.

What if $x = 2, y = 3$? $x = 1, y = 2$? $x = 0, y = {}^-2$? $x = {}^-2, y = 3$?
$x = \frac{1}{2}, y = {}^-1$? $x = 1, y = {}^-1$? $x = 0, y = 1$?
What if ...
Where, on the graph, are all the points for which the
inequality $2x + y > 1$ is true? **Discuss**.

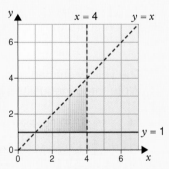

Worked Example
Shade the region where
$y < x$ $x < 4$ *and* $y \geqslant 1$.

Answer
First we must draw the lines $y = x$, $x = 4$ and $y = 1$.
The lines $y = x$ and $x = 4$ are dashed because the inequality signs
are $>$ and $<$.
The line $y = 1$ is solid because the inequality sign is $\geqslant$.
To find whether to shade above or below the line $y = x$, test a point.
Choose the point $(3, 2)$.
    $2 < 3$ so $y < x$ is true for this point.
Shade the region to the right of the line $y = x$.
The answer is shaded in orange.

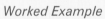

Worked Example
The blue shaded region is bounded by the line $y = 3$ and the curve
$y = x^2 + 1$.
Which of these fully describe the shaded region?
A $y > x^2 + 1$ *and* $y < 3$ B $y \leqslant x^2 + 1$ *and* $y \leqslant 3$
C $y \geqslant x^2 + 1$ *and* $y < 3$ D $y \geqslant x^2 + 1$ *and* $y \leqslant 3$

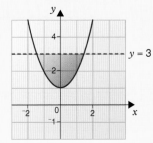

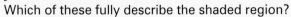

Answer

C because the shaded region has values greater than or equal to $x^2 + 1$. $y = x^2 + 1$ is included in the region because it is a solid line. The shaded region has values less than $y = 3$. The dotted line shows us that $y = 3$ is not included in the region.

It is best to test a point to see if $y < x^2 + 1$ or $y \geqslant x^2 + 1$.

Test $(1, 2)$: $2 \geqslant (1)^2$ **True** so $(1, 2)$ lies in the region where $y \geqslant x^2 + 1$

Exercise 10

1 Which four inequalities fully describe the orange shaded region?
Choose one from each row.

$x \leqslant 2, x \geqslant 2, x > 2, x < 2$
$x > {}^-3, x \geqslant {}^-3, x < {}^-3, x \leqslant {}^-3$
$y < 2, y \leqslant 2, y > 2, y \geqslant 2$
$y > 0, y \geqslant 0, y < 0, y \leqslant 0, x \geqslant 0$

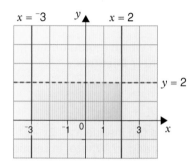

T **2** Use a copy of this.
    I am thinking of a point on the dotted grid.
    The coordinates of my point are (x, y).
    You have 3 clues to find which of the dots is my point.

[SATs Paper 2 Level 7]

a **First clue:** $x > \mathbf{0}$ **and** $y > \mathbf{0}$
    Which dots **cannot** represent my point?
    On the grid **cross them out** like this ✖

b **Second clue:** $x + y < \mathbf{4}$
    Which other dots **cannot** represent my point?
    This time, put a **square around them** like this ◉

c **Third clue:** $x > y$
    What are the coordinates of my point?

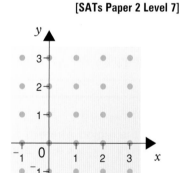

3 The yellow shaded region is formed by straight-line graphs.
Write three inequalities to describe the shaded region.

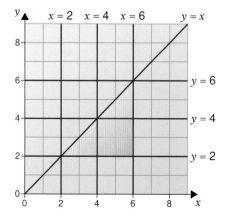

4 The darker shaded area is given by which inequality?

a

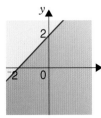

A $y \leqslant x + 2$
B $y < x + 2$
C $y \geqslant x + 2$
D $y > x + 2$

b

A $x - y \geqslant {}^-3$
B $x - y > {}^-3$
C $x - y \leqslant {}^-3$
D $x - y < {}^-3$

c

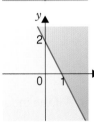

A $2x + y > 2$
B $2x + y \geqslant 2$
C $2x + y < 2$
D $2x + y \leqslant 2$

5 Shade the region in which these inequalities are true. Draw a separate graph for each.
Number both the x- and y-axes from $^-5$ to 5.
a $y \geqslant 2x + 1$ **b** $y \leqslant x - 4$ **c** $y > {}^-x + 3$ **d** $x + y < 1$

6 Shade the region in which both of the inequalities are true.
a $x < {}^-1$ and $y \leqslant 3$ **b** $x \geqslant 2$ and $y \geqslant 3$ **c** $x > 3$ and $y < x$
d $y < 3$ and $y \geqslant x + 1$ **e** $y \geqslant 2x - 1$ and $x \geqslant 1$ **f** $y > 3$ and $x - y > {}^-3$

7 The blue shaded region is bounded by the line $y = 3$
and the curve $y = x^2 - 1$.
Write down two of these inequalities that fully describe
the shaded region.

$y < x^2 - 1$ $x \leqslant 0$ $y \leqslant 3$ $y \leqslant 0$
$y \geqslant x^2 - 1$ $x \geqslant 0$ $y \geqslant 3$ $y \geqslant 0$

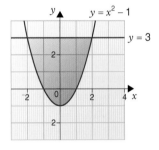

*8 A region is defined by four inequalities. Shade this region if the inequalities are
a $y \leqslant 4, y \geqslant {}^-3, x \leqslant 5, x \geqslant 0$ **b** $y \geqslant 0, x \geqslant {}^-3, y \leqslant 3, x \leqslant 1$.

*9 The region R is defined by three inequalities. Shade R in the following cases.
a $x \geqslant 0, y \geqslant 0, y \geqslant x + 3$ **b** $y \leqslant x, x < 3, y \geqslant {}^-3$ ***c** $y \geqslant {}^-3, x \geqslant {}^-1, y < 2x + 1$
*d $x + y \leqslant 3, x \geqslant 0, y \geqslant 0$ ***e** $x > {}^-3, y \geqslant {}^-1, x + y < 2$

*10 Draw a diagram to show where both inequalities are true.
a $y \geqslant x - 2$ and $y < x + 4$ **b** $y < 2x$ and $y \geqslant x + 1$ **c** $x + y > 2$ and $y > x - 3$
d $2x + y \geqslant 4$ and $2y \leqslant x + 2$ **e** $x - 2y < 4$ and $y \leqslant x$ **f** $2x - 3y \leqslant 6$ and $3x + 2y < 0$

*11 Write inequalities that fully describe the regions, R.

a

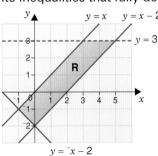

*b

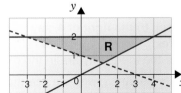

Review 1

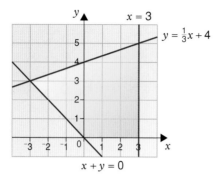

The yellow shaded region is formed by four straight lines.
Write four inequalities to describe the shaded region.

Review 2 The region S is defined by two inequalities. Shade this region S if the inequalities are
a $x \leqslant 5$ and $y \geqslant 1$ **b** $y \geqslant x - 1$ and $y < 3$ **c** $2x - y > 1$ and $x + y < 0$.

* **Review 3** Draw a diagram to show where all three inequalities are true.
a $x \geqslant {}^-2$ and $y \leqslant 4$ and $y > 2x - 3$
b $x \geqslant 0$ and $4y \leqslant x - 4$ and $x - y \leqslant 4$

Solving quadratic inequalities

The quadratic equation $x^2 = 16$ has two solutions, $y = 4$ and $y = {}^-4$.
These two points divide the number line into three regions.
The quadratic inequality $x^2 \geqslant 16$ is true for two of these regions.
We test a point in each region to check which two.

Worked Example
Solve the inequality $x^2 \geqslant 16$.

Answer
Find the end points of the regions by solving $x^2 = 16$.
If $x^2 = 16$, then $x = 4$ or $x = {}^-4$.

Since the inequality sign is $\geqslant$, the symbol ● is placed on these end points.

Now test a point in each of the three regions.

Choose a point to the left of $^-4$, say $^-5$. Substitute $^-5$ for x.
Is $({}^-5)^2 > 16$? Yes.
So the inequality is true in the region to the left of $^-4$.

Choose a point between $^-4$ and 4, say 3. Substitute 3 for x.
Is $(3)^2 > 16$? No.
Solutions for the inequality are not in the region between $^-4$ and 4.

Choose a point to the right of 4, say $x = 6$. Substitute 6 for x.
Is $6^2 > 16$? Yes.
Solutions for the inequality are in the region to the right of 4.

From this number line graph we see that the solutions for $x^2 \geqslant 16$ are $x \leqslant {}^-4$ **and** $x \geqslant 4$.

Exercise 11

1 Katie is solving the inequality $y^2 < 9$.   [SATs Paper 1 Level 8]
She says

'$y^2 < 9$ whenever y is less than 3.'

Katie is not correct. Explain why.

2 Solve these inequalities.

  a $x^2 \geqslant 9$ **b** $x^2 > 4$ **c** $n^2 > 36$ **d** $n^2 \geqslant 100$ **e** $a^2 < 25$
  f $a^2 < 100$ **g** $x^2 \leqslant 49$ **h** $x^2 > 64$ **i** $x^2 > 81$ **j** $n^2 \leqslant 4$

*3 Solve these inequalities.

  a $2x^2 \geqslant 8$ **b** $3x^2 \geqslant 48$ **c** $3x^2 < 12$ **d** $x^2 - 2 < 47$
  e $x^2 + 5 > 30$ **f** $x^2 - 4 \leqslant 60$ **g** $2x^2 - 5 < 13$ **h** $\frac{x^2}{4} \leqslant 9$
  i $3x^2 + 2 < 77$ **j** $7 + 2x^2 > 9$

Make x^2 the subject first before testing points.

Review 1   Solve these inequalities.
a $x^2 > 1$ **b** $x^2 \leqslant 16$ **c** $x^2 < 81$ **d** $x^2 \geqslant 144$

* Review 2   Solve these inequalities.
a $5x^2 > 80$ **b** $x^2 + 3 < 28$ **c** $3x^2 - 1 \geqslant 11$ **d** $\frac{x^2 - 1}{3} \leqslant 16$

Summary of key points

In algebra letters are used in **equations, formulae and functions**.

This is a **linear equation**. The unknown has a particular value.

$$2x - 4 = 9$$

A **formula** gives the relationship between variables which stand for something specific.

Example $E = \frac{1}{2}mv^2$ where E is energy, v is speed, m is mass.

A **function** gives the relationship between two variables, usually x and y.

Example $y = 2x - 7$

An **identity** is true for all values of the unknown.

Example $2(3x - 4) \equiv 6x - 8$ $\equiv$ means 'is identically equal to'.

We can choose any value of x and the left-hand side will equal the right-hand side.

If there is an unknown on both sides of the equation, we **transform** both sides in the same way.

Example $\frac{1}{3}(7p - 3) = \frac{1}{6}(4p + 12)$

           $\frac{1}{3}(7p - 3) \times 6 = \frac{1}{6}(4p + 12) \times 6$ multiplying both sides by 6 to get whole number values

           $\frac{1}{3} \times 6(7p - 3) = \frac{1}{6} \times 6(4p + 12)$

           $2(7p - 3) = 4p + 12$

           $14p - 6 = 4p + 12$ multiplying out the brackets

           $10p - 6 = 12$ subtracting $4p$ from both sides

           $10p = 18$ adding 6 to both sides

           $p = 1 \cdot 8$ dividing both sides by 10

C **Simultaneous equations** are equations which need to be solved together to find the value of the unknowns.

We can solve simultaneous equations in three ways.

1 adding and subtracting equations

Sometimes we can add or subtract the equations and one of the unknowns will be eliminated.

See page 132 for an example.

Sometimes we have to **multiply both sides of one or both equations** before we can solve by adding and subtracting.

See page 133 for an example.

2 substitution

See page 135 for an example.

3 drawing graphs

If each equation is graphed, the point where the graphs intersect gives the solution for the simultaneous equations.

See page 136 for an example.

D We can use simultaneous equations to **solve problems**.

E We can show solutions to **inequalities** on a number line.

Examples **a** $x > {}^-2$ **b** $0 \leqslant n < 5$

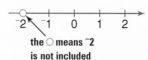
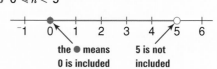

The integer solutions of **a** are $^-1, 0, 1, 2, \ldots$

The integer solutions of **b** are $0, 1, 2, 3, 4$.

F We solve **inequalities** in the same way we solve equations.

When working through the steps, if we multiply or divide both sides of the inequality by a negative number, we must change the direction of the inequality.

Examples **a** $2n + 3 \geqslant 7$ **b** $3(2 - x) \leqslant 15$

                $2n \geqslant 4$ subtract 3 from both sides $3 - 3x \leqslant 15$

                  $n \geqslant 2$ divide both sides by 2 ${}^-3x \leqslant 12$ subtract 3 from both sides

                                              $x \geqslant {}^-4$ divide both sides by $^-3$

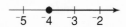

G A **linear inequality with two variables** can be shown on a set of axes.

See page 147 for an example.

H **Quadratic inequalities** have a term with an index 2.

We can solve quadratic inequalities and show the solutions on a number line.

See page 150 for an example.

Test yourself

1 Which of these are equations, which are formulae, which are functions?

   a $y = 2x + 3$ **b** $A = \pi r^2$ where r is the radius and A is the area of a circle

   c $a + 7 = 15$ **d** $S = \frac{D}{T}$ where S is speed, D is distance, T is time

   e $6(x + 5) = 20$ **f** $2y - x = 7$

2 **a** Show that $4(x - 3) \equiv 4x - 12$ is true for

     i $x = 2$ **ii** $x = 7$ **iii** $x = ^-4$.

   b By expanding the brackets show that it is true

   c What is the special name we give to $4(x - 3) \equiv 4x - 12$?

3 Solve these equations.
Show your working.

   a $7x - 3 = 1 + 5x$ **b** $2x - 3 = 5(x + 2)$ **c** $\frac{2}{3}(x - 1) = \frac{3}{4}(x + 2)$

   d $2(4x - 1) = 4(2x + 1) - 2(5x - 4)$ **e** $\frac{3y}{2y - 1} = 6$ **f** $\frac{16}{y + 3} = y + 3$

4 Use either the adding and subtracting method or the substitution method to
solve these simultaneous equations.

   a $5x - 6y = 8$ **b** $x - 2y - 1 = 0$
      $4x + 3y = ^-17$ $x + 4y = 4$

5 Is it possible to find a solution to these pairs of simultaneous equations? Explain.

   a $y = \frac{1}{2}x - 2$ **b** $y = ^-x - 2$ **c** $y - 2x = 6$
      $y = \frac{1}{2}x + 3$ $y = ^-x + 4$ $y = 2(x + 3)$

6 Solve these simultaneous equations.   **[SATs Paper 1 Level 7]**
      $y = 2x + 1$
      $3y = 4x + 6$
Show your working.

7 The diagram shows a triangle.   **[SATs Paper 2 Level 8]**
Side XY is of length $11b$.
Side XZ is of length $2a + 3b$.
Side YZ is of length a.

The triangle is isosceles, with XY = XZ.
The perimeter of the triangle is 91.
Use algebra to find the values of a and b.

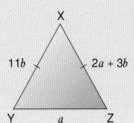

8

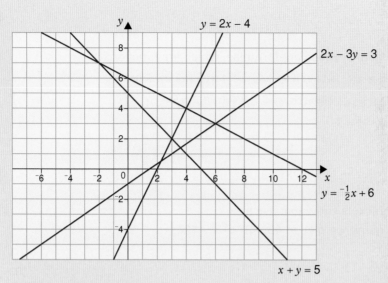

a Use the graph to find the solution of the simultaneous equations $2x - 3y = 3$ and $y = -\frac{1}{2}x + 6$.

b Rearrange $y = {}^-x + 5$ as $x +$ ___ $= 5$.

c Use the graph to find the values of x and y for which $y = {}^-x + 5$ and $y = -\frac{1}{2}x + 6$.

d Use the graph to write down two simultaneous equations which have the solution $x = 3, y = 2$.

9 On the same graph, draw the lines $x - y = 3$ and $y = 2x - 5$.
Use your graph to write down the solution of the simultaneous equations $y = 2x - 5$ and $x - y = 3$.

10 Avonlee Golf Club fees are £80 per year, with a special rate of £75 if paid before 1 May. One year, the 140 members paid a total of £11 100 in fees. Write down two simultaneous equations for e and l, where e is the number of members who paid before 1 May and l is the number who paid after this.
How many members paid their fees before 1 May?

11 Choose the correct inequality for each graph.

a

    A $n > 1$ **B** $n < 1$ **C** $n \geqslant 1$ **D** $n \leqslant 1$

b

    A ${}^-1 < n < 2$ **B** ${}^-1 \leqslant n \leqslant 2$ **C** ${}^-1 < n \leqslant 2$ **D** ${}^-1 \leqslant n < 2$

12 List all the integers for which ${}^-1 < n \leqslant 4$.

13 Show the following inequalities on a number line.
Write down all the integer solutions.
  a $x < 5$ **b** ${}^-1 \leqslant n < 5$ **c** $a \geqslant 0$

14 Write down the inequalities displayed on the number lines. Use n for the variable.

  a i ←———•———————   **ii** ———————○———→   **iii** ——○—————————•——
              3   ⁻4   1   5

  b i Show the inequality $^-3 < x < 2$ on a number line.
  ii Write down all the integer solutions for x if $^-3 < x < 2$.

15 Solve these inequalities.

  a $5a - 4 < 9$ **b** $^-5 < 2n + 1 \leqslant 11$ **c** $2(3x - 2) < 11$ **d** $\frac{1}{2}x \geqslant 1$ **e** $^-3x - 2 > 7$

16 Which four inequalities fully describe the pink shaded region?

  $x > 0$ $x \geqslant 0$ $x < 0$ $x \leqslant 0$

  $x > 6$ $x \geqslant 6$ $x < 6$ $x \leqslant 6$

  $y > 0$ $y \geqslant 0$ $y < 0$ $y \leqslant 0$

  $x + 2y < 8$ $x + 2y \leqslant 8$

  $x + 2y > 8$ $x + 2y \geqslant 8$

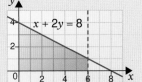

17 Solve these inequalities.

  a $x^2 \leqslant 4$ **b** $x^2 \geqslant 64$ **c** $5n^2 < 45$ **d** $\frac{n^2}{2} < 8$ **e** $3n^2 + 4 > 7$

18 Equations may have different numbers of solutions.   **[SATs Paper 1 Level 8]**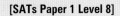

  For example: $x + 2 = 7$ has only one solution, $x = 5$ but $x + 1 + 2 = x + 3$ is true for all values of x.

  Use a copy of the table.
  Tick (✓) the correct box for each algebraic statement.

	Correct for **no** values of x	Correct for **one** value of x	Correct for **two** values of x	Correct for **all** values of x
$3x + 7 = 8$				
$3(x + 1) = 3x + 3$				
$x + 3 = x - 3$				
$5 + x = 5 - x$				
$x^2 = 9$				

T

7 Expressions and Formulae

You need to know

✓ expressions   page 114
✓ simplifying expressions   page 114
✓ substituting into expressions   page 114
✓ formulae   page 115

> ## Key vocabulary
>
> **algebraic expression,   collect like terms,   common factor,
> expand the product,   factorise,   formula,   formulae,   index law,
> simplest form,   subject of the formula,   take out common factors**

There's a skeleton in the cupboard

When a skeleton is found, forensic scientists can estimate the
height this person was from the length of various bones.

For **males**, the formulae are:

$$H = 3 \cdot 08h + 70 \cdot 45$$
$$H = 3 \cdot 7u + 70 \cdot 45$$
$$H = 2 \cdot 52t + 75 \cdot 79$$

For **females**, the formulae are:

$$H = 3 \cdot 36h + 57 \cdot 97$$
$$H = 4 \cdot 27u + 57 \cdot 76$$
$$H = 2 \cdot 90t + 59 \cdot 24$$

H = estimated height in centimetres
h = length of humerus bone (cm)
u = length of ulna bone (cm)
t = length of tibia bone (cm)

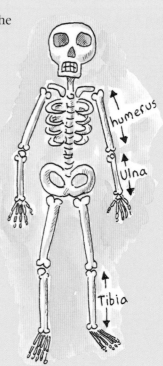

1 A forensic scientist found part of a female skeleton
with a tibia bone of 39·9 cm.
What was her estimated height?

* **2** Workout the length your humerus, ulna and tibia
should be for your height.

Writing and simplifying expressions

Remember

Before **collecting like terms** we must **expand brackets**.

Example 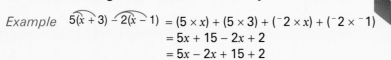 $5(x + 3) - 2(x - 1)$
$= (5 \times x) + (5 \times 3) + (^-2 \times x) + (^-2 \times ^-1)$
$= 5x + 15 - 2x + 2$
$= 5x - 2x + 15 + 2$
$= 3x + 17$ This is the expression written in its **simplest form**.

> To remind yourself about collecting like terms or multiplying out brackets go to page 114.

Example $3p - (^-p) = 3p - 1(^-p)$
$= 3p + ^-1 \times ^-p$
$= 3p + p$
$= \mathbf{4p}$

> A negative sign in front of a bracket can be written as $^-1 \times (...)$.

Note: Sometimes expanding brackets is called 'expanding the product'.

We sometimes use the **index laws** when simplifying expressions.

$$x^a \times x^b = x^{a+b} \qquad x^a \div x^b = x^{a-b} \qquad (x^a)^b = x^{ab}$$

> See page 41 for more index laws.

Examples   **a** $5a^3 \times 3a^2 = 5 \times 3 \times a^3 \times a^2$
$= 15 \times a^{3+2}$
$= \mathbf{15a^5}$

b $\dfrac{6p^5}{3p^2} = 2p^{5-2}$
$= \mathbf{2p^3}$

c $3x^2(2x^3 - 5x) = (3x^2 \times 2x^3) - (3x^2 \times 5x)$
$= 6x^{2+3} - 15x^{2+1}$
$= \mathbf{6x^5 - 15x^3}$

d $(2x^2)^4 = 2^4 \times x^{2 \times 4}$
$= \mathbf{16x^8}$

$$a^{-n} = \frac{1}{a^n} \qquad a^{\frac{1}{n}} = \sqrt[n]{x}$$

Example $(x^{-2})^3 = x^{-2 \times 3}$
$= x^{-6}$
$= \dfrac{1}{x^6}$

> See page 41 for more.

---

1 Show that this is a magic square.
The sum of the first row is:
$x + 4 + 4x + 4x + 5 = 9x + 9$

$x + 4$	$4x$	$4x + 5$
$6x + 4$	$3x + 3$	2
$2x + 1$	$2x + 6$	$5x + 2$

> **Remember**: each row, column and diagonal must have the same sum.

2 Simplify.   [SATs Paper 1 Level 6]
a $(3d + 5) + (d - 2)$ **b** $3m - (^-m)$

3 To cook roast lamb, allow 30 minutes per $\frac{1}{2}$ kg and then 25 minutes extra.
A leg of lamb weighs x kg.
Write an expression for the time needed to cook roast lamb.

Algebra

4 a Write expressions in their simplest forms for the area and perimeter of this rectangle.

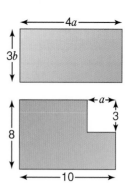

  b A different rectangle has area $15a^2$ and perimeter $16a$.
  What are the dimensions of this rectangle?

  c Chiraq and Isra both wrote expressions for the area of this shape.
  Chiraq wrote $80 - 3a$.
  Isra wrote $8(10 - a) + 5a$.
  Are they both correct? Justify your answer.

5 Write these without brackets and then simplify.
  a $(2n + 4) + (n - 3)$ **b** $n - 5 + 2(3 + 2n)$ **c** $3a^2 + a(3 + 2a)$
  d $x(2x + 5) - x$ **e** $6(x + 3) + 2(x - 1)$ **f** $4(m - 2) + 3(m + 4)$
  g $7b - (b + 2)$ **h** $5k - (^-k)$ **i** $3(n - 2) - 2(4 - 3n)$
  j $4(r + 2) - 3(r - 4)$ **k** $5(f - 2) - (3 - f)$ **l** $5n(n - 2) + 2n(1 - 2n)$
  m $3x(2x + 3y) + 2x(x - y)$ **n** $x(2 + x) - 2(2 - x)$ **o** $2x(2x - 3) - 3(4 - x)$

6 Simplify these using the index laws.
  a $m^4 \times m^2$ **b** $2b^3 \times 3b^2$ **c** $3n^4 \times 5n^3$ **d** $(6a^2)^2$ **e** $\dfrac{5y^4}{10y^2}$ **f** $\dfrac{24q^5}{8q^3}$

  g $6x^4 \div 3x^2$ **h** $k^5 \div k^8$ **i** $\dfrac{15x^2}{5x^4}$ **j** $\dfrac{12y^{12}}{8y^8}$ **k** $\dfrac{24a^2b^3}{20a^4b}$ **l** $\dfrac{48m^4n^3}{60mn^4p}$

7 You will need some square dot paper.
  On square dot paper you can join dots with two different length lines, a and b.
  The perimeter of the red shape is $4a + 3b$.

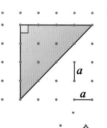

  a Write an expression for the perimeter of these shapes.
   i purple   **ii** green   **iii** blue

  b On square dot paper, draw shapes with these perimeters.
   i $4a + b$ **ii** $2(2a + b)$

  c What is the area of this triangle?
  Write it in terms of a.

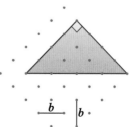

  d This is the same triangle and grid.
  What is the area of the triangle?
  Write it in terms of b.

  e Use you answers to **c** and **d** to explain why $2a^2 = b^2$.

8 Write an expression for the missing lengths in these shapes.

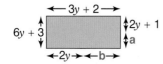

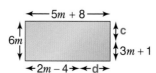

9 a Find a simplified expression for the perimeter and area of each of these rectangles.

b If the perimeters are the same, find the value of x.

10 a Alex, Kirsty and Ben each have a card with an expression on it.
Bens card is turned so we can't see the expression.
The mean of the three expressions is $3y$.
What is written on Ben's card?

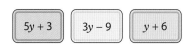

b Write three expressions which have a mean of $4y$.

c What is the mean of these three expressions?
Write your answer as simply as possible.

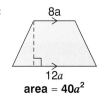

11 The answer is $6n - 3$. Make up a question that has this answer.

12 Write an expression for the red length in each of these.

a

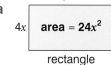

rectangle

b

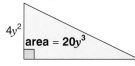

c

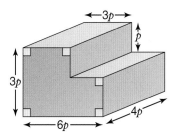

13 This solid is a prism with length $4p$.
The shaded shape is the cross-section.
Write an expression for the volume of the shape.
Simplify your expression.

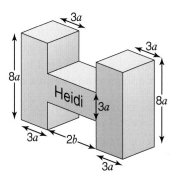

14 Heidi made a wooden H for her bedroom door.
She used three cuboids.
a Show that the area of the cross-section of the prism is $48a^2 + 6ab$.
∗b The volume of the prism is $6a$.
What is the depth of the prism?
Show your working.

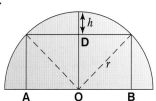

∗15 This diagram represents a rail tunnel with a semi-circular section.
O is the centre of the circle.
AB is the distance between the rails.
OD is the maximum height of a train able to use this tunnel.
a Find an expression for OD in terms of r and h.
b If AB = x, show that the relationship between x, h and r is
$x^2 + 4h^2 = 8rh$.
Hint use Pythagoras' theorem.

Algebra

Review 1 Simplify these.

a $2(a - 3) + 3a$ **b** $3t - (5 - t)$ **c** $4(q + 2r) - 3(2q + r)$

d $2p(p - 4) + p(6 - 2p)$ **e** $(4x^2)^3$ **f** $5n \times 2n^5$ **g** $\dfrac{8n^2}{12n^3p}$

Review 2 The perimeter of this rectangle is $3p + 2s$.
Write an expression, with p and s, for the fourth side.

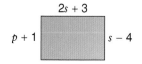

Review 3 Write an expression for the red length.

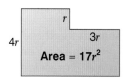

Review 4 A solid ramp is built by joining a solid
rectangular prism and a triangular prism as shown.

a Show that the area of the shaded cross-section of the
ramp is $n^2 + 4n + 4$.

b The width of the ramp is $2n$. Write an expression for
the volume of the ramp and expand the product.

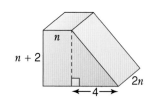

Expanding two brackets

Discussion

● We can write 23×64 as $(20 + 3)(60 + 4)$.
This can be worked out using the grid method.

	60	**+4**
20	1200	80
+3	180	12

Answer $= 1200 + 80 + 180 + 12$
         $= 1472$

We can use a grid diagram to expand $(a + b)(c + d)$
How might you continue? **Discuss**.

Use the same method to expand these.
     $(a + b)(a + b)$ $(a + b)(a - b)$ $(a + b)(c - d)$
Can any of the answers be simplified? **Discuss**
Can a and b be any values or expressions?

● Karl expanded $(3n - 4)(2n + 5)$ using a grid diagram.

	2n	**+5**
3n	$6n^2$	$15n$
⁻4	$⁻8n$	$⁻20$

$(3n - 4)(2n + 5) = 6n^2 + 7n - 20$

Jenni expanded $(3n - 4)(2n + 5)$ like this.

$$\left(3n - 4\right)\left(2n + 5\right) = 3n\left(2n + 5\right) - 4\left(2n + 5\right)$$
$$= 6n^2 + 15n - 8n - 20$$
$$= 6n^2 + 7n - 20$$

Which way do you like best? **Discuss**.

When we **expand two brackets** we multiply all the terms in the first bracket by all the terms in the second bracket.

Worked Example
Expand and simplify.　　**a** $(3x - 2)(x - 5)$　　**b** $(a + 2b)(3a - b)$

Answer

a

	x	-5
$3x$	$3x^2$	^-15x
$^-2$	^-2x	$+10$

$3x^2 - 15x$
$-2x + 10$
$\overline{3x^2 - 15x - 2x + 10}$

$(3x - 2)(x - 5) = 3x^2 - 15x - 2x + 10$
$\qquad\qquad\qquad\quad = \mathbf{3x^2 - 17x + 10}$

b $(a + 2b)(3a - b) = a(3a - b) + 2b(3a - b)$
$\qquad\qquad\qquad\qquad = 3a^2 - ab + 6ab - 2b^2$
$\qquad\qquad\qquad\qquad = \mathbf{3a^2 + 5ab - 2b^2}$

Note a and **b** show different ways of working out the answer. Either way can be used.

Exercise 2

1 Show that these identities are true by expanding the brackets.
   a $(m + n)(m + n) \equiv m^2 + 2mn + n^2$　　**b** $(p + q)(r + s) \equiv pr + ps + qr + qs$
   c $(a + b)(a - b) \equiv a^2 - b^2$　　**d** $(c - d)(e - f) \equiv ce - cf - de + df$

2 Expand and simplify if possible.
   a $(x + 3)(x + 7)$　　**b** $(n + 4)(n - 2)$　　**c** $(a - 3)(a - 7)$
   d $(2n + 3)(3n + 2)$　　**e** $(2x + 1)(3x + 5)$　　**f** $(4a + 3)(3a + 1)$
   g $(4n - 5)(3n + 1)$　　**h** $(2a - 3)(a + 7)$　　**i** $(2x + 3)(x - 4)$
   j $(5x - 1)(2x - 3)$　　**k** $(n - 7)(n + 4)$　　**l** $(x + 3)(2x - 5)$
   m $(a + b)(2a - b)$　　**n** $(3p + 2q)(p - 3q)$　　**o** $(c + 2d)(3e - 2f)$
   p $(x + 3y)(2w - 3z)$　　**q** $(n - 8)(3m - 4)$　　**r** $(2a + 4)(2a + 4)$
   s $(2p - q)(2p - q)$　　**t** $(5a + 2)(5 + 2a)$　　**u** $(2 - 3d)(d + 3) - d^2 + 3$
   v $(3 + 2x)(2 - 3x) + 6x^2$　　**w** $(3 - n)(2n + 1) - 7$　　**x** $(3x + 2)(3x - 4) - (2x + 4)$
   y $(2 + n)(3 - n) + n + 3$

> Use the identities in question 1 to help.

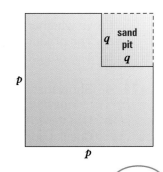

3 Multiply out and simplify these.
   a $(a + m)^2$　　**b** $(f - g)^2$　　**c** $(x + 3)^2$
   d $(2p + 2)^2$　　**e** $(a + 1)^2 - (a + 1) + 1$　　**f** $(3 + x)^2 - (x + 3) + 2$
   g $(y - 3)^2 - y^2 + 4$　　**h** $(m + 2)^2 + (2m - 5)^2$

4 Expand these using the identity $(a + b)(a - b) \equiv a^2 - b^2$.
   a $(x + 2)(x - 2)$　　**b** $(x + 3)(x - 3)$　　**c** $(a - 4)(a + 4)$　　**d** $(b + 1)(b - 1)$
   e $(n - 7)(n + 7)$　　**f** $(x - 9)(x + 9)$　　**g** $(a + 6)(a - 6)$　　**h** $(2x + 1)(2x - 1)$
   i $(5x - 2)(5x + 2)$　　**j** $(4a + 3)(4a - 3)$　　**k** $(2e - 3)(2e + 3)$　　**l** $(3n + 5)(3n - 5)$

5 Janie's backyard was a square lawn of side p metres.
   In one corner she built a square sand pit of side q metres.
   One expression for the area of lawn left after the sandpit was built
   is $p^2 - q^2$.
   a Show that this is true.
   b Write three other expressions for the area of lawn left, by
   dividing the shape in three different ways.
   c Show, by multiplying out brackets and simplifying, that all four
   expressions are equivalent.

6 a Use the identity $(a + b)(a - b) \equiv a^2 - b^2$ to factorise these into two brackets.
    i $y^2 - 9$ **ii** $4m^2 - n^2$ **iii** $16x^2 - 4y^2$
b Factorise $10^2 - 6^2$ into brackets.
    Use this to find the answer.

7 Rectangular paving tiles come in different sizes.
If the length of a tile is $3x - 1$ the width is $2x + 1$.
a Write an expression for the area of a tile.
b Multiply out and simplify your expression.

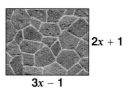

$2x + 1$

$3x - 1$

8 a Explain how you know that $(y + 3)^2$ is not equal to $y^2 + 9$.   **[SATs Paper 1 Level 8]**
b Multiply out and expand these expressions.
    i $(y + 2)(y + 5)$ **ii** $(y - 6)(y - 6)$ **iii** $(3y - 8)(2y + 5)$

*9 Use this diagram to show that $4a^2 - 3^2 = (2a + 3)(2a - 3)$.

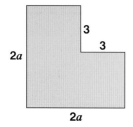

3
3
$2a$
$2a$

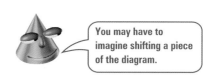

You may have to imagine shifting a piece of the diagram.

*10 a n and $n + 1$ are two consecutive whole numbers.
    The difference of their squares is 41.
    Find the two numbers.
    Use the diagram to help.
b Use a diagram to simplify $(y + 1)^2 - (y - 1)^2$.
c Find two consecutive odd numbers whose squares differ by 104.

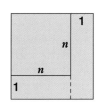

1
n
n
1

*11 Jamahl cut out four identical right-angled triangles from some gold paper.
He placed these on some backing card as shown to make a photo frame.
Each right-angled triangle has shorter sides of length a and
b and hypotenuse of length c.
a Show that the space for the photo is a square of side c.
b Show that the outer frame is a square of side $a + b$.
c Use the diagram to deduce Pythagoras' theorem $a^2 + b^2 = c^2$.

a c b

*12 Una had a sheet of lino of length p and width q.
She cut out a U-shaped piece to fit around a drain.
a By dividing the U-shaped piece into rectangles, write an
    expression for its area.
b By dividing it a different way, write another expression for
    its area.
c Show that the expressions you wrote in **a** and **b** are equivalent.

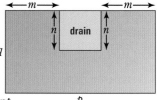

m m
n drain n
q
p

*13 Prove whether these statements are true or false.
a The product of two consecutive even numbers is always even.
b The product of two consecutive odd numbers is always even.

*14 a Show that
    $(n + 2)^2 = n^2 + 4n + 4$.
b Use this to calculate these.   **i** 22^2 **ii** 82^2 **iii** 802^2
    Check your results using a graphical calculator.

Review 1 Expand and simplify if possible.
a $(t-3)(t+4)$ **b** $(a-2)(a-1)$ **c** $(2a+3)(3a-4)$
d $(2f+3)(2f-3)$ **e** $(r-2)^2+5r$ **f** $(c+3)^2-(c3-2)$

Review 2 Sarah has a rectangular flower garden that measures l metres by w metres. It is surrounded on 3 sides by a 2 metre wide grass border.
a Show that the grass area is $(l+4)(w+2)-lw$.
b Write another expression for the grass area by dividing the shape into rectangles.
c Show that the two expressions are equal.

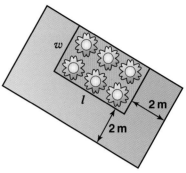

* **Review 3** The difference of the squares of two consecutive even numbers is 60. Use a diagram to help find the two numbers.

Factorising

Remember
Factorising an expression is the inverse of multiplying out a bracket.

Example $3x^2+2x = x(3x+2)$ x is the HCF of $3x^2$ and $2x$.

> See page 114 for more on common factors of algebraic expressions.

We take out the highest common factor possible.

Example $18x^3+6x^2$ could be factorised as
$$x^2(18x+6)$$
$$\text{or} \quad 6x^2(3x+1)$$

> Link to simplifying fractions by factorising, page 2.

$6x^2(3x+1)$ is completely factorised.

Sometimes we can **simplify fractions by factorising the numerator and/or denominator**.

Example $\dfrac{4m+8mn}{2m^2} = \dfrac{4m(1+2n)}{2m^2}$ Take out $4m$ as a common factor.

$$= \dfrac{{}^{2}4m(1+2n)}{{}^{1}2m \times m}$$

$$= \dfrac{2(1+2n)}{m}$$ Divide numerator and denominator by $2m$.

Example $\dfrac{5x-10}{x-2} = \dfrac{5(x-2)}{x-2}$

$$= \dfrac{5(x-2)^{1}}{x-2^{1}}$$

$$= 5$$

Example $\dfrac{6a^2+3a}{3a^2} = \dfrac{3a(2a+1)}{3a^2}$ Take out $3a$ as a common factor.

$$= \dfrac{{}^{1}3a(2a+1)}{{}^{1}3a \times a}$$ Divide numerator and denominator by $3a$.

$$= \dfrac{2a+1}{a}$$

Algebra

Discussion

Mr Chan gave these fractions to his class to simplify.

$$\frac{p+q}{q} \qquad \frac{xy-2}{y} \qquad \frac{ab+3a}{ab^2}$$

This shows what some pupils wrote.

Donna

$$\frac{p+q}{q} = p+1$$

$$\frac{xy-2}{y} = x-2$$

$$\frac{ab+3a}{ab^2} = \frac{3a}{b}$$

Alex

$$\frac{p+q}{q} = p$$

$$\frac{xy-2}{y} = \text{can't simplify}$$

$$\frac{ab+3a}{ab^2} = \frac{1+3}{b}$$

$$= \frac{4}{b}$$

Mandy

$$\frac{p+q}{q} = \text{can't simplify}$$

$$\frac{xy-2}{y} = \text{can't simplify}$$

$$\frac{ab+3a}{ab^2} = \frac{a(b+3)}{ab^2}$$

$$= \frac{b+3}{b^2}$$

Which ones are wrong? **Discuss**.

Exercise 3

For more practice at factorising see page 114.

1 Factorise these. The first three are started for you.

   a $3a + 3b = 3(\underline{\ \ } + \underline{\ \ })$ **b** $8y^2 - 4y = 4y(\underline{\ \ } - \underline{\ \ })$ **c** $m^3 + 2m^2 + 4m = m(\underline{\ \ } + \underline{\ \ } + \underline{\ \ })$

   d $5x + 10$ **e** $30 - 25w$ **f** $8t^3 + 4t - 4$ **g** $2n^2 + n$

   h $a^2 + 4a$ **i** $p^3 + p^2 + 6p$ **j** $5x^3 - x^2$ **k** $12m^3 + 4m^2$

   l $12p^2 + 18p$ **m** $25e - 15e^2$ **n** $3x^2 + x - 6x^3$ **o** $4y^3 + 8y - 12y^2$

   p $3p - 12p^2 + 30p^3$ **q** $10m^3 - 15m^2 - 25m$

2 a One of these expressions is **not** a correct factorisation of $12a + 24$. Which one is it?

      A $12(a+2)$ **B** $3(4a+8)$ **C** $2(6a+12)$ **D** $12(a+24)$ **E** $6(2a+4)$

   b Factorise $8y + 16$.

   c Factorise this expression as fully as possible. $8y^3 - 2y^2$

3 a Explain why $\dfrac{m}{m+n}$ will not simplify any further.

   b Explain why $\dfrac{3a+b}{b}$ is not equivalent to $3a$.

   c Explain why $\dfrac{xy-3}{x}$ does not cancel to give $y - 3$.

4 Only two of these are correct. Which two?

   a $\dfrac{a+b}{2a} = \dfrac{b}{2}$ **b** $\dfrac{5a+10}{5} = a+2$ **c** $\dfrac{4p+q}{q} = 4p$ **d** $\dfrac{pq-4}{4} = pq$

   e $\dfrac{mn-3}{mn} = {}^-3$ **f** $\dfrac{xy+x}{x} = y+1$ **g** $\dfrac{3a^2+a}{a^2} = 3+a$

5 Copy these and fill in the gaps to simplify.

   a $\dfrac{2a+4}{2} = \dfrac{2(\underline{\ \ }+\underline{\ \ })}{2}$ **b** $\dfrac{8y+12}{2y+3} = \dfrac{4(\underline{\ \ }+\underline{\ \ })}{2y+3}$

       $= \underline{\hspace{2cm}}$ $= \underline{\hspace{2cm}}$

   c $\dfrac{10m+5n}{6m+3n} = \dfrac{5(\underline{\ \ }+\underline{\ \ })}{3(\underline{\hspace{1cm}})}$ **d** $\dfrac{4p+2pq}{4p^2} = \dfrac{2p(\underline{\ \ }+\underline{\ \ })}{4p^2}$

       $= \underline{\hspace{2cm}}$ $= \underline{\hspace{2cm}}$

6 The missing numbers are found by adding the two numbers in the boxes below.

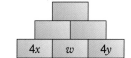

   a Prove that the number in the top box is even if x, y and w are integers.

   b What if $4x$ is replaced by $x - 1$? Is it possible to prove that the number in the top box is odd? or even?

   c What if $4y$ is replaced by y?

7 Simplify these by factorising the numerator, the denominator or both, then cancelling.

   a $\dfrac{5p + 25}{5}$ **b** $\dfrac{6x + 12}{3}$ **c** $\dfrac{10x - 15}{10}$ **d** $\dfrac{12x - 8}{8}$ **e** $\dfrac{8x + 16}{x + 2}$

   f $\dfrac{4x - 12}{x - 3}$ **g** $\dfrac{25 + 10x}{5 + 2x}$ **h** $\dfrac{49 - 21x}{7 - 3x}$ **i** $\dfrac{3x^2 + 3x}{x + 1}$ **j** $\dfrac{5x^2 + 10x}{5}$

   k $\dfrac{25x^2 + 10x}{5x + 2}$ **l** $\dfrac{8x^2 - 4x}{16x}$ **m** $\dfrac{16x^2 - 4x}{4x - 1}$ **n** $\dfrac{20x^2 - 15x}{4x - 3}$ **o** $\dfrac{10x - 4x^2}{5 - 2x}$

   p $\dfrac{27x - 9x^2}{3 - x}$ **q** $\dfrac{24x + 3x^2}{3x}$ **r** $\dfrac{81x^2 - 27x}{9x - 3}$ **s** $\dfrac{100x^2 - 20x}{5x - 1}$ **t** $\dfrac{10p + 5q}{12p + 6q}$

   u $\dfrac{24a - 16b}{15a - 10b}$ **v** $\dfrac{30x - 25y}{18x - 15y}$ **w** $\dfrac{24a - 16b}{18a - 12b}$

8 Simplify these.

   a $\dfrac{6ab + a}{a}$ **b** $\dfrac{4mn + 2m}{2m}$ **c** $\dfrac{3xy + 6x}{3x}$ **d** $\dfrac{5pq - 10p}{5p}$

   e $\dfrac{3cd - 6c^2e}{3cde}$ **f** $\dfrac{ab + 5ab^2}{a^2b}$ **g** $\dfrac{5m + 10mn}{4m^2}$ **h** $\dfrac{12wz^2 - 18w^2z}{6wz}$

   *__i__ $\dfrac{24a^2b - 18ab}{12ab - 6ab^2}$ *__j__ $\dfrac{17pq^2 - 51q^2r^3}{q^2r}$ *__k__ $\dfrac{12a^2b - 8ab}{4ab^2}$

9 **a** Factorise $b^2 - b$

   b What is the smallest value of $b^2 - b$ if b is an integer?

   c What is the smallest value if b can have values other than an integer value?

Use a spreadsheet to help.

*__10__ **a** Prove that $p^2 - p + 4$ is divisible by 2 for any integer value of p.

    b Prove that $a^3 - a + 9$ is divisible by 3 for any integer value of a.

For 10b and 11 you need to know the identity $a^2 - b^2 \equiv (a + b)(a - b)$.

*__11__ a and b are two integers.

The difference between a and b is 7.

The difference between a^2 and b^2 is also 7.

   a Using algebra, find a and b

   b Write down another pair of values a and b could have.

*__12__ **a** Show that $\dfrac{a^2 - b^2}{a - b}$ simplifies to $a + b$.   **[SATs Paper 2 Level 8]**

   b Simplify the expression $\dfrac{a^3b^2}{a^2b^2}$.   **c** Simplify the expression $\dfrac{a^3b^2 - a^2b^3}{a^2b^2}$.

   Show your working.

Review 1 Use a copy of this box. Write the letter beside the question above the correct answer in the box.

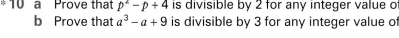

| $\dfrac{a + b}{b}$ | $\dfrac{4}{3}$ | $\dfrac{x}{2}$ | $a + 1$ | $\dfrac{a + 3}{3}$ | | 3 | $3ac^3$ | $a - 5$ | $\dfrac{a + 3}{3}$ | $3a + 5$ |

| $x + 4$ | $3ac^2$ | $\dfrac{x}{2}$ | $\dfrac{4}{3}$ | $\dfrac{1}{x}$ | $\dfrac{a}{b}$ | | $\dfrac{a}{b}$ | $\dfrac{a + b}{b}$ | $a - 5$ | 8 | $a + 1$ | | $x + 4$ | $\dfrac{1}{x}$ |

| $\dfrac{x}{2}$ | $\dfrac{y}{3x^2}$ | 8 | $a + 2$ | | $x + 4$ | $\dfrac{1}{x}$ | | $\dfrac{x}{2}$ | $a + 1$ | $\dfrac{a + 3}{3}$ |

continued on next page ... **165**

Algebra

Simplify each of these algebraic fractions.

A $\dfrac{4x + 16}{4}$ **B** $\dfrac{3y - 3}{y - 1}$ **C** $\dfrac{24b - 16}{3b - 2}$ **E** $\dfrac{a^2 + a}{a}$ **H** $\dfrac{2a + 4}{2}$

I $\dfrac{a^2 - 5a}{a}$ **K** $\dfrac{15a + 25}{5}$ **N** $\dfrac{2a^2 + 6a}{6a}$ **O** $\dfrac{8a + 4b}{6a + 3b}$ **S** $\dfrac{x - 1}{x^2 - x}$

T $\dfrac{a^2 - ab}{ab - b^2}$ **L** $\dfrac{3ac^3 + 3ac^2}{c + 1}$ **M** $\dfrac{6x^2 - 5xy}{12x - 10y}$ **U** $\dfrac{ay - 2y}{3ax^2 - 6x^2}$ **W** $\dfrac{a^2 + ab}{ab}$

Review 2   Rose started to prove that $x^3 - x + 4$ is divisible by 2 for any integer value of x.
She wrote:

$$x^3 - x + 4 = x(x^2 - 1) + 4$$
$$= x(\underline{\quad})(\underline{\quad}) + 4$$

Copy and finish Rose's proof.

> Hint: Think about which terms are divisible by 2.

*** Review 3** p and q are two integers. The sum of p and q is 15.
The difference between p^2 and q^2 is 45.
Using algebra find p and q.

Working with formulae

Sometimes we want to **change the subject of a formula**.

Example $v = \dfrac{s}{t}$ where v = average velocity in m/s, s = distance in m and t = time in seconds

I want t to be the subject.

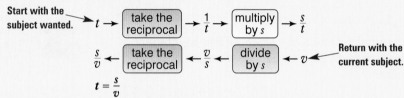

$$t = \frac{s}{v}$$

Worked Example
Make l the subject of the formula $S = \dfrac{n(a + l)}{2}$

Answer

Begin with l → [add a] → $a + l$ → [multiply by n] → $n(a + l)$ → [divide by 2] → $\dfrac{n(a + l)}{2}$

$\dfrac{2S}{n} - a$ ← [subtract a] ← $\dfrac{2S}{n}$ ← [divide by n] ← $2S$ ← [multiply by 2] ← Return with S

$$l = \frac{2S}{n} - a$$

Discussion

Freya says that she changes the subject of a formula by transforming both sides.
She started to change the subject of $A = \dfrac{bh}{2}$ to b like this.

$$A = \frac{b \times h}{2}$$
$$A \times 2 = \frac{b \times h}{2} \times 2$$

How might she continue? **Discuss**.
How could you use Freya's method to change the subject of these formulae?

$v = u + at$ **(subject u)** $c = 2\pi r$ **(subject r)**

Worked Example

Make b the subject of $a = 5\sqrt{\dfrac{c}{b}}$.

Answer

We can find b using inverse operations or by transforming both sides.

Using 'transforming both sides':

$$a = 5\sqrt{\dfrac{c}{b}}$$

$$\dfrac{a}{5} = \sqrt{\dfrac{c}{b}} \qquad \text{dividing both sides by 5}$$

$$\dfrac{a^2}{25} = \dfrac{c}{b} \qquad \text{squaring both sides}$$

$$\dfrac{25}{a^2} = \dfrac{b}{c} \qquad \text{taking the reciprocal of both sides}$$

$$\dfrac{25c}{a^2} = b \qquad \text{multiplying both sides by } c$$

$$\boldsymbol{b = \dfrac{25c}{a^2}}$$

Exercise 4

1 Make h the subject of these formulae.

  a $A = \frac{1}{2}bh$ **b** $A = \frac{1}{2}(a + b)h$ **c** $V = lbh$ **d** $V = \frac{1}{3}Ah$

2 A function is given by $y = \frac{x}{4} - 3$.

  Make x the subject of the function.

3 Make r the subject of
  a $d = 2r$ **b** $C = 2\pi r$.

4 Make l the subject of
  a $A = \pi rl$ **b** $P = 2(l + w)$.

5 Make m the subject of
  a $y = mx + c$ **b** $d = \dfrac{m}{v}$ **c** $a = \dfrac{F}{m}$.

6 **a** $R = \dfrac{V}{I}$.   Express I in terms of R and V.

  b The interest I, earned by a sum of money P, invested for T years at an interest rate of $R\%$ is given by $I = \dfrac{PRT}{100}$.   Express P in terms of I, R and T.

  c Make r the subject of the formula $D = \pi r + 5s$.

  d Make x the subject of $y = m(x + 3) + 2$.

  e $v^2 = u^2 + 2as$.   Express s in terms of v, u and a.

  f The area of a trapezium is given by the formula $A = \frac{1}{2}(a + b)h$ where a and b are the lengths of the parallel sides and h is the distance between these sides.
  Make b the subject of this formula.

  g Make R the subject of $A = P(\frac{100 + R}{100})$.

  h The area of metal needed to make a cylindrical tin of radius r and height h is given by $A = 2\pi r(r + h)$. Express h in terms of A and r.

7 **a** The subject of the equation below is p.   **[SATs Paper 1 Level 7]**
      $p = 2(e + f)$
    Rearrange the equation to make e the subject.

  b Rearrange the equation $r = \frac{1}{2}(c - d)$ to make d the subject.
    Show your working.

8 The formula for the surface area of a cylinder is
$$A = 2\pi r^2 + \pi rh.$$
   a Factorise the expression $2\pi r^2 + \pi rh$.
   b Make h the subject of $A = 2\pi r^2 + \pi rh$.

9 Make l the subject of
   a $n = \sqrt{l}$
   b $n = \sqrt{l - 2}$
   c $n = \dfrac{a}{\sqrt{l}}$
   d $n = \sqrt{\dfrac{a}{l}}$.

*10 $T = 2\pi\sqrt{\dfrac{l}{g}}$. Make l the subject.

Review

a $m = \dfrac{y}{x}$. Make y the subject of this formula.

b $m = \dfrac{y}{x}$. Express x in terms of m and y.

c $v = u + at$. Express t in terms of v, u and a.

d $F = \frac{9}{5}C + 32$ is a formula to convert temperatures given in degrees Celsius to degrees Fahrenheit. Make C the subject of this formula.

e The formula for the volume of a cylinder is $v = \pi r^2 h$. Make h the subject of the formula.

Discussion

● x is to be made the subject of the formula $a = x^2 + b$.
  Is $x = \sqrt{a - b}$ or $x = \overset{\pm}{\sqrt{}}\,a - b$? **Discuss**.

● x is to be made the subject of the formula $a = x^3 + b$.
  Is $x = \sqrt[3]{a - b}$ or $x = \pm\sqrt[3]{a - b}$? **Discuss**.

● $V = \pi r^2 h$ is the formula for the volume of a cylinder of radius r and height h.
  Eden made r the subject.

$$V = \pi r^2 h$$
$$\frac{V}{\pi h} = r^2 \quad \text{dividing both sides by } \pi h$$
$$r = \sqrt{\frac{V}{\pi h}}$$

Eden said 'I dont have to put $\pm\sqrt{\dfrac{V}{\pi h}}$ because ...'

How might Eden have finished his sentence? **Discuss**.

Exercise 5

1 Make r the subject of
   a $A = \pi r^2$
   b $V = \frac{1}{3}\pi r^2 h$
   if in each case r must be positive.

2 Make x the subject of the following. x may take negative or positive values.
   a $a = x^2$
   b $b = 5x^2$
   c $c = \dfrac{2}{x^2}$
   d $d = \dfrac{ax^2}{b}$

3 $I = \frac{1}{3}ml^2$. Make l the subject. l can have only positive values.

4 $E = \frac{1}{2}mv^2$. Make v the subject. v can have positive or negative values.

5 If a stone is dropped from the top of a lighthouse the distance, s, it falls in time, t, is given by $s = \frac{1}{2}gt^2$. Make t the subject of this formula.

6 $x = \frac{y^2}{4a}$.   Express y in terms of x and a. y may take positive or negative values.

7 a $A = \pi r^2$ is the formula for the area, A, of a circle of radius r.
Isobel began to rearrange $A = \pi r^2$ to make r the subject.

$$A = \pi r^2$$

$$\frac{A}{\pi} = r^2$$

$$\underline{} = r$$

What goes in the gap?

b Rearrange $V = \frac{4}{3}\pi r^3$ to make r the subject.

8 Make r the subject of $V = r^3$.

*9 The formula $\frac{1}{f} = \frac{1}{u} + \frac{1}{v}$ is used in a physics experiment.

Make u the subject of this formula.

Review

a $I = \frac{c}{d^2}$. Express d in terms of I and c. d may take positive values only.

b The surface area of a sphere of radius r is given by the formula $A = 4\pi r^2$.
Make r the subject of this formula.

c $y = \frac{x^2}{a^2}$.   Express x in terms of y and a. x may take positive or negative values.

d Make n the subject of $a = b\sqrt{n+1}$.

Substituting into expressions and formulae

Worked Example

$V = \frac{4}{3}\pi r^3$.

a Find V if $r = 2\cdot4$ cm.
b Find r if $V = 20\cdot6$ cm^3.

Use π on the calculator. Give the answer to 3 significant figures.

Answer

a $V = \frac{4}{3}\pi r^3$

$ = \frac{4}{3} \times \pi \times 2\cdot4^3$

Key ④ ÷ ③ × Shift π × 2·4 x^3 = to get **57·9 cm^3 (3 s.f.)**.

b We can substitute for V and then solve an equation to find V **or** we can make r the subject of the formula first.

$V = \frac{4}{3}\pi r^3$

$20\cdot6 = \frac{4}{3} \times \pi \times r^3$

$\frac{20\cdot6 \times 3}{4} = \pi \times r^3$ \qquad **dividing both sides by $\frac{4}{3}$**

$\frac{20\cdot6 \times 3}{4 \times \pi} = r^3$ \qquad **dividing both sides by π**

$r = \sqrt[3]{\frac{20\cdot6 \times 3}{4 \times \pi}}$

Key $\sqrt[3]{}$ ((20·6 × ③ ÷ ((④ × Shift π))) = to get **1·70 cm (3 s.f.)**.

Note There are other possible keying sequences.

Algebra

1   $s = vt$.   Find the value of
   a s when $v = 45$ and $t = 0.2$
   b v when $s = 120$ and $t = 3$
   c t when $s = 75$ and $v = 20$
   d s when $v = {}^-15$ and $t = \frac{1}{3}$.

When you are asked to find a value that is **not** the subject, you can make it the subject **or** solve an equation.

2   Work out the value of the expression $2y^2 - 3$ when
   a $y = 2$ **b** $y = 10$ **c** $y = {}^-1$ **d** $y = {}^-3$ *$*$**e** $y = 0.1$ *$*$**f** $y = 0.5$.

[T]

3   Use a copy of this.
   Work out the value of each expression when $p = 10$.
   Start at the red square and move to an adjacent square with a greater
   value. What colour square do you end in? You may not move diagonally.

   'Adjacent' means next to.

   start

$\dfrac{p^2}{2}$	$\dfrac{p^2}{4} + 50$	$\dfrac{3p^2}{2}$	$p^2 - 100$	$5p^2(p + 1)$
$p^2 - 75$	$\dfrac{p^2}{5}$	$4p^2 + 4$	$\dfrac{2p^2(p - 3)}{2}$	p^3
$p^2 + 8$	$\dfrac{(p - 8)^2}{2}$	$\dfrac{(p + 2)^2}{24}$	$(p - 4)^2$	$\dfrac{2p^2}{8}$

4   **a** y is an odd number.
      Draw an odd and an even box on your page.
      Which of these numbers must be odd and which
      must be even?
      Put each in one of your boxes.
         $2y$ y^2 $2y - 1$ $3y - 2$ $(y - 1)(y + 1)$
   b y is an odd number.
      Is the number $\dfrac{y + 1}{2}$ odd or even or is it not possible to tell? Explain your answer.

     odd
    even

5   **A** $\boxed{n - 3}$ **B** $\boxed{2n}$ **C** $\boxed{n^2}$ **D** $\boxed{\dfrac{n}{2}}$ **E** $\boxed{\dfrac{3}{n}}$

   a   Which of these expressions has the greatest value when n is between 1 and 2?
   b   Which expression has the greatest value when n is between 0 and 1?
   c   Which expression has the greatest value when n is negative?

6   Write true or false for each of these.   **[SATs Paper 1 Level 7]**
   a   **i**   When x is even, $(x - 2)^2$ is even.
      ii   When x is even, $(x - 2)^2$ is odd.
      Show how you know it is true for **all** even values of x.
   b   **i**   When x is even, $(x - 1)(x + 1)$ is even.
      ii   When x is even, $(x - 1)(x + 1)$ is odd.
      Show how you know it is true for **all** even values of x.

7   Use the formula $A = \pi r^2$ to find the radius of the circle which has an area of 7·6 cm^2.

8   The formula for the surface area, A in cm^2, of a cone is $A = \pi r l$
   r = radius in cm and l = slant height in cm.
   Find
   a   the surface area if $r = 7.57$ cm and $l = 2.46$ cm
   b   the slant height, l, if $A = 22$ cm^2 and $r = 3.18$ cm
   c   the radius if $A = 984$ cm^2 and $l = 26$ cm.

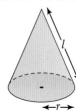

9 The sum of the first n odd numbers is given by $S = n^2$.
 a Find the sum of the first 20 odd numbers.
 b Find n if $S = 289$.

10 a Make x the subject of the formula $V = x^3$.
 b The volume of a cube is 80 mm^3. Find the length of an edge of this cube.

11 The surface area, A in units^2, of a sphere is given by
 the formula $A = 4\pi r^2$.
 a Find A if $r = 2 \cdot 36$ units.
 b Find r if $A = 18 \cdot 2$ units^2.

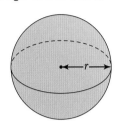

12 $s = \dfrac{v^2 - u^2}{2a}$
 s = distance in m, v = final velocity in m/s, u = initial velocity in m/s, a = acceleration in m/s^2.
 Find
 a s if $v = 6$ m/s, $u = 4 \cdot 2$ m/s, $a = 2$ m/s^2
 b the two values for v if $u = 24 \cdot 8$ m/s, $a = {}^-2$ m/s^2, $s = 14$ m
 c a if $v = 40$ m/s, $u = 25$ m/s, $s = 120$ m.

13 $T = 2\pi \sqrt{\dfrac{l}{g}}$

 T = time in seconds taken for a pendulum to make one complete swing,
 l = length of pendulum in m, g = acceleration due to gravity in m/s^2.
 a Find T if $l = 10$ m and $g = 9 \cdot 8$ m/s^2.
 b Find l if $T = 5 \cdot 2$ seconds and $g = 9 \cdot 8$ m/s^2.

14 The formula for the surface area of a cylinder is $A = 2\pi r(r + h)$.
 Find
 a A if $r = 2 \cdot 7$ cm and $h = 5 \cdot 62$ cm
 b h if $A = 226 \cdot 4$ mm^2 and $r = 4 \cdot 25$ mm

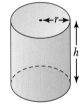

15 Work out the values of M and N for $a = 1 \cdot 6$ and $b = 0 \cdot 7$.
$$M = a - b + \frac{3\sqrt{a^2 + b^2}}{5} \qquad N = \frac{ab}{4} + \frac{a^3 - b^2}{7}$$

***16 a** Emily drew the graphs of two functions,
 $x \rightarrow (x + 1)^2$ and $x \rightarrow 13 - 3x^2$.
 Find the value of each function when $x = {}^-3$.
 b For what values of x does $x \rightarrow (x + 1)^2$ equal 1?
 c For what values of x does $x \rightarrow 13 - 3x^2$ equal 1?
 d Using your answers to **b** and **c**, write down the coordinates of one of the points where
 the functions meet.
 e The value of $x \rightarrow (x + 1)^2$ is 6·25 at the second point where the two functions meet.
 What are the coordinates of the second point where they meet?

***17** A life-ring is made from polystyrene.
 The volume of the ring is given by
 $V = \frac{1}{4}\pi^2(p + q)(q - p)^2$.
 If $p = 18 \cdot 62$ cm and $q = 21 \cdot 34$ cm, find V.

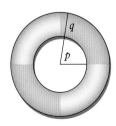

Algebra

*18 The volume of this prism is given by the expression
$2x^3 \sin b$.

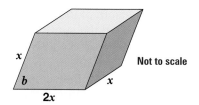

Not to scale

a   What value of b would make the volume of the prism $2x^3$?
b   The volume of the prism is 120 cm³ when $b = 60°$.
What is the value of x?

19 The formula $\dfrac{1}{R} = \dfrac{1}{R_1} + \dfrac{1}{R_2}$ is used in a physics experiment, where R, R_1 and R_2 are all resistances in ohms.

a   Find the value of R if $R_1 = 6 \cdot 3$ ohms and $R_2 = 7 \cdot 8$ ohms.
b   Find the value of R_1 if $R = 3 \cdot 6$ ohms and $R_2 = 4$ ohms.

Review 1 $v = u + at$ (v = final velocity in m/s, u = initial velocity in m/s, a = acceleration in m/s², t = time in seconds).
Find the value of

a v when $u = 6 \cdot 8$ m/s, $a = 15$ m/s², $t = 25$ s
b u when $v = 70$ m/s, $a = ^-10$ m/s², $t = 2$ s
c a when $v = 8 \cdot 8$ m/s, $u = 6$ m/s, $t = 0 \cdot 4$ s
d t when $v = 85$ m/s, $u = 100$ m/s, $a = ^-5$ m/s².

Review 2 $y = \dfrac{4a}{x^2}$ is the equation of a curve.

a   Find the value of y when $x = 2$ and $a = 1 \cdot 6$.
b   Find the values of x when $y = 0 \cdot 4$ and $a = 3 \cdot 6$.

Review 3   The volume of a sphere is given by $V = \dfrac{4}{3}\pi r^3$
(V = volume in cm³, r = radius of sphere in cm).

a   When a balloon is inflated its radius increases. What happens to the volume of a spherical balloon when the radius is doubled?
b   Calculate the volume when the radius is $3 \cdot 8$ cm.

Review 4 $x \rightarrow \dfrac{x}{2x - 3}$

a   For what value of x is the function equal to 3?
*b   For what value of x does the function have no value?

 Practical

You will need a spreadsheet package.

Ask your teacher for ICT worksheet '**Modelling with Formulae**'.

Finding formulae

Worked Example
When we put two cubes together we can count 10 square faces if we pick the shape up and turn it round.
Find a formula for the number of square faces we can count when n cubes are joined in a line.

2 cubes

3 cubes

4 cubes

Answer

It is sometimes best to draw a table.

Number of cubes (n)	2	3	4	5	6	...
Number of faces (f)	10	14	18	22	26	...

We can then draw (or make) the shapes, count the faces, and fill in the table.
The number of faces forms the sequence 10, 14, 18, 22, 26 ...
Each number is **4** more than the one before.
The formula will be:

$$F = 4 \times n + ?$$

We can find what **?** is by looking at the sequence.

| 1st term | $10 = 4 \times 2 + 2$ | |
| 2nd term | $14 = 4 \times 3 + 2$ | and so on. |

So the formula is $F = 4n + 2$.

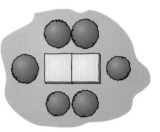

Link to finding the *n*th term of a sequence, page 115.

Exercise 7

1 Mel wants to put square paving stones in a row.
She wants to plant a shrub along each side of the paving stones.
If she has one paving stone she can plant four plants.
If she has two paving stones she can plant six plants.
Find a formula for the number of plants (p) that Mel can plant if she puts n paving stones in a row.
A table like this might help.

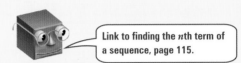

Number of paving stones (n)				
Number of plants (p)				

2 This network has 3 nodes ($\bullet$), 5 arcs (⌒) and 4 regions (A, B, C and D – we count the outside).
Look at these networks and count the number of nodes, arcs and regions.
Note Arcs must not cross.
All nodes do not have to be joined to every other node.

a 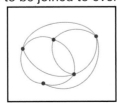   **b**   **c**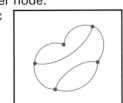

Draw some more diagrams.
Fill in this table.
Find a formula for the relationship between n, r and a.

	example	a	b	c
Number of nodes (n)	3	4		...
Number of regions (r)	4	3		...
Number of arcs (a)	5	5		...

Algebra

*3 Pies are cut into pieces by making straight cuts which intersect each other.

Jillian investigated the number of pieces for various numbers of cuts. She put her results on a table.

a Copy and fill in this table.

No. of cuts (c)	No. of pieces (n)
1	2
2	4
3	7
4	
5	

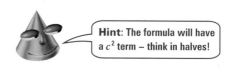

Hint: The formula will have a c^2 term – think in halves!

b Find a formula for the number of pieces, N, if there are c cuts.

c Use the formula you found to find the number of pieces for 15 cuts.

4 Kent started to derive the formula for the area of a trapezium $A = \frac{h}{2}(a + b)$ where a and b are the lengths of the parallel sides and h is the height.

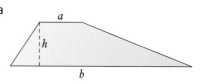

Divide the trapezium into a rectangle and two triangles.
Area of trapezium = area of A + area of B + area of C
$$= \frac{xh}{2} + ah + \frac{yh}{2}$$
Finish Kent's working to derive the formula.

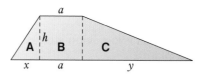

5 Derive the formula for the area, A of an annulus with outer radius, R, and inner radius, r.

6 Derive the formula for the perimeter, P of an semi circle with radius, r.

Review 1

1 square
1 region

2 squares
3 regions

3 squares
5 regions

a Copy and fill in the table.

Number of squares (s)	1	2	3	4	5	...
Number of regions (n)	1	3	5			...

b Find the formula for the number of regions, n, formed by s intersecting squares.

Review 2

Derive the formula for the area of the pink shaded region, in terms of r. The circles are congruent.

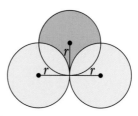

Summary of key points

 When **simplifying expressions** we often need to use the index laws. If the expression has brackets, we must first expand these,

$$x^a \times x^b = x^{a+b} \qquad x^a \div x^b = x^{a-b} \qquad (x^a)^b = x^{ab} \qquad a^{-n} = \frac{1}{a^n} \qquad a^{\frac{1}{n}} = \sqrt[n]{x}$$

Examples
$$4x - (x - 2) = 4x - 1(x - 2) \qquad\qquad (x^4)^{-3} = x^{4 \times -3}$$
$$= 4x - 1 \times x + {}^-1 \times {}^-2 \qquad\qquad = x^{-12}$$
$$= 4x - x + 2 \qquad\qquad\qquad = \frac{1}{x^{12}}$$
$$= 3x + 2$$

 When we **expand two brackets**, we must multiply all terms in the first bracket by all terms in the second bracket.

Examples

	x	-3	
$2x$	$2x^2$	^-6x	$2x^2 - 6x$
4	$4x$	$^-12$	$4x - 12$
			$2x^2 - 2x - 12$

or
$$(2x + 4)(x - 3) = 2x(x - 3) + 4(x - 3)$$
$$= 2x \times x + 2x \times {}^-3 + 4 \times x + 4 \times {}^-3$$
$$= 2x^2 - 6x + 4x - 12 \qquad \textbf{Write like terms together.}$$
$$= 2x^2 - 2x - 12 \qquad \textbf{Collect like terms.}$$

 When **factorising an expression**, we take out the highest common factor possible. Sometimes we need to factorise to simplify expressions.

Examples
$$\frac{6a + 9ab}{3ab} = \frac{{}^1 3a(2 + 3b)}{{}_1 3ab} \qquad\qquad \frac{4x^2 - 6x}{2x^2} = \frac{2x(2x - 3)}{2x^2}$$
$$= \frac{2 + 3b}{b} \qquad\qquad\qquad = \frac{{}^1 2x(2x - 3)}{{}_1 2x \times x}$$
$$= \frac{2x - 3}{x}$$

 When using formulae, we sometimes need to **change the subject of the formula**. We can use a flow chart or transforming both sides.

Example Make a the subject of $v^2 = u^2 + 2as$.

Start with the subject you want. $\quad a \rightarrow \boxed{\text{multiply by } 2s} \rightarrow 2as \rightarrow \boxed{\text{add } u^2} \rightarrow u^2 + 2as$

$\frac{v^2 - u^2}{2s} \leftarrow \boxed{\text{divide by } 2s} \leftarrow v^2 - u^2 \leftarrow \boxed{\text{subtract } u^2} \leftarrow v^2$ **Return with the current subject.**

$$a = \frac{v^2 - u^2}{2s}$$

Making u the subject, using transforming both sides,
$$v^2 = u^2 + 2as$$
$$v^2 - 2as = u^2 \qquad \textbf{Subtract 2as from both sides.}$$
$$\sqrt{v^2 - 2as} = u \qquad \textbf{Take the square root of both sides.}$$
$$u = \sqrt{v^2 - 2as}$$

When **substituting into expressions or formulae** we sometimes need to use a calculator.

Example $V = \pi r^2 h$ is the formula for the volume of a cylinder.

If $r = 1.5$ cm and $h = 6.2$ cm, to find V

key (Shift) (π) ($\times$) (1.5) (x^2) ($\times$) (6.2) (=) to get **43.8 (3 s.f.)**.

If $V = 15.8$ cm^3 and $h = 3.2$ cm, to find r

key ($\sqrt{}$) (() (15.8) (÷) (() (Shift) (π) ($\times$) (3.2) ()) ()) (=)
to get 1.25 (3 s.f.).

$$r = \sqrt{\frac{V}{\pi h}}$$

F We can **derive a formula** using algebra.
See page 172 for an example.

Test yourself

1 Simplify these.
   a $(4x + 7) + (x - 3)$ **b** $5y - (^-y + 2)$

2 Rachel collected information about the number of people living in households.
   She displayed the information on a frequency chart but then spilt some ink on it.

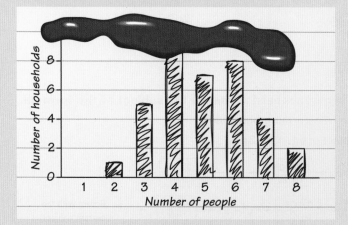

Call the number of households with 4 people x.
   a Show that the total number of people in all the households is $144 + 4x$.
   b Write an expression for the total number of households.
   c The mean number of people per household is 5.
   What is the value of x?
   Show your working.

3 A number grid is inside a large triangle.
The small triangles are numbered consecutively.
The diagram shows the first 4 rows.

[SATs Paper 1 Level 6]

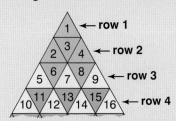

a An expression for the **last** number in row n is n^2.

 ← row n

Write an expression for the **last but one** number in row n.

 ← row n

b An expression for the first number in row n is $n^2 - 2n + 2$.
Calculate the value of the first number in row 10.

c What goes in the gap in the table?

first number in row n	$n^2 - 2n + 2$
second number in row n	

d What goes in the gap in this table? [Level 7]

centre number in row n	$n^2 - n + 1$
centre number in row ___	$(n + 1)^2 - (n + 1) + 1$

e Multiply out and simplify the expression $(n + 1)^2 - (n + 1) + 1$.
Show your working.

4 Multiply out. Simplify if possible.

   a $(x - 3)(x + 4)$ **b** $(y - 1)(y - 5)$ **c** $(a + 2n)(a + 3n)$ **d** $(c - 3x)(2c + x)$
   e $(2n - 3y)(5n - y)$ **f** $(2 - 3d)(d + 3)$ **g** $(3x - 2)(2 + x)$ **h** $(5x + 2n)(a - 3b)$

5 Multiply out and simplify these.

   a $(x + y)^2$ **b** $(m - 2)^2 + m^2$ **c** $(p + 4)^2 - (2p - 3)^2$

6 Factorise these.

   a $4x + 4y$ **b** $6m^2 + 3m$ **c** $28 - 14y$ **d** $4p^2 - p + 6p^3$ **e** $36p - 24p^2$

7 Simplify these, if possible, by factorising first.
If it cannot be simplified, explain why not.

   a $\frac{3c + 6}{3}$ **b** $\frac{4x - 6}{2x - 3}$ **c** $\frac{x + 2y}{y}$ **d** $\frac{6m + 3m^2}{6m^2}$ **e** $\frac{15xy - 10xy^2}{5x^2y}$

8 This diagram shows a square of side x cm from which a square of side y cm has
been removed.

   a Use the diagram to show that $x^2 - y^2 = (x - y)(x + y)$.
   b Use **a** to factorise $a^2 - b^2$.
   c Now factorise $4a^2 - b^2$.
   d Use **a** to write $8^2 - 4^2$ in brackets.
      Then find the value of $8^2 - 4^2$.

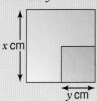

Algebra

9 **a** Make y the subject of the formula $x = \frac{a}{y}$.

    b Make P the subject of the formula $k = PVT$.

    c Make l the subject of the formula $S = \frac{n(a + l)}{2}$.

    d $A = 2\pi rh + \pi r^2$. Make h the subject of this formula.

    e $s = \frac{1}{2}(a + b + c)$. Express b in terms of s, a and c.

10 $A = lb$ where A is the area of a rectangle in units^2, l is length in units, b is breadth in units. Find the value of

    a A if $l = 4$ cm and $b = 2.5$ cm

    b l if $A = 7.5$ cm^2 and $b = 1.5$ cm

    c b if $A = 63.8$ m^2 and $l = 7.2$ m.

11 $v^2 = u^2 + 2as$ where v is final velocity in m/s, u is initial velocity in m/s, a is acceleration in m/s^2, and s is distance in m. Find the two values for v if $u = 25$ m/s, $a = 8.2$ m/s^2 and $s = {}^-20$ m.

12 The energy, E in joules, is given by $E = \frac{1}{2}mv^2$.
    m is the mass in kg and v is the speed in m/s.
    Make v the subject of the formula.

13 The formula $\frac{1}{u} + \frac{1}{v} = \frac{1}{f}$ is used in a physics experiment on light.

    a Find the value of f if $u = 6.2$ and $v = 10$.

    b Find the value of u if $f = 4.6$ and $v = 8.2$.

14 Cubes have been put together like 'this'.
    When we pick the shape up and turn it we can count the number of faces.

1 cube
6 faces

3 cubes
14 faces

5 cubes

    Find a formula for the number of square faces we can count when n cubes are used for a shape.

15 The equation of the curve shown is $y = \pm\sqrt{\dfrac{x^3}{4 - x}}$.   **[SATs Paper 2 Level 8]**

    a When $x = 2.5$ calculate the positive value of y.
        Show all the digits on your calculator display.

    b When $x = 2.5$ give both values of y correct to 3 significant figures.

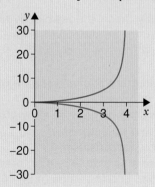

8 Sequences and Functions

You need to know

✓ sequences   page 115
      – sequences in practical situations
✓ functions   page 116

Key vocabulary

curve, first/second difference, identity function, inverse mapping, inverse function, maximum/minimum point, maximum/minimum value, quadratic function, quadratic sequence, self-inverse, $T(n)$

Don't tell Fibonaccis

The Fibonacci sequence is generated by this rule.
   first terms   1, 1   **rule** add the two previous terms together

   1, 1, 2, 3, 5, 8, 13, ...

1   Write down the first 20 terms of the Fibonacci sequence.

2   Look at every third number. What sort of number is it?
    What about every fourth/fifth number?

3   Count the petals on this daisy.
    Is it a Fibonacci number?
    Count the number of petals on some other flowers.

4   Fibonacci numbers occur often in nature.
    To find out more, here is a web site you could look at.
          www.mcs.surrey.ac.uk/Personal/R.Knott/Fibonacci

Generating sequences

Sequences can be generated using

1 a term-to-term definition

Example   **first term** $m + n$ **term-to-term rule** add $m + 2n$
This rule gives the sequence

$$m + n, \quad 2m + 3n, \quad 3m + 5n, \quad 4m + 7n, \quad 5m + 9n, \ldots$$
$$\quad\quad +m + 2n \quad\quad +m + 2n \quad\quad +m + 2n \quad\quad +m + 2n$$

Example $T(2) = T(1) + 2a, \quad T(3) = T(2) + 2a \ldots T(n) = T(n - 1) + 2a.$
If $T(1) = a + 2b$ this rule gives the sequence

$$a + 2b, \ 3a + 2b, \ 5a + 2b, \ 7a + 2b, \ldots$$

> The term-to-term rule is 'add $2a$ to the previous term'.

2 a rule for the nth term

Example $T(n) = 3n + n(n - 1)$
We substitute $n = 1, 2, 3, 4, \ldots$ into the expression given.
$$T(\mathbf{1}) = 3 \times \mathbf{1} + \mathbf{1}(\mathbf{1} - 1) = 3$$
$$T(\mathbf{2}) = 3 \times \mathbf{2} + \mathbf{2}(\mathbf{2} - 1) = 8$$
$$T(\mathbf{3}) = 3 \times \mathbf{3} + \mathbf{3}(\mathbf{3} - 1) = 15$$
$$T(\mathbf{4}) = 3 \times \mathbf{4} + \mathbf{4}(\mathbf{4} - 1) = 24$$
$$\vdots$$
$$T(\mathbf{20}) = 3 \times \mathbf{20} + \mathbf{20}(\mathbf{20} - 1) = 440$$

Discussion

Sometimes sequences 'converge' to a number.

Example   The sequence with term-to-term rule 'divide by 3 add 4' can be represented
by $x \rightarrow \frac{x}{3} + 4$.
Choose a first term, say 1.
Using a graphical calculator enter

    [1] [EXE]
    [Ans] [÷] [3] [+] [4] [EXE]
    [EXE] [EXE] [EXE] ...

If you keep pressing [EXE] , what number does $x \rightarrow \frac{x}{3} + 4$ converge to? **Discuss**.

What happens if you change the first term? **Discuss**.

Solve the equation $x = \frac{x}{3} + 4$.

What do you notice about the solution? **Discuss**.

Either on graph paper or using a graphical calculator, draw the graphs of $y = x$ and $y = \frac{x}{3} + 4$.
Write down the coordinates of the point of intersection of these graphs.
What do you notice about the x-coordinate? **Discuss**.

What if you changed 3 and 4 to 2 and 5? **Discuss**.

Use a graphical calculator for question 4.

1 Write down the first ten terms of these sequences.
  a first term 256 **term-to-term rule** divide by 4
  b first terms 1, 2 **term-to-term rule** add the two previous terms
  c $T(n) = n - \frac{1}{2}$ **d** $T(n) = n(n + 1)$ **e** $T(n) = 100 - 3n$
  f $T(n) = \frac{n}{2} + 3$ **g** $T(n) = 3n + n(n - 1)(n - 2)$

2 Write down the first six terms of the sequences given by these rules.
  a $T(1) = p + q,\ T(2) = p + T(1) \dots T(n) = p + T(n - 1)$
  b $T(1) = a + b,\ T(2) = T(1) + 3a + 2b \dots T(n) = T(n - 1) + 3a + 2b$

3 a Paula made up a sequence that had every fourth number an integer.
    What might the term-to-term rule for Paula's sequence be?
  b Kent made up a sequence that had every fifth number a multiple of 4.
    What might the rule for the nth term be?
  c Manzoor made up a sequence that began as 1, 2, 4, ...
    Write down two ways he could continue this sequence and the rule for each.
  d Lilly made up a sequence with rule $T(n) = pn + q$. It increases in 4s.
    What will the value of p be?

4 a A sequence has the rule 'divide by 5, add 4' or $x \rightarrow \frac{x}{5} + 4$. Choose a starting number.
    Use a graphical calculator to generate some terms of the sequence.
    Does the sequence 'converge' to a number? If so, what number?
  b Solve the equation $x = \frac{x}{5} + 4$.
  c Draw the graphs of $y = x$ and $y = \frac{x}{5} + 4$.
    What is the x-coordinate of the point of intersection?
  d What do you notice about your answers to **a**, **b** and **c**?
  e Explore what happens when you change the numbers 5 and 4 to other numbers.
  ∗f Make a general statement about what happens when you generate a sequence from
    the rule 'divide by a then add b'.

Review 1   Write down the first six terms of these sequences.
a first term 4 **term-to-term rule** double then subtract 5
b $T(n) = 2n - 3$
c $T(1) = a - b,\ T(2) = T(1) + a + 2b \dots T(n) = T(n - 1) + a + 2b$

Review 2
a Use the rule $T(2) = \frac{T(1)}{4} + 3$ to find the first five terms of the
    sequence with
    i $T(1) = 2$
    ii $T(1) = 8$.
b What do the sequences in **a** converge to?
c Use a copy of this grid and plot your values for $T(1)$ to $T(5)$ for
    both sequences. Use crosses for **i** and dots for **ii**.
d The rule for the sequences can also be written as $x \rightarrow \frac{x}{4} + 3$.
    Solve the equation $y = \frac{x}{4} + 3$.
    Compare your answer to your answer in **b**.
e Draw the graphs of $y = x$ and $y = \frac{x}{4} + 3$.
    What is the x-coordinate of the point of intersection?

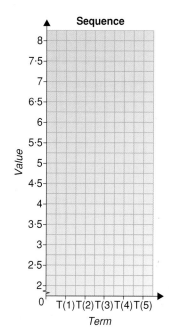

Quadratic sequences

1, 4, 9, 16, 25, ... is the sequence of square numbers.
The rule for this sequence is $T(n) = n^2$.
Sequences that have a rule with a squared term as their highest power are called **quadratic sequences**.

Example $T(n) = 2n^2 + 1$

Position	1	2	3	4	5
Term T(n)	$2 \times 1^2 + 1 = 3$	$2 \times 2^2 + 1 = 9$	$2 \times 3^2 + 1 = 19$	$2 \times 4^2 + 1 = 33$	$2 \times 5^2 + 1 = 51$

The sequence is 3, 9, 19, 33, 51, ...

Practical

Ask your teacher for the **Quadratic Sequences** ICT Worksheet.

Investigation

Quadratic sequences

The first few terms of the sequence of square numbers, $T(n) = n^2$, are given, together with the first and second differences.

```
1    4    9    16    25    36    49
  3    5    7     9    11    13         first difference
     2    2     2     2     2           second difference
```

The sequence ascends by successive odd numbers.
The second differences are all the same, 2.

Investigate the first and second differences of these sequences.

$T(n) = 2n + 2$	$T(n) = 5n - 4$	$T(n) = n^2 + 1$
$T(n) = 2n^2$	$T(n) = n^2 + n + 1$	$T(n) = n^2 + 3n - 1$
$T(n) = 3n^2 + 3$	$T(n) = \frac{n(n+1)}{2}$	$T(n) = 5 - 2n^2$

The first two sequences are **linear** sequences.

Comment on
- first differences of linear sequences
- second differences of quadratic sequences
- the value of the second differences and the number multiplying n^2 in a quadratic sequence.

Rules of the form $T(n) = an^2 + bn + c$ generate **quadratic sequences**.
The second differences are constant.

Example $T(n) = 2n^2 + 1$

For this rule $a = 2$, $b = 0$, $c = 1$.

3   9   19   33   51

6   10   14   18   ← 1st differences

→ 4   4   4   ← 2nd differences

constant difference

For a **linear** sequence,
the **first differences**
are all the same.

We can use differences to predict the next terms of a sequence.

Worked Example
Predict the next term of the sequence 6, 7, 11, 18, 28, ...

Answer

6   7   11   18   28   41

1   4   7   10   13   ← First differences

3   3   3   3   ← Second differences

Write another 3 in the second differences row then add this three to the 10 in the first differences row to get 13; then add 13 to 28 to get 41 as the next number of the sequence.

Exercise 2

1 Write down the first six terms of the sequences given by these rules.
  a $T(n) = n^2 + 1$ **b** $T(n) = 2n^2 + 2$ **c** $T(n) = 5 - 3n^2$ **d** $T(n) = n^2 + 2n + 4$

2 For each of the sequences you found in question **1**, find the second difference.

3 Which of these sequences are quadratic? How can you tell?
  a 4, 12, 20, 28, 36, ...   **b** 3, 6, 11, 18, 27, ...   **c** 2, 11, 26, 47, 74, ...
  d 96, 89, 82, 75, 68, ...   **e** 5, 11, 21, 35, 53, 75, ...

4 Which of the sequences in question **3** are linear? Explain how you know.

5 Find the next two terms of these sequences by finding first and second differences.
  a 2, 5, 10, 17, 26, ...   **b** 4, 8, 15, 25, 38, ...   **c** 3, 8, 15, 24, 35, ...
  d 5, 11, 21, 35, 53, ...   **e** 1, 8, 21, 40, 65, ...   **f** 1, 7, 17, 31, 49, ...
  g ⁻3, 3, 13, 27, 45, ...   **h** 4, 13, 28, 49, 76, ...   **i** 0, 9, 24, 45, 72, ...
  j ⁻6, 6, 26, 54, 90, ...

6 Write down the first six terms of these quadratic sequences.

	First term	First 1st difference	2nd difference
a	12	3	2
***b**	50	⁻8	4
***c**	5	12	⁻3

Algebra

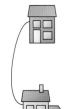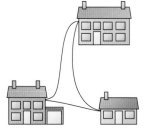

7

Wendy investigated the number of phone connections that can be made between various numbers of houses. She put her results on a table.

a Use differences to find the missing numbers in this table.

b How many connections would be needed for 9 houses?

c In a room there are 4 people. Each person shakes hands with every other person. How many handshakes will there be?

No. of houses	No. of connections
1	0
2	1
3	3
4	6
5	10
6	
7	

d What is the connection between the number of handshakes between any number of people and the number of connections needed between houses?

e In a county cricket competition there are 14 teams. Each team plays every other team. How many games will there be?

Review 1 Write down the first six terms of the sequences given by
a $T(n) = n^2 - 2$ **b** $T(n) = 6 - 2n^2$.

Review 2 Which of these sequences are quadratic? Explain how you know.
a 4, 6, 10, 16, 24, ...   **b** 17, 11, 5, ⁻1, ⁻7, ...
c 4, 2, ⁻2, ⁻8, ⁻16, ...

Review 3 Find the next two terms in these sequences by finding first and second differences.
a 0, 2, 6, 12, 20, ...   **b** ⁻3, ⁻1, 4, 12, 23, ...

Review 4 Cubes are drawn on isometric paper as shown.
a Use a copy of this and draw the next set of cubes in the sequence. How many dots are joined to draw it?
b Complete this table.

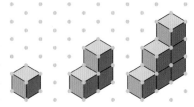

Number of cubes on base	Number of dots
1	7
2	13
3	20
4	
5	
6	

Use differences to find the number of dots for 5 and 6 cubes on the base.

c How many dots are joined for 9 cubes on the base?

Finding rules for the *n*th term of linear and quadratic sequences

Remember

To find the **rule for the *n*th term of a linear sequence**, find the constant **difference** between consecutive terms.

Worked Example

Find the *n*th term of 46, 39, 32, 25, ...

Answer

Term	46	39	32	25
Difference		⁻7	⁻7	⁻7

The difference between consecutive terms is ⁻**7**, so the *n*th term is of the form $T(n) = {}^-7n + b$ or $b - 7n$.

$T(1) = 46$ $b - 7 \times 1 = 46$
          so $b - 7 = 46$
                    $b = 53$

$T(n) = 53 - 7n$

Check by testing a few more terms

$n = 2$ $53 - 7n = 53 - 7 \times 2$ $n = 3$ $53 - 7n = 53 - 7 \times 3$ $n = 4$ $53 - 7n = 53 - 7 \times 4$
                  $= 39$ ✓ $= 32$ ✓ $= 25$ ✓

To find the **rule for the *n*th term of a quadratic sequence**, first look to see if it is obvious.

Example   The rule for the *n*th term of the sequence 2, 5, 10, 17, 26, 37, 50, ... is $T(n) = n^2 + 1$. Each term is one more than a square number.

The following investigation leads you through finding the rule for the *n*th term of any quadratic sequence.

Investigation

Rules for quadratic sequences

● Robbie wanted to prove that for a sequence given by $T(n) = an^2 + bn + c$, the second difference is always $2a$.
He began like this.

$T(1) = a \times 1^2 + b \times 1 + c$ $T(2) = a \times 2^2 + b \times 2 + c$
        $= a + b + c$ $= 4a + 2b + c$
$T(3) = a \times 3^2 + b \times 3 + c$ $T(4) = a \times 4^2 + b \times 4 + c$
        $= 9a + 3b + c$ $= 16a + 4b + c$

Terms	$a + b + c$	$4a + 2b + c$	$9a + 3b + c$	$16a + 4b + c$	___
1st differences		$3a + b$	$5a + b$	___	___
2nd differences			___	___	___

Work out the missing expressions.
Will the first expression in each row be the same for all quadratic sequences?

Algebra

- How could Robbie use the above table to find the rule for the quadratic sequence 3, 4, 7, 12, ...? **Investigate**.

  Hint: Find a first, using a difference table. Then find b, using the first difference between the first two terms, then find c using the first term.

  Test your method for these sequences.
  6, 16, 32, 54, ...   0, 8, 22, 42, ...   5, 11, 21, 35, ...

We can find the **rule for a quadratic sequence** using a difference table.
The rule will be of the form $T(n) = an^2 + bn + c$ and $T(1) = a + b + c$.

Worked Example

Find the rule for the nth term of the sequence 3, 10, 21, 36, ...

Answer

Terms		3		10		21		36
1st differences			7		11		15	
2nd differences				4		4		

The rule will be of the form $T(n) = an^2 + bn + c$

2nd difference = $2a$
$$4 = 2a$$
$$a = \mathbf{2}$$

$T(1) = 2 + b + c$ because $T(1) = a + b + c$
$T(2) = 2 \times 2^2 + 2b + c = 8 + 2b + c$ because $T(2) = a \times 2^2 + 2b + c$
1st difference = $8 + 2b + c - (2 + b + c)$ $T(2) - T(1)$
$$= 6 + b$$
$$6 + b = 7$$
$$b = \mathbf{1}$$

$T(1) = a + b + c$
$3 = 2 + 1 + c$ substituting for $T(1)$, a and b
$c = \mathbf{0}$

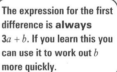

The expression for the first difference is **always** $3a + b$. If you learn this you can use it to work out b more quickly.

The rule for the sequence 3, 10, 21, 36... is $T(n) = \mathbf{2n^2 + n}$.

Worked Example

Ruth investigated how rolls of paper could be stacked. She drew the following diagrams.

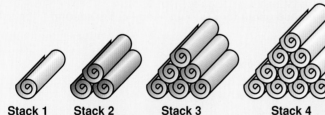

Stack 1   **Stack 2**   **Stack 3**   **Stack 4**

Ruth put her results in a table.
a Copy and finish Ruth's table.
b Draw a difference table for the sequence for the number of rolls.
c Find the rule for the nth term of this sequence.
d Use the rule you found to find the number of rolls in stack 24.

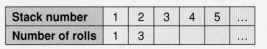

Stack number	1	2	3	4	5	...
Number of rolls	1	3				...

Answer

a

Stack number	1	2	3	4	5	...
Number of rolls	1	3	6	10	15	...

b

Terms	1		3		6		10		15
1st differences		2		3		4		5	
2nd differences			1		1		1		

c $2a = 1$

$\quad a = \frac{1}{2}$

$T(1) = a + b + c$
$\quad\quad = \frac{1}{2} + b + c$

$T(2) = a \times 2^2 + b \times 2 + c$
$\quad\quad = \frac{1}{2} \times 4 + 2b + c$
$\quad\quad = 2 + 2b + c$

1st difference, $T(2) - T(1) = 2 + 2b + c - (\frac{1}{2} + b + c)$

$\quad\quad\quad\quad\quad 2 = 1\frac{1}{2} + b$ **2 is the 1st difference**

$\quad\quad\quad\quad\quad b = \frac{1}{2}$

$T(1) = a + b + c$
$\quad 1 = \frac{1}{2} + \frac{1}{2} + c$
$\quad c = \mathbf{0}$

The rule for the nth term is $T(n) = \frac{1}{2}n^2 + \frac{1}{2}n$ or $T(n) = \frac{n^2}{2} + \frac{n}{2}$.

d $T(24) = \frac{1}{2} \times 24^2 + \frac{1}{2} \times 24$
$\quad\quad\quad = 300$

There are **300 rolls** in stack 24.

Exercise 3

1 Find the rule for the nth term of these **linear** sequences.

  a 8, 14, 20, 26, 32, ...   **b** 58, 49, 40, 31, 22, ...

  c 0·8, 1·6, 2·4, 3·2, 4, ...   **d** 2·2, 3·4, 4·6, 5·8, 7, ...

  e $1\frac{1}{4}, 1\frac{3}{4}, 2\frac{1}{4}, 2\frac{3}{4}, 3\frac{1}{4}$, ...   **f** ⁻7, ⁻11, ⁻15, ⁻19, ⁻23, ...

> See the **Remember** on page 185 to remind yourself if you need to.

2 Write down the quadratic rule for the nth term of these sequences by comparing them to
*the sequence 1, 4, 9, 16, 25, ... which has the rule $T(n) = n^2$.

  a 0, 3, 8, 15, 24, ...   **b** 3, 6, 11, 18, 27, ...   **c** 2, 8, 18, 32, 50, ...

  d 1, 7, 17, 31, 49, ...   ***e** 4, 10, 20, 34, 52, ...   ***f** 98, 94, 88, 80, ...

3 Use a difference table to work out the rule for the nth term of these sequences.

  a 10, 40, 90, 160, ...   **b** 4, 7, 12, 19, ...   **c** 4, 16, 36, 64, ...

  d ⁻9, ⁻6, ⁻1, 6, ...   **e** 21, 24, 29, 36, ...   **f** 5, 20, 45, 80, ...

  g 3, 9, 19, 33, 51, ...   **h** ⁻1, 0, 3, 8, 15, ...   **i** ⁻2, 7, 22, 43, 70, ...

  j 4, 11, 22, 37, 56, ...   **k** 2, 3, 6, 11, 18, ...   **l** 10, 27, 52, 85, 126, ...

  m 3, 16, 39, 72, 115, ...   **n** 2, 13, 32, 59, 94, ...   **o** 3, 19, 45, 81, 127, ...

Algebra

*4 Hong investigated the growth of bacteria at low temperature. She took readings each 24 hours. She recorded her results in a table.

No. of 24-hour periods	1	2	3	4	5	6
No. of bacteria (million)	3	10	21	36	55	78

a   Use Hong's results to find the rule for the number of bacteria after n 24-hour periods.
b   How many bacteria would there be after fifteen 24-hour periods if the bacteria continued to grow at the same rate?

*5

  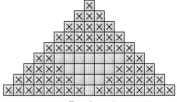

Design 1
8 cross-stitches

Design 2
21 cross-stitches

Design 3

Design 4
65 cross-stitches

Emma designed this sequence of cross-stitch patterns for table cloths.
a   Draw design 3.
b   Write down the sequence for the number of cross-stitches.
c   Using a difference table, find the rule for the number of cross-stitches in design n.

Review 1   Find the rule for the nth term of these *linear* sequences.
a   ⁻1, 2, 5, 8, 11, ...
b $2\frac{1}{2}$, 3, $3\frac{1}{2}$, 4, $4\frac{1}{2}$, ...
c   ⁻4, ⁻3·5, ⁻3, ⁻2·5, ⁻2, ...
d $\frac{1}{4}$, $1\frac{1}{2}$, $2\frac{3}{4}$, 4, $5\frac{1}{4}$, ...

Review 2   Write down the rule for the nth term of these quadratic sequences by comparing them with 1, 4, 9, 16, 25, ...
a   ⁻2, 1, 6, 13, 22, ...   b   3, 12, 27, 48, 75, ...   c   1, 10, 25, 46, 73, ...

Hint: Compare c with b to help.

Review 3   Use a difference table to work out the rule for $T(n)$ for these sequences.
a   11, 14, 19, 26, 35, ...   *b   20, 80, 180, 320, 500, ...   c   5, 11, 21, 35, 53, ...
d   3, 8, 15, 24, 35, ...   e   3, ⁻2, ⁻11, ⁻24, ⁻41, ...

Review 4   Debbie investigated the design of some stone monuments.
a   Draw monument 4 in this pattern.

 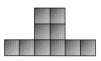 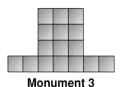

Monument 1

Monument 2

Monument 3

b   Copy and complete the table.
c   Write down the rule for the nth term of the sequence.
d   Use this rule to find the number of squares needed for monument 6.

Monument	Number of square stones
1	6
2	10
3	
4	

Discussion

Beatrice was exploring the rule for the sequence

7, 16, 29, 46, ...

She wrote

	$T(1)$	$T(2)$	$T(3)$	$T(4)$
	7	16	29	46

1st difference   9   13   17

↑   ↑   ↑

$4 \times 2 + 1$ $4 \times 3 + 1$ $4 \times 4 + 1$

So $T(n) = T(n-1) + 4n + 1$
Is she correct? **Discuss**.

Marty said he had written the rule as

$T(n) = T(n-1) + 9 + 4(n-2)$.

Is he correct? **Discuss**.

Beatrice's and Marty's rules are term-to-term rules.

How could you use the term-to-term rule to write the rule for the nth term? **Discuss**.

How could you write the rule for the nth term of this sequence as a term-to-term rule and a rule for the nth term? **Discuss**.

3, 6, 11, 18, 27, ...

Exercise 4

1 a Write down the nth term of the sequence

4, 7, 12, 19, 28, ...

as a term-to-term rule.

b Use term to term rule to write the rule for the nth term.

c What is the nth term as a term-to-term rule if $T(1) = 2$?

d What is the nth term as a term-to-term rule if $T(1) = 8$?

See the **Discussion** above for an example.

Review

a Write down the nth term of the sequence

6, 13, 24, 39, ...

as a term-to-term rule.

b Use the term to term rule to write the rule for the nth term.

c What is the nth term as a term-to-term rule if $T(1) = 5$?

Fraction sequences

Worked Example

Predict the next three terms of each of these sequences and explain the rule.

a $\frac{1}{3}, \frac{2}{4}, \frac{3}{5}, \frac{4}{6} \dots$ **b** $\frac{1}{1}, \frac{1}{2}, \frac{2}{3}, \frac{3}{5}, \frac{5}{8} \dots$

Answer

a The sequence for the numerators is 1, 2, 3, 4, ...

The sequence for the denominators is 3, 4, 5, 6, ...

Each denominator is two more than the numerator.

The next three terms could be $\frac{5}{7}, \frac{6}{8}, \frac{7}{9}$.

An expression for the nth term is $\frac{n}{n+2}$.

b The sequences for the numerators and denominators follow the Fibonacci rule 'add the two previous terms'.

The next three terms could be $\frac{8}{13}, \frac{13}{21}, \frac{21}{34} \dots$

Algebra

Discussion

● Consider the sequence $\frac{1}{2}, \frac{2}{3}, \frac{3}{4}, \frac{4}{5}, \frac{5}{6}, \frac{6}{7}, \ldots$. Is $T(n) = \frac{n+1}{n}$ a rule for this sequence? What is a possible rule? **Discuss**.

● **Discuss** possible rules for the following sequences.

$\frac{3}{2}, \frac{4}{3}, \frac{5}{4}, \frac{6}{5}, \ldots$ $2, \frac{3}{2}, \frac{4}{3}, \frac{5}{4}, \ldots$ $0, \frac{1}{2}, \frac{2}{3}, \frac{3}{4}, \ldots$

$1, \frac{1}{2}, \frac{1}{3}, \frac{1}{4}, \ldots$ $3, \frac{3}{2}, 1, \frac{3}{4}, \frac{3}{5}, \ldots$ $1, \frac{2}{3}, \frac{3}{5}, \frac{4}{7}, \ldots$

Exercise 5

1 Predict the next three terms of each of these sequences and explain the rule.

 a $\frac{1}{2}, \frac{1}{4}, \frac{1}{8}, \frac{1}{16}, \ldots$ **b** $\frac{1}{4}, \frac{2}{5}, \frac{3}{6}, \frac{4}{7}, \ldots$ **c** $\frac{1}{2}, \frac{2}{3}, \frac{3}{4}, \frac{4}{5}, \ldots$

 d $\frac{1}{1}, \frac{2}{1}, \frac{3}{2}, \frac{5}{3}, \frac{8}{5}, \ldots$ **e** $\frac{1}{5}, \frac{2}{5}, \frac{3}{10}, \frac{5}{15}, \ldots$

2 Write down the first five terms of the sequences given by these rules. Leave your answers as fractions.

 a $T(n) = \frac{1}{n}$ **b** $T(n) = \frac{n}{n+4}$ **c** $T(n) = \frac{n+1}{3n-2}$ **d** $T(n) = \frac{n}{n^2+1}$

 e $T(n) = \frac{n}{2n^2-1}$ **f** $T(n) = \frac{n^2+1}{2n^2+1}$

3 a What happens to the sequence given by $T(n) = \frac{n}{n+1}$ as n gets very large?

 b What if $T(n) = \frac{n}{n^2+1}$?

* **4** A spider is at the bottom of a 2-metre pipe. Each hour it moves up the pipe. In the first hour it moves halfway up the pipe. Each hour after that it moves half of the remaining distance up the pipe.

 a Write down the first eight terms of the sequence for the total distance moved by the spider after each hour.

 b How long will it take the spider to reach the top of the pipe?

Review 1 Predict the next three terms of each of these sequences and explain the rule.

a $\frac{1}{3}, \frac{2}{9}, \frac{3}{27}, \frac{4}{81}, \ldots$ **b** $\frac{16}{5}, \frac{14}{6}, \frac{12}{7}, \frac{10}{8}, \ldots$

Review 2 Write down the first five terms of the sequences given by these rules. Leave your answers as fractions.

a $T(n) = \frac{n}{2n-1}$ **b** $T(n) = \frac{2n}{n^2-2}$

***Review 3** A ball is dropped from a height of 2 metres. On each bounce the ball bounces to half of the height of the previous bounce.

a Find the total distance travelled by the ball before it starts falling for its third time.

b Find the total distance travelled before it starts falling for its

   i fourth time

   ii fifth time

   iii sixth time.

c Estimate how far the ball will travel altogether.

d Following the pattern, will the ball ever stop bouncing?

Spatial patterns

In the following exercise you will explore rules for **spatial patterns**.

1   The triangular numbers can be drawn as follows.

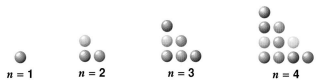

$n = 1$ $n = 2$ $n = 3$ $n = 4$

a   What goes in the gaps?

Position of triangular numbers (n)	1	2	3	4	5
Triangular number	1	3	6	10	15
Made up from rows of	1	1+2	1+2+3		

b   Write a rule for $T(n)$ based on what you discovered.
$T(n) = 1 +$ _____ $+$ _____ $+$ _____ $+$ _____ $+ ... +$ ___

c   Rosie discovered that by repeating any triangular number she could make a rectangle.

She wrote
$T(3) + T(3) = 3 \times 4$
$T(n) + T(n) =$ _____
Write an expression for the gap in terms of n.
Now use this to write an expression for the nth triangular number.
$T(n) =$ _____

This means an expression with n in it.

d   Use your answers to **b** and **c** to find the sum of the first 50 whole numbers.
$1 + 2 + 3 + ... + 50$

2   Roseanna drew a square pattern of dots.
She divided it into two triangular numbers.
Roseanna deduced that the nth triangular number could be given by
$T(n) = n^2 - T(n - 1)$ $T(n)$ is the nth triangular number and
$T(n - 1)$ the preceding one.

a   Show how she might have deduced this.
b   Show that this formula is true for the 2nd up to the 8th triangular numbers.

3   Ronan divided a square pattern like this.
a   What can you deduce about the sum of the first four odd numbers and the square of 4?
b   Generalise a formula for the sum of the first n odd numbers.

Algebra

*4 Dipesh divided a square pattern like this.
What might he deduce from this?
Explain your answer fully and if possible write a formula for the deduction.

*5 A particular type of bacteria starts as a single cell and grows by doubling.

| **Stage 1** | **Stage 2** | **Stage 3** | **Stage 4** |
| 1 cell | 2 cells | 4 cells | 8 cells |

   a How many cells will be present at stage 6?
   b Write a rule for the number of cells present at stage n.
      Hint Consider powers of 2.

*6 Maggie drew this growth pattern of squares.
Each new pattern has a square added on all sides of existing squares.

| **pattern 1** | **pattern 2** | **pattern 3** | **pattern 4** |
| 1 square | 5 squares | 13 squares | 25 squares |

Investigate this pattern by drawing more of them, writing down the sequence for the number of squares and trying to find a rule for it.

*7 Gemma's Aunty Lucy was bored one afternoon so she started working out the maximum number of crossings for a certain number of knitting needles.

| **1 needle** | **2 needles** | **3 needles** | **4 needles** |
| 0 crossing | 1 crossing | 2 crossings | 6 crossings |

Predict how the sequence might continue. Explain your answer.

Review 1 Hans drew this square pattern.
He wrote $1 + 3 + 5 + 3 + 1 = 3^2 + 2^2$.
a Explain how this relates to the diagram.
b Deduce the answer to $1 + 3 + 5 + 7 + 5 + 3 + 1$ by drawing a
    similar square pattern.
c Deduce the answer to $1 + 3 + \dots 19 + \dots 3 + 1$.
d Find the answer to $1 + 3 + \dots (2n - 1) + \dots 3 + 1$.

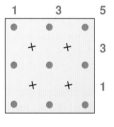

Review 2 This growth pattern of squares is made from matches.

| **pattern 1** | **pattern 2** | **pattern 3** | **pattern 4** |

Investigate this pattern.
a Link the number of matches to the pattern number in a table.
b Use your table to find the rule connecting the pattern number, n, with the number of matches, $T(n)$.

Functions

Remember

These are all ways of expressing a **function**.

$$y = \frac{x}{2} - 4, \qquad x \rightarrow \frac{x}{2} - 4, \qquad x \rightarrow \boxed{\text{divide by 2}} \rightarrow \boxed{\text{subtract 4}} \rightarrow y$$

For any given value of x, we can work out a value for y.

The input (x) and output (y) of a function can be shown on a table, a mapping diagram or on a graph.

Example $\quad x \rightarrow 2x - 3$

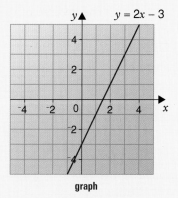

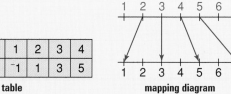

x	⁻1	0	1	2	3	4
y	⁻5	⁻3	⁻1	1	3	5

table   mapping diagram   graph

If $x \rightarrow x$ the function is called the **identity function**, because it maps every number onto itself. The number is unchanged.

$$x \rightarrow x$$

To find the input given the output we work backwards, doing the **inverse operations**.

Example

$$? \rightarrow \boxed{\text{add 4}} \rightarrow \boxed{\text{multiply by 2}} \rightarrow 14$$

$$3 \leftarrow \boxed{\text{subtract 4}} \leftarrow \boxed{\text{divide by 2}} \leftarrow 14 \leftarrow \quad \text{Start with the output and work backwards doing inverse operations.}$$

The input was **3**.

Every linear function has an **inverse**. This reverses the direction of the mapping.

Example

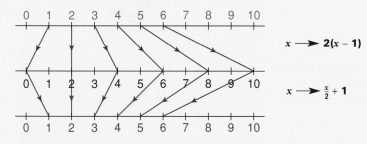

$$x \rightarrow 2(x - 1)$$

$$x \rightarrow \frac{x}{2} + 1$$

The inverse of $x \rightarrow 2(x - 1)$ is $x \rightarrow \frac{x}{2} + 1$.

To find the inverse of a function we can use inverse function machines.

Example   To find the inverse of $x \rightarrow \frac{1}{2}x + 3$ draw a function and inverse function machine.

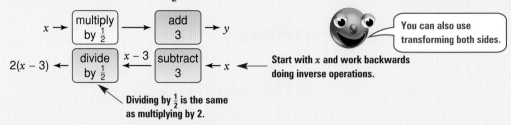

Start with x and work backwards
doing inverse operations.

Dividing by $\frac{1}{2}$ is the same
as multiplying by 2.

You can also use
transforming both sides.

The inverse of $x \rightarrow \frac{1}{2}x + 3$ is $x \rightarrow 2(x - 3)$.

Functions of the form $x \rightarrow c - x$, where c is a constant, are **self-inverses**.

Example

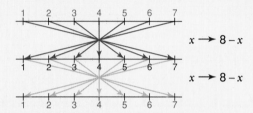

$x \rightarrow 8 - x$

$x \rightarrow 8 - x$

○ ○

Investigation

Graphs of linear and quadratic functions

You will need a graphical calculator or graph plotter.

A Plot each function and its inverse on the same set of axes .

   a $y = 3x$ and $y = \frac{x}{3}$ **b** $y = 2x$ and $y = \frac{x}{2}$ **c** $y = x + 1$ and $y = x - 1$

   d $y = x - 3$ and $y = x + 3$ **e** $y = 2x + 1$ and $y = \frac{x}{2} - \frac{1}{2}$ **f** $y = 2(x + 3)$ and $y = \frac{x}{2} - 3$

   g $y = \frac{1}{2}x + 1$ and $y = 2(x - 1)$

What do you notice about each pair?
Investigate other graphs of functions and their inverses.

B Draw the graphs of these quadratic functions.

$$x = x^2$$
$$y = x^2 - 2$$
$$y = x(x - 1)$$
$$y = {}^-x^2 + 1$$
$$y = {}^-x(x + 2)$$

This is linked to graphs
of quadratic functions.

Is the graph of a quadratic function
    – straight or curved?
    – symmetrical? If so, describe in general terms the line of symmetry.

Each quadratic function has a turning point.
Identify the turning point for each of the graphs you drew.
What can you say about the value of the y-coordinate at
the turning point?

turning
point

Investigate graphs of other quadratic functions.

Exercise 7

1 a Write down the output for each of these function machines.

b Match each function machine with a function from the box.

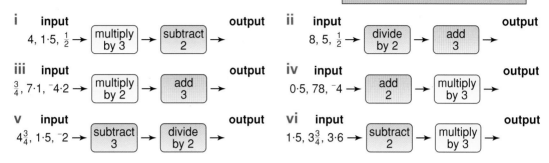

A $y = \frac{x-3}{2}$ D $y = \frac{x}{2} + 3$

B $y = 3(x - 2)$ E $y = 3x - 2$

C $y = 3(x + 2)$ F $y = 2x + 3$

i input → multiply by 3 → subtract 2 → output
4, 1·5, $\frac{1}{2}$

ii input → divide by 2 → add 3 → output
8, 5, $\frac{1}{2}$

iii input → multiply by 2 → add 3 → output
$\frac{3}{4}$, 7·1, ⁻4·2

iv input → add 2 → multiply by 3 → output
0·5, 78, ⁻4

v input → subtract 3 → divide by 2 → output
$4\frac{3}{4}$, 1·5, ⁻2

vi input → subtract 2 → multiply by 3 → output
1·5, $3\frac{3}{4}$, 3·6

T

2 Use a copy of this.
Fill in the missing input and output numbers.

a x → multiply by 2 → add 4 → y

Input	Output
0·75	
$\frac{7}{8}$	
	8·4

b x → divide by 3 → add 2 → y

Input	Output
$\frac{3}{4}$	
⁻0·12	
	4·6

c input (x)   output

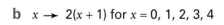

?
? → add 2 → multiply by 3 → 6·3
?   0·24
0·45 $\frac{6}{15}$
                               ?

d

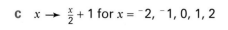

___, ⁻3, ___, $\frac{1}{2}$, 0·4 → subtract 1 → multiply by 2 → 1·2, ___, $1\frac{3}{4}$, ___, ___

T

3 Use a copy of these mapping diagrams.
Fill them in for the functions given.

a $y = 2x + 1$ for $x = 0, 1, 2, 3, 4$

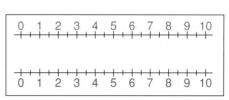

b x → $2(x + 1)$ for $x = 0, 1, 2, 3, 4$

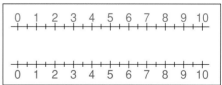

c x → $\frac{x}{2} + 1$ for $x = ⁻2, ⁻1, 0, 1, 2$

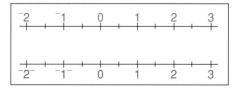

4 Marcia wants to know what the identity function is. How could you explain it to her?
Use a diagram to help.

5 Find the inverse function for each of these.
  a $x \rightarrow 5x - 2$ **b** $x \rightarrow 3(x - 4)$ **c** $x \rightarrow \frac{x + 3}{5}$ **d** $x \rightarrow \frac{1}{4}x + 2$
  e $x \rightarrow \frac{1}{2}x - 3$ **f** $x \rightarrow \frac{3(x - 4)}{2}$ ***g** $x \rightarrow x^2$ ***h** $x \rightarrow x^2 + 1$

6 Find the inverse of each of these functions.
  Draw a mapping diagram for the function and its inverse.
  What do you notice?
  a $x \rightarrow 12 - x$ **b** $x \rightarrow 8 - x$ **c** $x \rightarrow 4 \cdot 5 - x$

7 **a** Using paper, a graphical calculator or ICT package draw the graph of $y = 4x$.
  b What is the inverse function for $y = 4x$?
  c Draw the inverse function on the same grid as you used in **a**.
  d Write down the equation of the line that $y = 4x$ and its inverse are symmetrical about.

8 Repeat question **7** for these.
  a $y = x + 3$ **b** $y = 2(x + 1)$ **c** $y = \frac{1}{2}x + \frac{1}{2}$

9 Sajid thought of a number.
  He carried out these operations on the number.

  When he did them in one order he got 89.
  When he did them in the other order he got 114.
  a Draw a function machine for each of Sajid's calculations.
  b Write a function for each of the function machines.
  c What is the number Sajid thought of?
  d Explain why doing the operations in a different order gives a different result.
  *e The difference between the answers is 25.
    Prove that the difference will **always** be 25, no matter what the number is.

*10 To answer these questions you need to have done **part B** of the **Investigation** on page 194. If you haven't done it, do it now.
  Write true or false for these.
  a The graph of a quadratic function is a curve symmetrical about the vertical line through the turning point.
  b Graphs of functions with a $^-x^2$ term do not have a turning point.
  c The value of the y-coordinate at the turning point is always the minimum value of the function.
  d The value of the y-coordinate at the turning point is always the maximum or minimum value of the function.

Review 1   Use a copy of these mapping diagrams. Fill them in for the functions given.
  a $y = 3x - 2$ for $x = 0, 1, 2, 3$

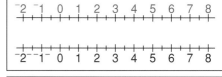

  b $x \rightarrow \frac{3x + 1}{2}$ for $x = ^-2, ^-1, 0, 1, 2$

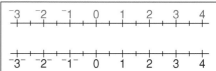

Review 2 Find the inverse function for each of these.

a $x \rightarrow 2(x-3)$ b $x \rightarrow \frac{1}{3}x - 4$ c $x \rightarrow \frac{2x+5}{3}$ *d $x \rightarrow \frac{1}{x} + 2$

Review 3

a Using paper, a graphical calculator or ICT package draw the graph of $y = {}^{-}2x$.

b What is the inverse function of $y = {}^{-}2x$?

c Draw the inverse function on the same grid as **a**.

d Write down the equation of the line that $y = {}^{-}2x$ and its inverse are symmetrical about.

e Repeat **a** to **d** for $y = 2x + 1$.

f What do you notice about the line of symmetry?

Review 4 This shows the input and output guesses Matthew made in a game of 'what is my rule?'

If the guess was correct it has a tick. If wrong, a cross.

Find the rule for each game.

Game 1

Input	Output	
3	2	✓
7	10	✓
12	18	✗
5	6	✓
6	12	✗
11	18	✓

Game 2

Input	Output	
5	17	✓
1	5	✓
3	9	✗
7	11	✗
3	11	✓
10	32	✓

* **Review 5** Write true or false for each of these statements.

a The quadratic function $y = x^2$ has no maximum turning point.

b The quadratic function $y = x^2 + 2$ has a minimum turning point when $y = 2$.

c The line of symmetry of a quadratic function is halfway between the x-intercepts.

Summary of key points

A Sequences can be generated using

1 a term-to-term definition

Example   **first term** $x + 2y$ **term-to-term rule** add x

generates the sequence $x + 2y, 2x + 2y, 3x + 2y, 4x + 2y, \ldots$

This could also be written

$T(1) = x + 2y, \quad T(2) = T(1) + x, \quad T(3) = T(2) + x, \ldots$

2 a rule for the nth term.

Example $T(n) = 2n + (n-1)(n-2)$

Substituting $n = 1, 2, 3, 4, \ldots$ gives the sequence

2, 4, 8, 14, 22, 32, ...

Sometimes sequences **converge to a number**.

Example   The sequence with the rule 'divide by 2, add 3'

converges to 6.

B A **quadratic sequence** has a rule with a squared term as its highest power.

Example $n^2 + 3n + 2$

Position	1	2	3	4	5
Term T(n)	$1^2 + 3 \times 1 + 2 = 6$	$2^2 + 3 \times 2 + 2 = 12$	$3^2 + 3 \times 3 + 2 = 20$	$4^2 + 3 \times 4 + 2 = 30$	$5^2 + 3 \times 5 + 2 = 42$

The 2nd difference is constant in a quadratic sequence.

It can be used to predict the next terms of a sequence.

Example

6,   12,   20,   30,   42,   56   42 + 14 = 56

1st differences ⟶   6   8   10   12   14   12 + 2 = 14

2nd differences ⟶   2   2   2   2   ⟵ constant 2nd difference

 C The **rule for the nth term** of a quadratic sequence can sometimes be found by comparing with $T(n) = n^2$. $T(n) = n^2$ is the sequence 1, 4, 9, 16, 25, ...

Example 2, 8, 18, 32, 50, ... has rule $T(n) = 2n^2$.

Each term is twice a square number.

The rule for any quadratic sequence can be found using the **1st and 2nd differences**.

Example   4,   11,   22,   37,   56,

     7   11   15   19   ⟵ 1st differences

       4   4   4   ⟵ 2nd differences

The rule will be in the form $T(n) = an^2 + bn + c$.

$2a = 4$ 2nd difference

$a = 2$

> To find a, use the 2nd difference, 2a. To find b, use the difference between $T(1)$ and $T(2)$ which is $3a + b$. To find c, use $a + b + c =$ first term.

$T(2) - T(1) = 3a + b$

$7 = 3 \times 2 + b$ substituting $T(2) - T(2) = 2$ and $a = 2$

$7 = 6 + b$

$b = 1$

$T(1) = a + b + c$

$4 = 2 + 1 + c$ substituting $a = 2$ and $b = 1$

$c = 1$

The rule for the sequence 4, 11, 22, 37, 56, ... is

$T(n) = 2n^2 + n + 1$.

 D The terms in some sequences are all **fractions**.

Example $\frac{1}{1}, \frac{2}{2}, \frac{3}{3}, \frac{4}{5}, \frac{5}{8}, ...$ ⟵ add 1 to the previous numerator

                           ⟵ add the two previous denominators

Functions

If $x \rightarrow x$ the function is called the **identity function**.

It maps every number onto itself.

If we are given the output of a function machine, we work
backwards doing inverse operations to find the input.

Example

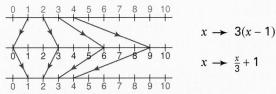

The input was 3.

The **inverse of a function** can be found by doing the inverse operations in the
reverse order.

Example

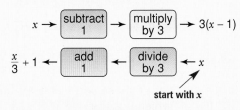

$x \rightarrow 3(x-1)$

$x \rightarrow \frac{x}{3} + 1$

The direction of the mapping is reversed.

We can draw a function and inverse function machine or transform both
sides to find the inverse of $x \rightarrow 3(x-1)$.

Inverse function machine

$x \rightarrow$ [subtract 1] $\rightarrow$ [multiply by 3] $\rightarrow 3(x-1)$

$\frac{x}{3}+1 \leftarrow$ [add 1] $\leftarrow$ [divide by 3] $\leftarrow x$

start with x

Transform both sides

$x \rightarrow 3(x-1)$

$\frac{x}{3} \rightarrow \frac{3(x-1)}{3}$

$\frac{x}{3} \rightarrow x-1$

$\frac{x}{3}+\mathbf{1} \rightarrow x-1+\mathbf{1}$

$x \rightarrow \frac{x}{3}+1$

The graph of an inverse function is a reflection of the graph of the function in the
mirror line $y=x$.

 The **graph of a quadratic function** is a symmetrical curve.

The line of symmetry is the vertical line through the turning point.

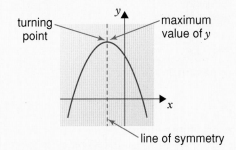

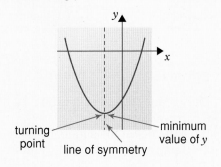

Algebra

Test yourself

1 Write down the first ten terms of these sequences.
  a **1st term** 10   **rule** subtract 0·5
  b **1st terms** 2, 2   **rule** add the two previous terms together
  c $T(n) = n + \frac{1}{2}$ **d** $T(n) = n + (n - 2)(n - 3)$

2 A sequence has the rule 'to find the next number subtract ☐'.
What could be the first term and the rule for the sequence if every fifth number is
an integer?

3 Write down the first six terms of the sequence given by
$T(1) = m + 2n, \quad T(2) = m - n + T(1), \dots T(n) = m - n + T(n - 1).$

4 **a** A sequence has the rule $x \rightarrow \frac{x}{3} + 6$.
    Use a graphical calculator to generate some terms of the sequence.
    Does the sequence 'converge' to a number? If so, what number?
  b Solve the equation $x = \frac{x}{3} + 6$. What do you notice about your answer?

5 Which of these sequences are quadratic? How can you tell?
  a 4, 5, 7, 10, 14, ...   **b** ⁻2, 1, 4, 7, 10, ...   **c** ⁻2, 1, 6, 13, 22, ...

6 Write down the first six terms of these sequences
  a $T(n) = n^2 + 2$ **b** $T(n) = 2n^2 - 3$

7 For each of the sequences in question **6**, find the second difference.

8 Find the next two terms of these quadratic sequences by finding the first and
second difference.
  a 2, 7, 14, 23, 34, ...   **b** 3, 9, 19, 33, 51, ...   **c** ⁻4, ⁻1, 4, 11, 20, ...

9 Write down the quadratic rule for the nth term of these sequences by comparing
them to the sequence 1, 4, 9, 16, 25, ... which has the rule $T(n) = n^2$.
  a 0, 3, 8, 15, 24, ...   **b** 4, 16, 36, 64, 100, ...

10 Use a difference table to work out the rule for the nth term of these sequences.
  a 4, 7, 12, 19, 28, ...   **b** 0, 4, 10, 18, 28, ...   **c** 6, 14, 28, 48, 74, ...

11

pattern 1   **pattern 2**   **pattern 3**

Mary made these designs on tiles.
  a Draw pattern 4.
  b Write down the sequence for the number of white squares.
  c Using a difference table, find the rule for the number of white squares in pattern n.

12 Predict the next three terms of these sequences and explain the rule.
  a $\frac{1}{1}, \frac{2}{4}, \frac{3}{9}, \frac{4}{16}, \dots$ **b** $\frac{1}{1}, \frac{1}{2}, \frac{2}{3}, \frac{3}{5}, \frac{5}{8}, \dots$

13 Write down the first five terms of the sequences given by these rules.
Leave your answers as fractions.
a $T(n) = \frac{n}{n+2}$ **b** $T(n) = \frac{n+1}{n^2}$ **c** $T(n) = \frac{n^2+1}{n+2}$

14 What happens to the terms of the sequence $T(n) = \frac{n}{n+2}$ as n gets very large.

15 **a** Each term of a number sequence is made by adding 1 to the numerator and 3 to the denominator of the previous term.
It starts $\frac{1}{2}, \frac{2}{5}, \frac{3}{8}, \frac{4}{11}, \frac{5}{14}, \ldots$
Write an expression for the nth term of the sequence.
b The nth term of a different sequence is $\frac{n}{2n^2-1}$.
The first term of the sequence is $\frac{1}{1}$.
Write down the next three terms.

16 **a** Write down the missing inputs and outputs for each of these function machines.
b Match each function machine with a function from the box.

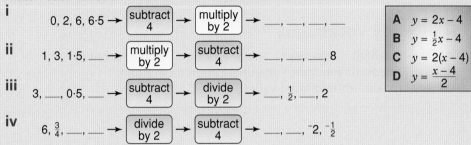

i $0, 2, 6, 6\cdot5 \rightarrow$ subtract 4 $\rightarrow$ multiply by 2 $\rightarrow$ ___, ___, ___, ___

ii $1, 3, 1\cdot5, \_\_ \rightarrow$ multiply by 2 $\rightarrow$ subtract 4 $\rightarrow$ ___, ___, ___, 8

iii $3, \_\_, 0\cdot5, \_\_ \rightarrow$ subtract 4 $\rightarrow$ divide by 2 $\rightarrow$ ___, $\frac{1}{2}$, ___, 2

iv $6, \frac{3}{4}, \_\_, \_\_ \rightarrow$ divide by 2 $\rightarrow$ subtract 4 $\rightarrow$ ___, ___, $^-2$, $\frac{-1}{2}$

A $y = 2x - 4$
B $y = \frac{1}{2}x - 4$
C $y = 2(x - 4)$
D $y = \frac{x-4}{2}$

17 Use a copy of this mapping diagram.
Fill it in for
$x \rightarrow \frac{x}{3} - 1$ for $x = ^-3, ^-1, 0, 2, 3$.

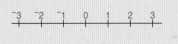

18 Find the inverse functions for these.
a $x \rightarrow 4(x - 1)$ **b** $x \rightarrow \frac{1}{2}x - 3$ **c** $x \rightarrow \frac{2(x+1)}{3}$ ***d** $x \rightarrow x^2 - 1$

19 **a** Find the inverse functions for these.
i $x \rightarrow 6 - x$ **ii** $x \rightarrow x$ **iii** $y = 2x$
b What is the function in **ii** called?
c Draw the graph of $y = 2x$ and the graph of its inverse.
Write down the equation of the line that $y = 2x$ and its inverse are symmetrical about.

20 **a** I think of a number then I carry out these operations on my number.

$\rightarrow$ multiply by 3 $\rightarrow$ add 5 $\rightarrow$

When I carry out the operations in one order the answer is 42.
When I carry out the operations in the other order, the answer is 32.
What is my number? Show your working.
b The difference between my two answers is 10.
Prove that the difference will always be 10, no matter what my number is.

21 Write true or false for each of these statments.
a The graph of a quadratic function is a curve with no line of symmetry.
b The quadratic function $y = -x^2$ has a maximum turning point.
c The value of the y-coordinate at the turning point of $y = x^2$ is the minimum value of the function.

You need to know

✓ graphs   page 116
    – graphs of real-life situations

Key vocabulary

**gradient,   intercept,   linear function,   cubic function,
quadratic function,   maximum/minimum point,
maximum/minimum value**

Dealing in Dollars

American Dollars to 1 GBP

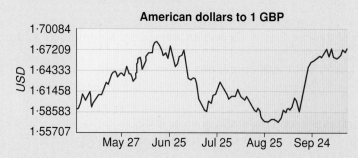

GBP means Great Britain Pounds

This graph showing how many American dollars you got for £1 versus time, came from the Internet.
Find some other real-life graphs in newspapers, magazines or the Internet.
Make a poster of your graphs explaining what each is about.
Are any of them misleading?

Linear graphs

Remember

$y = mx + c$ represents a **straight-line graph**.

   m is the gradient and c is the y-intercept.

   If m is positive the line slopes

   If m is negative the line slopes ⟍ ⊓   ⊓ **for negative**

Sometimes the equation of a straight line is not given in the form $y = mx + c$.
If we want to find m and c we must **rearrange it into the form** $y = mx + c$.

Example   We can rearrange $3y + 4x = 7$.

$$3y + 4x = 7$$
$$3y = 7 - 4x \qquad \text{subtracting } 4x \text{ from both sides}$$
$$y = \frac{7}{3} - \frac{4}{3}x \qquad \text{dividing both sides by 3}$$
$$y = \frac{4}{3}x + \frac{7}{3} \qquad \text{putting in } y = mx + c \text{ order}$$

For this equation $m = \frac{-4}{3},\ c = \frac{7}{3}$.

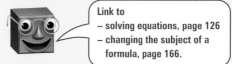

Link to
– solving equations, page 126
– changing the subject of a formula, page 166.

All points that lie on the line $3y + 4x = 7$ satisfy the equation.

For any straight line, the change in y is proportional to the change in x.
The gradient, m, of a straight line joining (x_1, y_1) to (x_2, y_2) is:

$$m = \frac{y_2 - y_1}{x_2 - x_1} = \frac{\text{change in } y}{\text{change in } x}$$

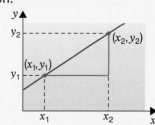

Investigation

Gradients of parallel and perpendicular lines

You will need graph paper, graph plotting software or a graphical calculator.

● Graph these lines on the same set of axes.

   $y = 2x + 2 \qquad y = 2x - 1 \qquad y = 2x + 4 \qquad y = 2x$

   Describe the similarities and differences.

   What if the lines were

   $y = {}^{-}3x + 1 \qquad y = {}^{-}3x - 2 \qquad y = {}^{-}3x \qquad y = {}^{-}3x + 4?$

   What can you say about m_1 and m_2 when $y = m_1x + c_1$ and $y = m_2x + c_2$ are parallel?
   Investigate.

● Graph each of these pairs on the same set of axes.

   $y = 2x + 1$ and $y = {}^{-}\frac{1}{2}x + 1$ $\qquad\qquad$ $y = 3x + 2$ and $y = {}^{-}\frac{1}{3}x$

   $y = x - 1$ and $y = {}^{-}x + 2$ $\qquad\qquad\qquad$ $y = {}^{-}4x - 6$ and $y = \frac{x}{4} + 1$

   $y = {}^{-}2x$ and $y = \frac{x}{2} - 2$

   What can you say about m_1 and m_2 when $y = m_1x + c_1$ and $y = m_2x + c_2$ are perpendicular?
   Investigate.

Algebra

Two lines, $y = m_1x + c_1$ and $y = m_2x + c_2$, are **parallel**.
Then

    Gradient, m_1, of $y = m_1x + c_1 = \frac{a}{1}$

    Gradient, m_2, of $y = m_2x + c_2 = \frac{b}{1}$

Since the lines are parallel a is the same length as b.

$m_1 = m_2$.

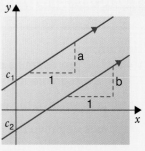

Two lines, $y = m_1x + c_1$ and $y = m_2x + c_2$, are **perpendicular**.
Then

    Gradient, m_1, of $y = m_1x + c_1 = \frac{p}{1}$

    Gradient, m_2, of $y = m_2x + c_2 = \frac{-1}{q}$

Since the lines are perpendicular p is the same length as q.

$m_1 = \frac{^-1}{m_2}$

      and $m_1 m_2 = {}^-1$

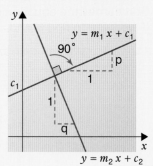

Exercise 1

1 The mid-points of each side, E, F, G and H, join   **[SATs Paper 1 Level 6]**
to make a different square.
  a Write the equation of the straight line
     through E and H.
  b Is $y = {}^-x$ the equation of the straight line
     through E and G?
     Explain how you know.

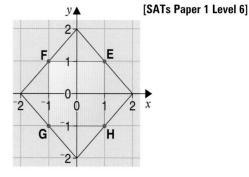

2 **a** Rearrange these equations into the form $y = mx + c$.
       $2y - 3x = 4$ $2y + 3x + 8 = 0$
  b Draw the graph of each on the same set of axes.
  c Write down the coordinates of the point where they meet.

3 Here are six different equations.

A $y = 2x - 3$	**B** $y = 5$	**C** $x = {}^-3$	**D** $x + y = 13$	**E** $y = 3x + 1$	**F** $y = x^2$

I draw the graphs of these equations.
  a Which graph goes through the point (0, 1)?
  b Which graph is parallel to the y-axis?
  c Which graph is not a straight line?
  d Which two graphs pass through the point (3, 10)?
  e Are any of the lines parallel? Explain your answer.
  f Are any of the lines perpendicular? Explain your answer.

4 Find the gradient of each of these lines.

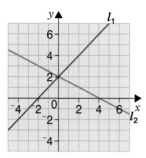
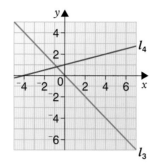
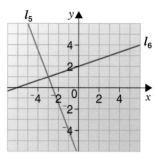

5 If the value of y decreases as x increases, will the gradient, m, be negative or positive?

6 Write down the equations of lines **a** to **f**.

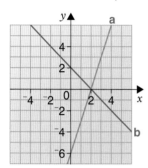
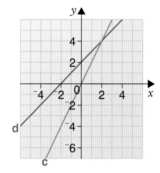
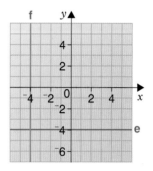

7 Six equations of straight lines are given in the box.
   a Which three are parallel?
   b Which two are perpendicular?

> $y = 3x - 2$
> $y = ^-3x - 2$
> $y = \frac{x}{3} - 2$
> $y = 3x + 4$
> $y = \frac{3}{x} + 2$
> $3y = 9x - 4$

8 a Write all the equations in the form $y = mx + c$.

   A $y + 3 = 2x$ **B** $2y = x + 4$ **C** $4x - 2y = 7$
   D $6x + 3y = 2$ **E** $\frac{1}{2}y = x - 5$ **F** $x + 4y = 3$

   b Which of the lines in **a** have a gradient of 2?
   c Which of the lines in **a** are   **i** parallel   **ii** perpendicular?
   d Which of these lines cut the y-axis at the same point?

   A $y = 3x + 2$ **B** $x + 2y = 4$ **C** $x + y + 2 = 0$
   D $6x - y + 2 = 0$ **E** $3x + 2y = 6$

   e Does the point (2, 3) lie on the line $2y = x + 4$?
   f Draw the graph of $4x - 2y = 8$.

9 a Write down the coordinates of the point that lies on both
   the straight lines $y = 3x + 1$ and $y = 5x - 3$.
   Show your working.
   b Explain how you can tell there is no point that lies on
   both straight lines $y = \frac{1}{2}x + 4$ and $y = \frac{1}{2}x + 6$.

> Link to simultaneous equations.

Algebra

10 Match these sketches with equations from the box.

a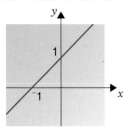

b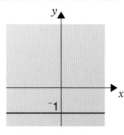

c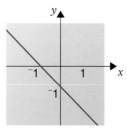

d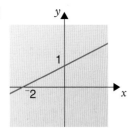

e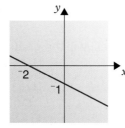

A $y = {}^-1$
B $y + x + 1 = 0$
C $2y - x = 2$
D $3y = 3x + 3$
E $2y + x + 2 = 0$

11 Two straight lines, $y = m_1 x + 4$ and $y = m_2 x - 3$ are
a parallel. If $m_1 = {}^-2$, what does m_2 equal?
b perpendicular. If $m_1 = {}^-2$, what does m_2 equal?

12 Write true or false for these.
a The equations for the lines that form two opposite sides of a parallelogram could be
$y = 2x - 3$ and $y = 2x + 2$.
b The equations of the lines in **a** could be for two opposite sides of a trapezium.
c These equations could be for the lines that form the four sides of a rhombus.
$y = 2x + 3 \qquad y = 2x + 6 \qquad y = 2x - 3 \qquad y = 2x - 6$
d Four of these equations are for lines that form the four sides of a rectangle.
Which four are they?

$y = 2x + 3 \qquad y = 2x - 1 \qquad y = \frac{1}{2}x - 3 \qquad y = {}^-\frac{1}{2}x \qquad y = {}^-\frac{1}{2}x - 3$

$y = \frac{1}{2}x \qquad y = 3x + \frac{1}{2} \qquad y = x + \frac{1}{2} \qquad y = \frac{{}^-2x}{3} + 2$

****13** This is a graphical calculator screen.
a Suggest possible equations for these lines.
b Find the shortest distance between the lines.

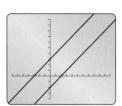

Review 1 $2x + 3y = 3$ is an equation for a line.
a Rearrange the equation into the form $y = mx + c$.
b Draw a graph of the equation.
c Write an equation for a line parallel to the one you've drawn.
d Write an equation for a line perpendicular to the one you have drawn that goes through the point (0, 1).

Review 2 Match each line with an equation from the box.

a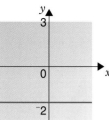

b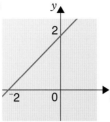

c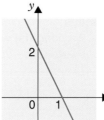

d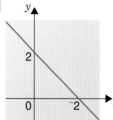

e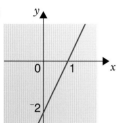

f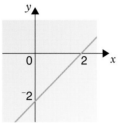

g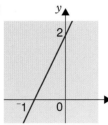

A	$y = ^-2$
B	$y = x - 2$
C	$y = 2x$
D	$y = 2x + 2$
E	$y = 2x - 2$
F	$y = ^-2x + 2$
G	$y = x + 2$
H	$y = ^-x + 2$

Review 3 The lines $y = 2x + 3$, $y = \frac{-x}{2} - 2$, $y = 3 - \frac{x}{2}$ and $y = 2x - 2$ form a quadrilateral. Is the quadrilateral a

A trapezium   **B** rectangle   **C** parallelogram   **D** rhombus?

 Practical

You will need a graphical calculator.

Use your graphical calculator to draw a rectangle and a square.

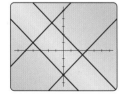   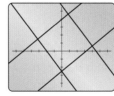

On your graphical calculator draw some other shapes made from parallel and perpendicular lines.
Write down the equations you used to draw them.

Graphs of quadratic, cubic and reciprocal functions

Investigations

A Quadratic graphs

You will need graph paper, a graph plotting package or a graphical calculator.
Use a graphical calculator if possible.

If you do not have a graphical calculator or a computer graphics package available, draw the graphs by plotting points.

For instance, for $y = x^2 - 9$, taking the whole number values of x from $^-4$ to 4 and using the calculator to find the y-values we get:

x	$^-4$	$^-3$	$^-2$	$^-1$	0	1	2	3	4
y	7	0	$^-5$	$^-8$	$^-9$	$^-8$	$^-5$	0	7

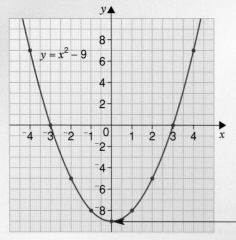

Include positive and negative values for x in your table.

A quadratic graph always has a minimum or maximum point.

● **Investigate** the following groups of quadratic functions.

Group 1 $y = x^2$, $y = x^2 + 2$, $y = x^2 - 2$, $y = x^2 + 5$, $y = x^2 - 5$
What is the relationship between $y = x^2$ and $y = x^2 + a$?

Group 2 $y = x^2$, $y = 2x^2$, $y = 4x^2$, $y = \frac{1}{2}x^2$, $y = \frac{1}{4}x^2$
What can you say about $y = x^2$ and $y = bx^2$?

Group 3 $y = x^2$, $y = (x + 2)^2$, $y = (x - 2)^2$, $y = (x + 5)^2$, $y = (x - 5)^2$
What is the relationship between $y = x^2$ and $y = (x + c)^2$?

Group 4 $y = x^2$, $y = (x - 3)^2 + 2$, $y = (x + 3)^2 + 2$, $y = (x - 2)^2 + 3$, $y = (x + 2)^2 - 3$,
$y = (x - 1)^2 - 3$, $y = (x + 4)^2 - 2$

What is the relationship between $y = x^2$ and $y = (x + c)^2 + a$?
Where is the vertex (the turning point) of the graph of $y = (x + c)^2 + a$?
What is the equation of the line of symmetry?

● Quadratic functions have the form $y = ax^2 + bx + c$.
Use a graphical calculator to investigate graphs for different values of a, b and c.

Examples $y = x^2$, $y = x^2 + 3x - 4$, $y = x^2 + 4x + 3$, $y = x^2 - 3x + 4$, $y = ^-x^2 + 3x + 4$,
$y = ^-x^2 - 3x - 4$, ...

- Plot the graph of $y = (x - 3)(x + 2)$.
  Where does the graph cut the x-axis?
  Draw the graphs of
  $$y = (x + 4)(x + 1), \quad y = (x - 5)(x - 4) \quad \text{and} \quad y = x(x - 5).$$
  What is the relationship between $y = (x - a)(x - b)$ and the x-intercepts? **Investigate**

* ● Plot the graph of $y = 25 - x^2$.
  Where does the graph cut the x-axis?
  Solve the equation $0 = 25 - x^2$. What do you notice?
  Draw the graphs of
  $$y = 16 - x^2, \quad y = 4 - x^2 \quad \text{and} \quad y = 36 - x^2.$$
  Where does each cross the x-axis?
  Solve the equations
  $$16 - x^2 = 0, \quad 4 - x^2 = 0 \quad \text{and} \quad 36 - x^2 = 0.$$
  What do you notice? **Investigate**.

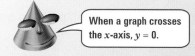

When a graph crosses the x-axis, $y = 0$.

B Cubic graphs

You will need graph paper, a graphical calculator or a graph plotting package.

- **Investigate** each of the groups of graphs given below. Use a graphical calculator or a computer graphics package. If you have neither of these available, draw the graphs by plotting points.

For instance, for $y = {}^-x^3 - 2x^2 + 5x + 6$, taking whole number values of x from $^-4$ to 4 and using the calculator to find the y-values we get:

x	$^-4$	$^-3$	$^-2$	$^-1$	0	1	2	3	4
y	18	0	$^-4$	0	6	8	0	$^-24$	$^-70$

Since the y-value corresponding to $x = 4$ is very large we will not include the point $(4, {}^-70)$ on our graph.

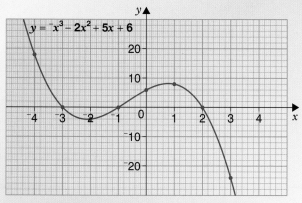

Group 1 $y = x^3,\ y = 2x^3,\ y = 4x^3,\ y = \frac{1}{2}x^3,\ y = \frac{1}{4}x^3$

Group 2 $y = x^3,\ y = {}^-x^3,\ y = 3x^3,\ y = {}^-3x^3$

Group 3 $y = x^3,\ y = x^3 + 2,\ y = x^3 - 4,\ y = {}^-x^3 + 3,\ y = {}^-x^3 - 3$

Group 4 $y = x^3,\ y = (x - 2)^3,\ y = (x + 2)^3,\ y = (4 - x)^3,\ y = (3 - x)^3$

Group 5 $y = (x + 1)(x - 2)(x - 3),\ y = (x - 2)(x + 2)(x + 4),\ y = (x + 2)^2(x - 3),\ y = (x - 3)^2(x + 2)$

***Group 6** $y = x^3 + x^2 - 6x + 3,\ y = x^3 - x^2 - 2x,\ y = 8 - 12x + 6x^2 - x^3,\ y = {}^-x^3 - 2x^2 + 5x + 6$

Algebra

- Plot the graph of $y = (x - 2)(x + 4)(x - 5)$
  Where does the graph cut the x-axis?
  Draw the graphs of
  $$y = x(x - 4)(x + 3) \text{ and } y = (x - 5)(x + 3)(x + 6).$$
  What is the relationship between $y = (x - a)(x - b)(x - c)$ and the x-intercepts? **Investigate**.

C Reciprocal graphs

You will need graph paper, a graph plotting package or a graphical calculator.

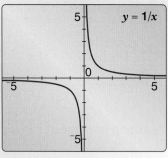

Use a graphical calculator or a computer graphics package to get this graph of $y = \frac{1}{x}$ on the screen. **Discuss** the following questions.

Can x take all values?
Can y take all values?
Between which values of x does y increase as x increases?
Between which values of x does y decrease as x increases?

On the same screen, display the following graphs. $y = \frac{1}{x}, \quad y = \frac{2}{x}, \quad y = \frac{3}{x}, \quad y = \frac{4}{x}$
Explore relationships between these graphs.

What if 1, 2, 3 and 4 were replaced with $^-1$, $^-2$, $^-3$, $^-4$?

What if 1, 2, 3 and 4 were replaced with 0·1, 0·2, 0·3, 0·4?

What if x was replaced with $x - 1$?

What if x was replaced with $x + 1$?

What if ...

The graphs of **quadratic functions** are called **parabolas**.
We can draw many **parabolas** by knowing the shape of $y = x^2$.

$y = x^2 + a$	is the graph of $y = x^2$ moved up a units.
$y = x^2 - a$	is the graph of $y = x^2$ moved down a units.
$y = (x - a)^2$	is the graph of $y = x^2$ moved to the right a units.
$y = (x + a)^2$	is the graph of $y = x^2$ moved to the left a units.
$y = (x - a)^2 + b$	is the graph of $y = x^2$ moved to the right a units and up b units.
$y = (x + a)^2 - b$	is the graph of $y = x^2$ moved to the left a units and down b units.

Examples

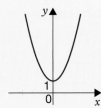

$y = x^2 + 1$
($y = x^2$ is moved up
1 unit)

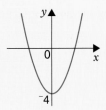

$y = x^2 - 4$
($y = x^2$ is moved down
4 units)

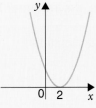

$y = (x - 2)^2$
($y = x^2$ is moved
2 units to the right)

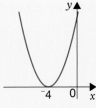

$y = (x + 4)^2$
($y = x^2$ is moved
4 units to the left)

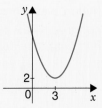

$y = (x - 3)^2 + 2$
($y = x^2$ is moved
3 units to the right
and 2 units up)

 Parabolas always have a maximum or minimum point – sometimes called the turning point.

$y = ax^2$ has the same vertex as $y = x^2$ but is 'wider' or 'thinner'. If a is between $^-1$ and 1 then the graph is 'wider' than $y = x^2$. For other values of a, the graph is thinner than $y = x^2$.

Example $y = \frac{1}{2}x^2$, $y = x^2$ and $y = 3x^2$ are shown sketched on the same set of axes.

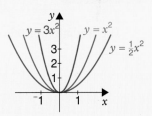

$y = {}^-x^2$ is the reflection of $y = x^2$ in the x-axis. Whenever the x^2 term is negative, the graph will be upside down.

Example The graph of $y = {}^-x^2 + 3$ is sketched as shown.

Maximum point. The maximum value of the function is 3.

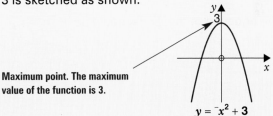

$y = {}^-x^2 + 3$

A **cubic** is the graph of an expression which has x^3 as its highest power. Some examples are shown.

$y = x^3$

$y = {}^-x^3$

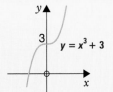

$y = x^3 + 3$

$y = (x - 4)^3$

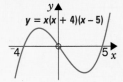

$y = x(x + 4)(x - 5)$

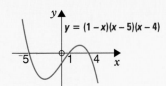

$y = (1 - x)(x - 5)(x - 4)$

Algebra

$\frac{1}{x}$, $\frac{3}{2x}$, $\frac{-4}{x}$ are all **reciprocal functions** of x. The graphs of $y = \frac{1}{x}$, $y = \frac{3}{2x}$, $y = \frac{-4}{x}$ are called hyperbolas. These graphs have two axes of symmetry. They always consist of two separate congruent curves.

In a reciprocal function, the x is on the denominator.

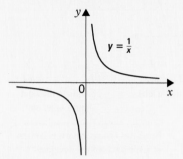

Exercise 2

1 The equation of this graph could be

   A $y = (x - 4)^2$
   B $y = x^2 + 4$
   C $y = x^2 - 4$
   D $y = (x + 4)^2$.

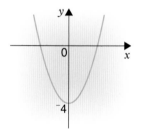

2 The equation of this graph could be

   A $y = (x + 2)^2$
   B $y = x^2 + 2$
   C $y = x^2 - 2$
   D $y = (x - 2)^2$.

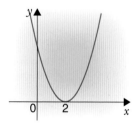

3 Match each of these graphs with an equation from the given list.

a 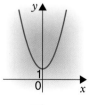   **b** 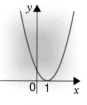   **c**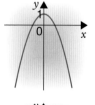

d   **e**   **f**

   A $y = 1 - x^2$
   B $y = x^2$
   C $y = {}^-1 - x^2$
   D $y = (x + 1)^2$
   E $y = (x - 1)^2$
   F $y = (x + 1)^2 - 1$
   G $y = x^2 - 1$
   H $y = x^2 + 1$
   I $y = (x - 1)^2 + 1$

4 The diagram shows the graph of the equation $y = 4 - x^2$.

[SATs Paper 1 Level 7]

What are the coordinates of the points where the graph of this equation meets the graph of equation $y = 2x + 1$?

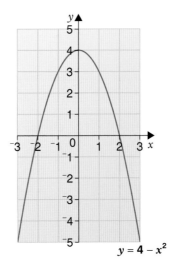

$y = 4 - x^2$

5 This is a sketch of the curve $y = (x - 4)^2$.
  a What are the coordinates of point A?
  b What is the equation of the line of symmetry?

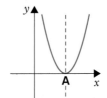

6 a Which of the following could be the graph of $y = x^2 - 7x + 10$? Explain your choice.

A

B

C

D

b Which of the following could be the graph of $y = {}^-2x^3$? Give reasons for your choice.

A

B

C

D

c Which of the following could be the graph of $y = 4x - x^2$? Explain your choice.

A

B

C

D

d Which of the following could be the graph of $y = \frac{3}{x}$? Explain your choice.

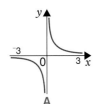
A

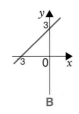
B

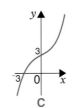
C

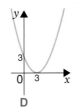
D

Algebra

7 Match the graphs with the equations.

a

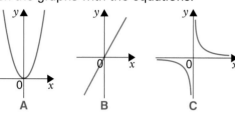

A	**B**
C	**D**

1 $y = 2x$
2 $y = 2x^2$
3 $y = 2x^3$
4 $y = \frac{2}{x}$

b

A	**B**
C	**D**

1 $y = 2 - x$
2 $y = x^2 - 2$
3 $y = \frac{-2}{x}$
4 $y = 2 - x^2$

c

A	**B**
C	**D**

1 $y = x + 4$
2 $y = x^2 - 4x + 4$
3 $y = x^3 - 4x^2 + 4x$
4 $y = 4x^3$

T **8** Use a copy of this graph of $y = x^2$.
On the same set of axes *sketch* the graphs of
$y = \frac{1}{2}x^2$, $y = 3x^2$ and $y = -x^2$.
Label your graphs clearly.

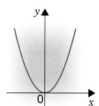

Sketch means draw a rough graph showing significant features.

T **9** Use a copy of this graph of $y = x^2$.
On the same set of axes *sketch* the graphs of
$y = x^2 + 2$ and $y = (x + 2)^2$.
Label your graphs clearly.

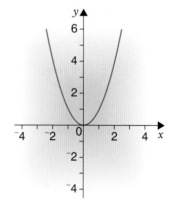

---

10 a Copy and complete the table for $y = x^2 - 5$.
Draw the graph of $y = x^2 - 5$.

x	-4	-3	-2	-1	0	1	2	3	4
x^2				1					16
-5				-5					-5
y				-4					11

* b Copy and complete the table for $y = x^3 - 6x$.
Plot the graph of $y = x^3 - 6x$. You may like to
plot a few more points around the turning
points.

x	-3	-2	-1	0	1	2	3
x^3		-8			1		
-6x		12			-6		
y		4			-5		

11 This is the graph of $y = (x - 2)^2 + 1$.
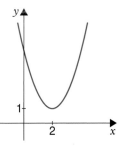
  a This graph crosses the y-axis at
        A 4   **B** 1   **C** 5   **D** $^-$2.
  b The coordinates of the minimum point are
        A (2, 0)   **B** (0, 1)   **C** (2, 1)   **D** can't tell.
  c Write down the equation of the line of symmetry.

12 **a** Use a copy of this table.
    Fill it in to find the values of y if $y = 2x^2 - x + 3$.
  b Plot the values of x and y on a set of axes.
    Draw and label the graph of $y = 2x^2 - x + 3$.
  ∗c Is the point (0, 3) the lowest point? Explain.
  ∗d What is the equation of the line of symmetry?

x	$^-2$	$^-1$	0	1	2	3	4
x^2							
$2x^2$							
^-x							
$+3$							
y							

13 The coordinates of the vertex of the graph of $y = (x - 3)^2 + 4$ are
  A ($^-3$, 4)   **B** ($^-4$, 3)   **C** ($^-3$, $^-4$)   **D** (3, 4)

14 Jeff got this on his graphical calculator screen.
Suggest a possible equation for this curve.
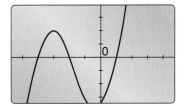

15 What might the equations of these
calculator drawn graphs be?
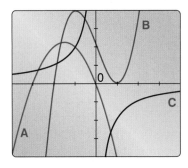

16 Where will the following graphs cut the x-axis?
  a $y = (x - 2)(x + 3)$ **b** $y = (x + 4)(x + 5)$ **c** $y = (x - 5)(x + 3)$
  d $y = (x + 6)(x - 7)$ **e** $y = (x - 4)(x - 3)$ **f** $y = (x - 6)(x + 1)$
  g $y = (x - 3)(x + 2)(x - 1)$ **h** $y = x(x + 6)(x - 4)$ **i** $y = (x - 8)(x + 12)(x - 1)$

17 The diagram shows a sketch of the curve $y = 16 - x^2$ **[SATs Paper 2 Level 8]**

  a What are the coordinates of points A, B and C?

  The curve $y = 16 - x^2$ is reflected in the line $y = 12$.
  b B_1 is the reflection of B.
    What are the coordinates of B_1?
  c What is the equation of the new curve?

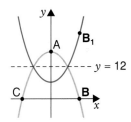

Algebra

Review 1   Match each graph with an equation from the box.

a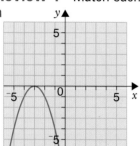
b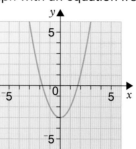
c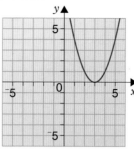

A $y = x^2 + 3$
B $y = x^2 - 3$
C $y = {}^-x^2 + 3$
D $y = (x - 3)^2$
E $y = (x + 3)^2$
F $y = (x + 3)^2 + 3$
G $y = (x - 3)^2 + 3$
H $y = {}^-(x + 3)^2$

d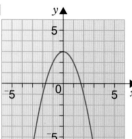
e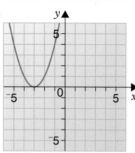
f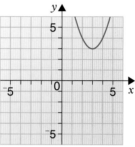

Review 2   Jacqui graphed two equations using her graphical calculator.
The inner parabola is the graph of $y = 3x^2$. Is the equation for the outer parabola
$y = x^2$ or $y = 6x^2$?

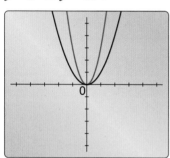

Review 3   Which of the following could be the graph of $y = x^3 - 2x^2 - x + 2$?
Explain your choice.

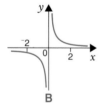

Review 4   Which of the graphs drawn in **Review 3** could have the equation
a $y = \dfrac{2}{x}$
b $y = x^2 - 4x + 4$
c $y = x^3$?

* **Review 5**   This is the graph of $y = {}^-(x - 2)^2 - 1$.
a   What are the coordinates of the maximum point?
b   Where does the parabola cross the y-axis?
c   What is the equation of the line of symmetry?

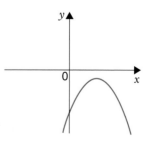

Practical

You will need a graphical calculator.

- Plot the graph of $y = 4x^2$ on your graphical calculator.
  Use the 'Trace' function on your calculator to read coordinates as you trace along the curve. Describe the relationship between the x- and y-values.

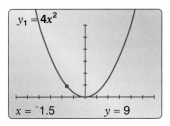

$x = {}^-1.5$ $y = 9$

- Create these designs on your screen.

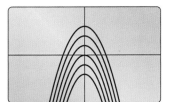
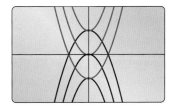

Drawing, sketching and interpreting real-life graphs

Remember
If we are asked to **draw** a real-life graph we must
- construct a table of values using a formula or relationship
- choose suitable scales for the axes
- plot the points accurately
- draw a line through the points if it is sensible to do so
- give the graph a title and label the axes.

A **sketch** shows the **relationship between variables**.

Examples

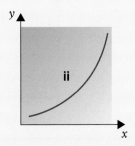

When x is large y is large.
As x increases in equal steps y increases by increasing amounts.

ii = increasing rate of increase.

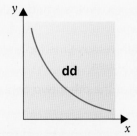

When x is large y is moving to 0.
As x increases in equal steps y decreases by decreasing amounts.

dd = decreasing rate of decrease.

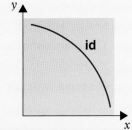

When x is large y becomes 0.
As x increases in equal steps y decreases by increasing amounts.

id = increasing rate of decrease.

Algebra

Discussion

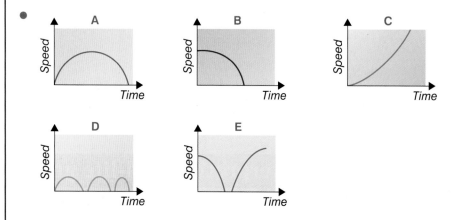

These are the graphs of three cycle journeys.

Graph A shows that at first the speed was constant but quite slow, then the speed was constant but quite fast, then the cyclist slowed down to a constant speed. The speed in the last section of the journey was not as slow as in the first part.

Describe the cycle journeys represented by graphs **B** and **C**. **Discuss**.

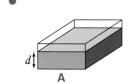
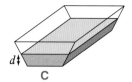
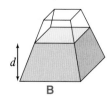

One of these graphs could represent a car in a traffic jam. Which one?
What might the other graphs represent? **Discuss**.

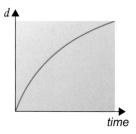

The above containers are filled with a liquid at the rate of 200 mℓ per second.

This graph shows the depth versus time for one of the containers. Which one? **Discuss**.

Discuss possible graphs for the other two containers.

Remember
Variables which are in **direct proportion** always give a **straight-line** graph.

Discussion

Bron did an experiment in science to test if the voltage in a circuit was directly proportional to the current in the circuit.
This table shows her results.

Voltage (volts)	3	6	9	12	15	18	21
Current (amps)	0·14	0·32	0·45	0·64	0·74	1·2	1·16

She plotted these results on a graph.

Do the points lie in an *exact* straight line?
Do the points lie in an *approximate* straight line?
Why might this be? **Discuss.**

Do you think that voltage and current **are** directly proportional? **Discuss.**

What might explain the point at (18, 1·2)? **Discuss.**

If the current in the same circuit is 2 amps, how could you work out the voltage? **Discuss.**

Write a formula for the relationship between voltage (*V*) and current (*A*).

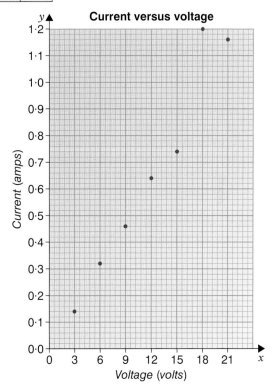

Current versus voltage

Exercise 3

1 Karen and Susan were two of the runners in a 400 m race.
   a At what times were Karen and Susan level with each other?
   b Who finished first, Karen or Susan?
   c Who had the faster speed during the first 10 seconds?
   d Who was leading after 50 seconds? About how far ahead was she?
   e How might an announcer have spoken about Karen's and Susan's progress during the race? Write a short report on this.

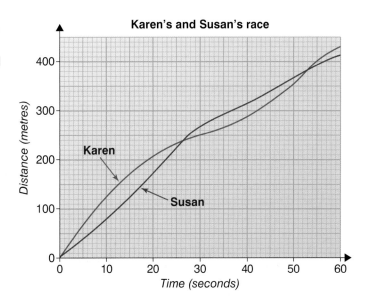

Karen's and Susan's race

Algebra

2 A beaker of liquid was used as part of an experiment. This graph shows the level of the liquid during this experiment.

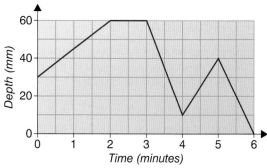

  a What was the depth of liquid at the beginning of this 6-minute experiment?

  b What depth of liquid was in the container at the end of 6 minutes?

  c Describe what was happening to the depth of liquid during the 6 minutes.

3 A car travels 10 km on every litre of petrol.

  a Copy and fill in this table.

Number of litres	1	2	3	4	5
Distance travelled					

  b Is the distance travelled directly proportional to the number of litres used? Justify your answer using ratio.

  c Draw a graph of distance travelled versus number of litres used.

  d Explain why your graph is or isn't a straight line.

  e Write the relationship between distance travelled (d) and litres used (l) as an equation.

  f How much petrol would be needed to travel 88 km?

T

4 Use a copy of this grid.
Plot a distance–time graph for two sisters' car journeys.

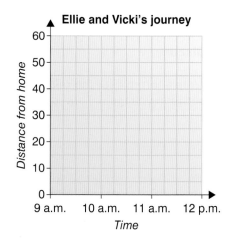

  A Ellie left home at 9 a.m. She travelled to her aunt's at a constant speed of 30 km/h. She arrived there at 9:30 a.m. She stayed for 30 minutes.
  She travelled home, arriving at 11:15 a.m.

  B Vicki left home at 9:30 a.m. She travelled at a constant speed until 10 a.m. She was then 35 km from home. She then slowed down and travelled the next 20 km in 45 minutes. She travelled home at a constant speed of 55 km/h.

5 Match these graphs and statements.

  a Distance travelled versus time taken graph for someone cycling up a steep hill going slower and slower, resting at the top then cycling down faster and faster.

  b Sales versus time graph for a magazine that is launched. Soon after, an advertising campaign increases sales but then they drop.

  c Population versus time for an island with a constant rate of emigration.

  d Distance versus time graph for a train that accelerates from rest then maintains a constant speed.

  e The number of bubbles made versus time in an experiment where the number of bubbles is initially constant then slows down rapidly to no bubbles.

  f Distance versus time for a runner who accelerates out of the blocks then gets cramp and slows down and eventually stops.

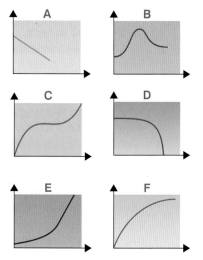

T

6 Jasmine did an experiment in science to work out the acceleration of an object when different forces were applied to it.

This table shows her results.

F (Newtons)	2	4	6	8	10	12	14	16	18	20
a (m/s²)	0·5	0·9	1·4	2·1	3·2	3	3·4	4·1	4·5	4·9

a Use a copy of this grid.
Draw a graph of acceleration versus force.
b Give a possible reason why the points don't lie in an exact straight line.
c Do you think that force and acceleration are directly proportional? Explain.
d Jasmine made an experimental error when collecting data for one of the points. Which point do you think it was?

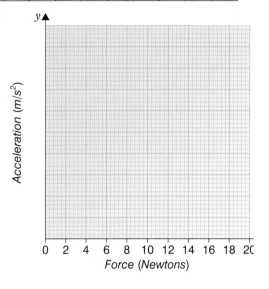

7 The simplified graph shows the flight details of an aeroplane travelling from London to Madrid, via Brussels.   [SATs Paper 2 Level 7]

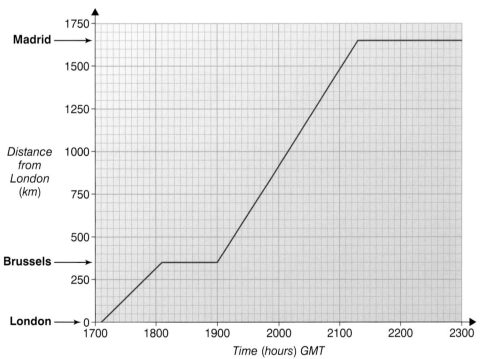

a What is the aeroplane's average speed from London to Brussels?
b How can you tell from the graph, **without calculating**, that the aeroplane's average speed from Brussels to Madrid is **greater** than its average speed from London to Brussels?

Algebra

c A different aeroplane flies from Madrid to London, via Brussels. The flight details are shown below:
Use a copy of the graph.
On your graph, show the aeroplane's journey from Madrid to London, via Brussels.
(Do not change the labels on the graph.)

Assume constant speed for each part of the journey.

Madrid	depart	1800
Brussels	arrive	2000
	depart	2112
London	arrive	2218

d At what time are the two aeroplanes the same distance from London?

8 **a** 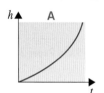 **b**   **c**   **d**

Water is poured into these containers at a rate of 150 mℓ per second. The graphs below show how the height of the water changes with time.
Match the containers with the graphs.

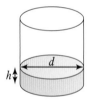  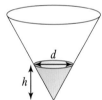

9 **a** 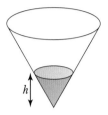 **b** 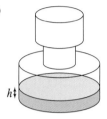 **c**  **d**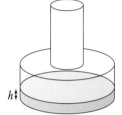

These containers are being filled with liquid. The graphs below show how the diameter of the surface of the liquid changes as the height of the liquid increases.
which graph belongs to which container?

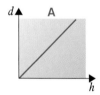 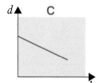 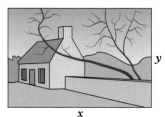

10 For an exhibition of posters, the miniature posters were to have an area of 60 cm².
The width of one of these miniature posters is x cm and the height is y cm².
 a Show that $y = \frac{60}{x}$.
 b Copy and complete the table for $y = \frac{60}{x}$.

x	5	10	15	20	25	30
y					2·4	

 c Draw the graph of $y = \frac{60}{x}$ for values of x from 5 to 30.
 d One of the miniature posters in the exhibition had a width of 7·5 cm. Use your graph to find the height of this poster.

11 i Berryfields Orchard pack their cherries into wooden boxes which have square ends, as shown. These boxes are twice as long as they are wide.

    a If the width of one of these boxes is x centimetres, show that the volume is given by $V = 2x^3$ cubic centimetres.

    b Copy and complete the table for $V = 2x^3$.

x	5	10	15	20	25
V	250				31 250

    c Draw the graph of $V = 2x^3$ for values of x from 5 to 25.

    d One of these cherry boxes has a volume of 20 000 cubic centimetres. Use your graph to find the approximate width of this box.

ii The cherry boxes do not have lids.

    a Show that the area of wood used in one of these boxes is $8x^2$ square centimetres.

    b Copy and complete the table for $A = 8x^2$.

x	5	10	15	20	25
A			1800		

    c Draw the graph of $A = 8x^2$ for values of x from 5 to 25.

    d 2500 cm^2 of wood is needed to make one of these boxes. Use your graph to find the approximate width of this box.

12 The nth term of an infinite sequence is $\frac{n}{n^2 + 2}$.

The first term of the sequence is $\frac{1}{3}$.

Which of these graphs shows the graph of value of term versus term number for the sequence?

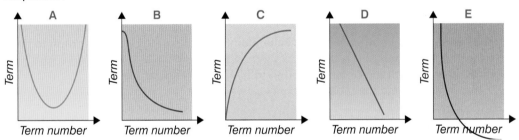

Review 1

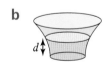
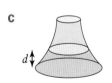

These containers are being filled with liquid at a constant rate. The graphs below show how the depth of the liquid (d) changes as the volume of liquid (V) increases. Match the containers to the graphs.

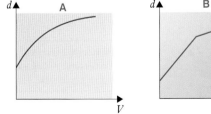
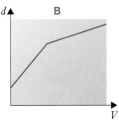
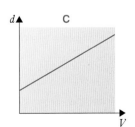
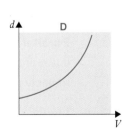

Algebra

Review 2

Before Felicity began an aerobics exercise programme, she did a 3-minute fitness test. The green line shows her pulse rate during this test and for 4 minutes afterwards.

After six months on the aerobics programme, she did the same fitness test again. The blue line shows her pulse rate during and immediately after this test.

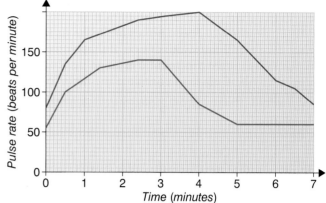

a What is the difference in Felicity's 'at rest' pulse rate before and after she started the aerobics programme?

b Felicity's pulse rate increased faster before she started the programme. How do we know this from the graph?

c Write a short report comparing Felicity's pulse rate before and after the aerobics programme.

Review 3

i Plastic trays filled with foam on which rings are displayed in a jewellers', have square bases. The dimensions of these trays are shown in the diagram.

a Show that the area of plastic needed to make one of these trays is $A = 9x^2$.

b Copy and complete the table for $A = 9x^2$.

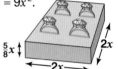

x	4	8	12	16	20
A	144			2304	

c Draw the graph of $A = 9x^2$ for values of x from 4 to 20. Choose sensible scales for the x- and A-axes.

d Use the graph to find the approximate area of a plastic needed to make a tray which is 15 cm long.

e 3000 square centimetres of plastic is needed to make a tray. Use your graph to find the approximate length of this tray.

ii a Show that the volume of foam needed to fill the tray is given by $V = \frac{5x^3}{2}$.

b Copy and complete the table for $V = \frac{5x^3}{2}$.

x	4	8	12	16	20
V		1280			

c Draw the graph of $V = \frac{5x^3}{2}$ for values of x from 4 to 20.

d The volume of foam in a tray is about 7000 cm^3. Use your graph to find the approximate dimensions of the tray.

Summary of key points

 A A **straight-line graph has equation** $y = mx + c$, where m is the gradient and c is the y-intercept.

For any straight line joining the points (x_1, y_1) and (x_2, y_2),

the gradient $\quad m = \dfrac{y_2 - y_1}{x_2 - x_1} = \dfrac{\text{change in } y}{\text{change in } x}$

Parallel lines have the same gradient. $\quad m_1 = m_2$.

If two lines are **perpendicular**, their gradients multiply to give $^-1$. $\quad m_1 m_2 = ^-1$ or $m_1 = \dfrac{^-1}{m_2}$.

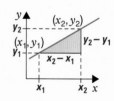

Equations of straight lines must be written in the form $y = mx + c$ before comparing gradients.

Examples $y = 3x - 2$ and $y = 3x + \frac{1}{2}$ are parallel.

$y = 4x + 1$ and $y = ^-\frac{1}{4}x - 3$ are perpendicular. $4 \times ^-\frac{1}{4} = ^-1$

B The graph of a **quadratic function is a parabola**.

The graph of $y = x^2$ is:

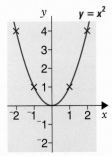

A quadratic function has an x^2 term.

$y = x^2 + a$ is $y = x^2$ moved up a units.

$y = x^2 - a$ is $y = x^2$ moved down a units.

$y = (x - a)^2$ is $y = x^2$ moved right a units.

$y = (x + a)^2$ is $y = x^2$ moved left a units.

$y = (x - a)^2 + b$ is $y = x^2$ moved right a units and up b units.

$y = (x + a)^2 - b$ is $y = x^2$ moved left a units and down b units.

$y = ax^2$ has the same vertex as $y = x^2$. If $a > 1$, the parabola will be thinner than $y = x^2$. If $a < 1$, the parabola will be wider than $y = x^2$.

$y = {}^-x^2$ is a reflection of $y = x^2$ in the x-axis.

See page 211 for examples of parabolas.

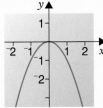

C A **cubic function** has x^3 as its highest power.

Examples

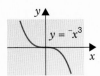

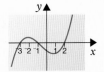

$$y = (x - 2)(x + 1)(x + 3)$$
$$y = x^3 + 2x^2 - 5x - 6$$

D $\frac{1}{x}$, $\frac{-1}{2}x$, $\frac{3}{x}$ are all **reciprocal functions**.

The graph of a reciprocal function is called a hyperbola. These graphs consist of two separate congruent curves which have two axes of symmetry.

The x is on the denominator.

Examples

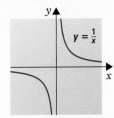

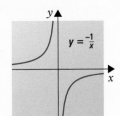

E We can **draw, sketch and interpret real-life graphs**.

When variables are in direct proportion, the graph is a straight line.

Algebra

Test yourself

1 a Find the gradients of the lines l_1, l_2 and l_3 shown on the grid.

   b Complete these equations for the three lines.

     l_1 $y = \_\_\_ x$ l_2 $y = \_\_\_ x + \_\_\_$

     l_3 $y = \_\_\_ x - \_\_\_$

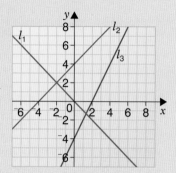

2 Here are six different equations.

    $y = 2x + 4$ $2y = 5x$ $y = {}^-4$

    $y = -\tfrac{1}{2}x - 2$ $y = x^2 - 2$ $y = \tfrac{5}{2}x + 3$

   I draw the graphs of each.

   a Which line goes through the point (0, 0)?   **b** Which two lines are perpendicular?

   c Which graph is **not** a straight line?   **d** Which line passes through (6, 18)?

   e Which lines are parallel?   **f** Which line has a gradient of 0?

3 Rearrange these line equations into the form $y = mx + c$.

   a $2x + y = 3$ **b** $x + 2y - 6 = 0$ **c** $3x - y = 1$

4 Write down the coordinates of the point that lies on both the straight lines $y = 4x - 2$ and $y = 3x - 5$.

   Show your working.

5 This is a graphical calculator screen.

   Match the lines with a possible equation from the box.

A $y = {}^-x - 2$	**B** $y = 3$
C $y = x + 2$	**D** $y = \tfrac{2}{3}x - 2$
E $x = {}^-5$	

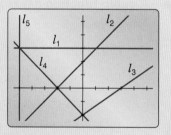

6 The equations of the three sides of a triangle are shown. The diagram is not drawn to scale.

   The line PQ, which has equation $y = 2x - 4$, is parallel to one of the sides of this triangle.

   Which one?

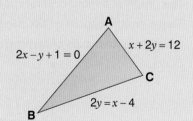

7 Match these equations and graphs.

   a $y = x^2 - 3$ **b** $y = (x - 3)^2$ **c** $y = (x + 3)^2$ **d** $y = x^2 + 3$

   e $y = {}^-x^2 + 3$ **f** $y = (x - 3)^2 + 3$ **g** $y = 3x^2$ **h** $y = x(x - 3)(x - 3)$

 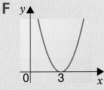

8 At Christmas time, a shop wraps small gifts and places them in boxes. These boxes are made from cardboard and shaped as shown.

   a Show that the area of cardboard needed for one of these boxes is given by $A = 9x^2$.

   b Draw the graph of $A = 9x^2$ for values of x between 1 and 10.

   c One of these boxes is made from 780 cm² of cardboard. Use your graph to find the approximate length of this box.

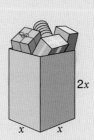

9 Use a copy of this graph of $y = x^2$.

   a On the same set of axes sketch

     i $y = x^2 - 4$ **ii** $y = (x - 4)^2$.

   b What are the equations of the lines of symmetry of the two graphs in **a**?

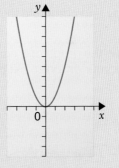

10 a Copy and complete this table for $y = x^3 - 4x$.

   b Plot the graph of $y = x^3 - 4x$.

   c $y = x^3 - 4x$ can also be written as $y = x(x - 2)(x + 2)$. Give the coordinates of the points where the graph cuts the y-axis and the x-axis.

x	$^-3$	$^-2$	$^-1$	0	1	2	3
x^3							
$-4x$							
y							

11 Match the graphs with the equations.

   a 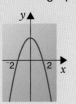   **b**   **c**   **d**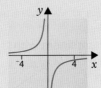

   A $y = \dfrac{^-4}{x}$

   B $y = 4 - x^2$

   C $y = x^3 + 4$

   D $y = 4 - x$

12 Write a possible explanation for the shapes of each of these graphs.

a
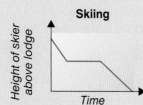
Skiing

b
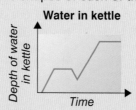
Water in kettle

c
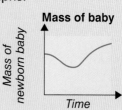
Mass of baby

13 A DJ can control the sound level of the CDs he plays at a disco.
The sketch graph below is a graph of the sound level against the time whilst one CD was played.

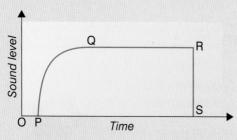

a Describe how the sound level changed between P and Q on the graph.
b Give one possible reason for the third part, RS, of the sketch graph.

14 In a kite-flying competition, all kites must have an area of 4 m². If a and b are the lengths of the diagonals of the kites, the relationship between a and b is $a = \frac{8}{b}$.

a Copy and complete this table.

b	1	2	3	4	5	6	7	8
a			2·7				1·1	

b Draw the graph of a against b.
c Use your *graph* to estimate the value of a when $b = 4.5$.

15 A
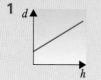
B

C

D

These containers are being filled with a liquid at the rate of 200 mℓ per second.

a These graphs show how the height of the liquid, h, is increasing with time.
Match the containers with the graphs.

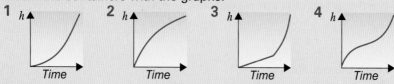

1 h / Time **2** h / Time **3** h / Time **4** h / Time

∗b These graphs show how the diameter of the surface of the liquid, d, changes as the height increases.
Match the containers with the graphs.

1 d / h **2** d / h **3** d / h **4** d / h

Lines and angles

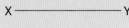

This is a **line segment**, XY.
It has **finite** length.

This line has **infinite length**.

We always use three letters to name an angle if there is more than one angle at the vertex.

Example The shaded angle is ∠RQS.

$a = b$
Vertically opposite angles are equal.

$x + y + z = 180°$
Angles on a straight line add to 180°.

$c + d + e = 360°$
Angles at a point add to 360°.

Angles made with parallel lines

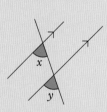

$x = y$
Corresponding angles on parallel lines are equal.

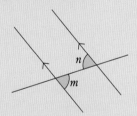

$m = n$
Alternate angles on parallel lines are equal.

Parallel lines are marked with arrows.

Angles in triangles

The interior angles of a triangle add to 180°.

Example $x + 52° + 74° = 180°$
$x = 180° - 52° - 74°$
$= 54°$

The exterior angle of a triangle is equal to the sum of the two opposite interior angles.

Example $g = 35° + 40°$
$= 75°$

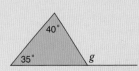

We often use **geometrical reasoning** to find an unknown angle or to prove something.
The steps and reasons must be written down clearly one by one.

Example

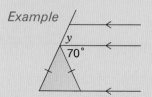

To find y, name other angles first.

$x = 70°$ (alternate angles on parallel lines are equal)

$z = 40°$ (angle sum of isosceles triangle $= 180°$)

$y = 180° - 70° - 40°$ (angles on a straight line add to $180°$)

  $= 70°$

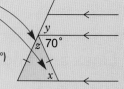

Practice Questions 8, 16, 34, 37, 44

Angles in polygons

The **sum of the interior angles of a polygon** with n sides is $(n - 2) \times 180°$.
For example, the hexagon on the right can be divided into four triangles as shown.
The sum of the angles in each triangle is $180°$.
So the sum of the angles in the four triangles $= 4 \times 180°$ $(n - 2) \times 180°$
                                                  $= 720°$

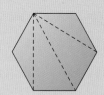

The **sum of the exterior angles of any polygon is 360°.**

Practice Questions 11, 42, 50

2-D shapes

Properties of triangles

A triangle is a 3-sided polygon.

right-angled
1 right angle

isosceles
2 equal sides
2 base angles equal

equilateral
3 equal sides
3 equal angles

scalene
no 2 sides are equal
no 2 angles are equal

Properties of quadrilaterals

A quadrilateral is a 4-sided polygon.
Some of the properties of the special quadrilaterals are shown in this table.

	Square	Rhombus	Rectangle	Parallelogram	Kite	Trapezium	Arrowhead
one pair of opposite sides parallel	✓	✓	✓	✓		✓	
two pairs of opposite sides parallel	✓	✓	✓	✓			
all sides equal	✓	✓					
opposite sides equal	✓	✓	✓	✓			
all angles equal	✓		✓				
opposite angles equal	✓	✓	✓	✓	1 pair		
diagonals equal	✓		✓				
diagonals bisect each other	✓	✓	✓	✓			
diagonals perpendicular	✓	✓	✓		✓		✓
diagonals bisect the angles	✓	✓					

A **polygon** is a closed 2-D shape made from line segments.
A **regular polygon** has all its sides and all its angles equal.

Congruence

Congruent shapes are the same size and the same shape.

Corresponding lengths are equal and corresponding angles are equal.

Example   ABC and DEF are congruent triangles.

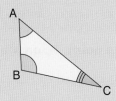
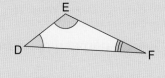

AB = DE
BC = EF
AC = DF

Practice Questions 3, 15, 28, 31

Circles

A **circle** is a set of points equidistant from its centre.
The **parts of a circle** are shown in this diagram.

circumference – distance around the outside
radius – distance from center to circumference
arc – part of the circumference
sector – area made by an arc and two radii
tangent – a straight line which just touches the circle at one point
chord – a straight line which intersects the circle at two points and
cuts the circle into two regions called **segments**
diameter – a line which passes through the centre of the circle and
divides it into two **semicircles**

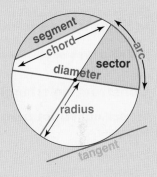

Practice Questions 13, 14

Constructions

We use compasses and a ruler to construct the **perpendicular bisector of a line segment**,
AB.

Open the compasses to a little
more than half the length of
AB. With compass point first
on A and then on B, draw arcs
to meet at P and Q.

Draw the line through P
and Q.   R is the point
which bisects AB.

We use compasses and a ruler to construct the **bisector of an angle P**.

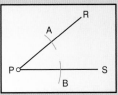
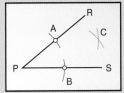
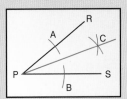

Open out the compasses to a length
less than PR or PS. With compass
point on P, draw arcs as shown.

With compass point first on A,
then B, draw arcs to meet at C.

Draw the line from P through C.
This line, PC, is the bisector of angle P.

These diagrams show the construction of a **perpendicular from a point, A, to a line segment, BC**.

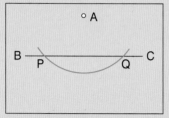

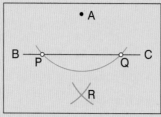

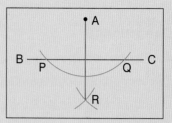

Open out the compasses. With the point on A, draw an arc to cross BC at P and Q.

With the point firstly on P, then on Q, draw two arcs to meet at R.

Join A and R.   AR is the perpendicular from A to the line segment BC.

These diagrams show the construction of a **perpendicular from a point, P, on a line segment, BC**.

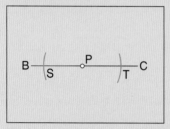

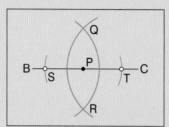

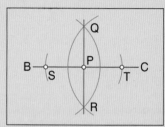

Open out the compasses to less than half the length of BC. With the point on P, draw arcs, one on each side of P. Label where they cross BC as S and T.

Open out the compasses a little more. With compass point first on S and then on T, draw arcs so they cut at Q and R.

Draw the line through Q and R.   QR is the perpendicular from P on the line segment BC.

Constructing triangles

We can **construct triangles and quadrilaterals** using a set square and ruler or compasses and ruler.

Examples

To construct this triangle:
1 Draw PR 2·6 cm long.
2 Draw an angle of 85° at R.
3 Draw RQ 2·8 cm long.
4 Join P to Q.

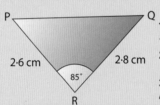

2·6 cm   2·8 cm

85°

To construct this triangle:
1 Draw AB 2·8 cm long.
2 Open compasses to 2·5 cm and with point on A draw an arc.
3 Draw an arc from B, 2 cm long.
4 Complete the triangle.

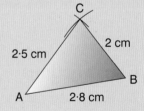

2·5 cm   2 cm

2·8 cm

We can **construct a right-angled triangle** given the right-angle, the length of the longest side and one other side (RHS).

The longest side is called the **hypotenuse**.

4 cm

3 cm

A ———————— B

Draw a line 3 cm long.

A ——————⊙—— B

Extend AB.
Construct a perpendicular at B.

A ——————⊙ B

Open compasses out to 4 cm. With the point on A, draw an arc that crosses the perpendicular. Join A to the point of intersection. Label this point C.

Practice Questions 30, 46

Locus

A **locus** is a set of points that satisfy a rule or set of rules. The locus of a moving object is the path that it follows.

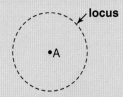

The locus of a point which is always the same distance from a fixed point is a circle.

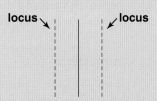

The locus of a point which is always the same distance from a straight line is two parallel lines.

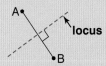

The locus of a point which is always an equal distance from two fixed points is the perpendicular bisector of the line joining the points.

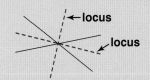

The locus of a point which is always the same distance from two intersecting lines is the line which bisects the angle between the lines.

Practice Questions 1, 22, 51

3-D shapes

3-D stands for three dimensional.

3-D shapes have length, width and height.

When we slice a shape, the face that is made is called a **cross-section**.

Example   A vertical slice through this pyramid gives a triangle.

We can represent 3-D shapes using **plans and elevations**.

Example

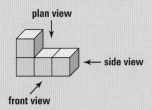

plan view

side view

front view

plan view

front view   side view

A 2-D shape that can be folded to make a 3-D shape is called a **net**.

Example   This net folds to make a cuboid.

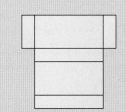

Practice Questions 5, 6, 29, 38, 49

Transformations and coordinates

Coordinates

The **mid-point** of the line segment joining $P(x_1, y_1)$ to $Q(x_2, y_2)$ is given by $\left(\frac{x_1 + x_2}{2}, \frac{y_1 + y_2}{2}\right)$.

Example   The mid-point, M, of the line segment joining PQ is $\left(\frac{2+8}{2}, \frac{3+9}{2}\right) = (5, 6)$.

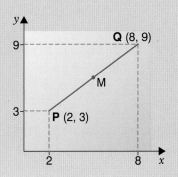

Symmetry

If one half of a shape can be reflected in a line to the other half, the shape has **reflection symmetry**.

If a shape fits onto itself **more than once** during a complete turn, it has **rotation symmetry**. The **order** of rotation symmetry is the number of times a shape fits onto itself exactly during that complete turn.

Some 3-D shapes are **symmetrical**.

Example   The shaded shape is called a **plane of symmetry**. It divides the shape into two congruent pieces.

Transformations

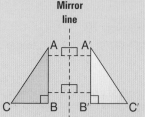

Mirror line

When we reflect, rotate or translate a shape, the image is always **congruent** to the object. Congruent shapes are exactly the same size and shape.

Example   Triangle ABC has been reflected to triangle A'B'C'.
AB = A'B', BC = B'C', AC = A'C', ∠A = ∠A', ∠B = ∠B', ∠C = ∠C'
The triangles are **congruent**.

A shape **tessellates** if, when it is reflected, translated or rotated, it completely fills a space leaving no gaps.

Enlargement is a transformation of a plane (2-D) shape in which points such as A, B, C and D are mapped onto A', B', C' and D' by the same scale factor and from a fixed centre of enlargement. The distance of A', B', C' and D' from the centre are found by multiplying each of the distances of A, B, C and D from the centre by the **scale factor**.

Example   scale factor = 3
OA' = 3 × OA
OB' = 3 × OB
OC' = 3 × OC
OD' = 3 × OD

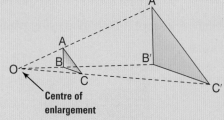

Centre of enlargement

Practice Questions 4, 7, 20, 21, 23, 33, 34, 39, 40, 47

Scale drawings

A **scale drawing** represents something in real life.
It has a scale which tells us what each unit on the drawing represents.

Example   A scale of 1 mm to 5 cm means that 1 mm on the scale drawing represents 5 cm or 50 mm in real life.
1 mm to 5 cm = 1 mm to 50 mm
= 1 : 50

Scale: 1 mm represents 5 cm

The bike is 38 mm long in the scale drawing.
In real-life it is
38 × 50 = 1900 mm
= 1.9 m

Practice Questions 9, 27, 32, 41

Measures

You need to know these **metric conversions**.

length	mass	capacity (volume)	area	time
1 km = 1000 m	1 kg = 1000 g	1 ℓ = 1000 mℓ	1 ha = 10 000m^2	1 minute = 60 seconds
1 m = 100 cm	1 tonne = 1000 kg	1 ℓ = 100 cℓ	(hectare)	1 hour = 60 minutes
1 m = 1000 mm		1 cℓ = 10 mℓ	1 m^2 = 10 000 cm^2	1 day = 24 hours
1 cm = 10 mm		1 ℓ = 1000 cm^3	1 cm^2 = 100 mm^2	1 year = 12 months or
		1 mℓ = 1 cm^3		52 weeks and 1 day
		1 m^3 = 1000 ℓ		1 year = 365 days or
		1 cm^3 = 1000 mm^3		366 in a leap year
		1 m^3 = 1 000 000 cm^3		1 decade = 10 years

Examples   650 cm^3 = (650 ÷ 1000) ℓ 3 m^3 = (3 × 10 000) cm^2
          = **0·65 ℓ**   = **30 000 cm^2**

These are some rough **metric and imperial equivalents**.

length	mass	capacity
5 miles ≈ 8 kilometres	1 kg ≈ 2·2 lb	1 pint ≈ 600 mℓ
1 yard ≈ 3 feet ≈ 1 m	1 oz ≈ 30 g	1 gallon ≈ 4·5 ℓ
1 inch ≈ 2·5 cm		1 litre ≈ 1·75 pints

≈ means 'approximately equal to'.

Bearings are angles measured in a clockwise direction from North and always have three digits.

Examples

The bearing of A from B is 075°.

The bearing of C from B is 210°.

Practice Questions 2, 10, 18, 24, 25, 26

Perimeter, area and volume

Perimeter is the distance around the outside of a shape. It is measured in mm, cm, m or km.
Area is the amount of space covered by a shape. It is measured in mm^2, cm^2, m^2, km^2 or ha.
Surface area is the total area of the faces making up a 3-D shape.

Area of a triangle = $\frac{1}{2}bh$

Area of a parallelogram = bh

Area of a trapezium
= $\frac{1}{2}(a + b) \times h$

Surface area of a cuboid
= $2lw + 2wh + 2lh$

Area of a circle = πr^2
Circumference of a circle = $2\pi r = \pi d$

There is a π key on your calculator.

Volume is the amount of space taken up by a solid.
It is measured in mm^3, cm^3, m^3 or ℓ.

    Volume of a cuboid = lwh

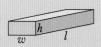

    Volume of a prism = area of cross-section × length

Example   volume = area of triangle × 10 cm

$$= \frac{1}{2} \times 5 \times 4 \times 10$$
$$= 100 \text{ cm}^3$$

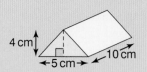

Practice Questions 12, 17, 19, 36, 43, 45, 48

Practice Questions

1 Which of these could be the locus of a ball thrown into the air?

  A   **B**   **C**   **D**

2 **a** How many minutes in $2\frac{3}{4}$ hours?

    b How many hours and minutes in 565 minutes? Round sensibly.

    c Change 126 months to years and months.

    d Change 5·6 ha to m^2.

    e Change 13·96 cm^3 to ℓ.

    f Change 0·62 ℓ to cm^3.

    g Change 355 cm^3 to cℓ.

    h Change 6·5 m^2 to cm^2.

    i Change 560 000 mm^3 to cm^3.

    j Change 0·328 m^3 to cm^3.

    k Change 3·7 cm^2 to mm^2.

3 Write true or false for each of these.

    a A kite can be split into two congruent triangles.

    b A parallelogram can be split into two congruent triangles.

    c In any triangle the longest side is opposite the smallest angle.

    d A rhombus can be split into four equilateral triangles.

4 This shape has 3 lines of symmetry.
What is the size of angle y?
Show your working.

5 This open tray was used for
displaying punnets of
strawberries.

    Sketch a net for the tray.

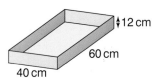

6 Match each shape with the correct plan and front and side elevations.

a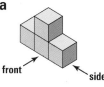

b

c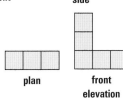

d

A

plan   front elevation   side elevation

B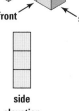

plan   front elevation   side elevation

C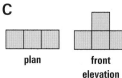

plan   front elevation   side elevation

D

plan   front elevation   side elevation

7 Use isometric (triangle dotty) paper to tessellate this shape. Describe how you did this using some of the words reflection, rotation or translation.

8 Find the angles marked with letters.

a
$36°$ a

b
$24°$ b $85°$

c
$52°$ c

d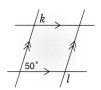
$88°$ d e $136°$

e
g $130°$ f $20°$

f
h $46°$

g
j i $95°$

h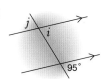
k $50°$ l

i
m $72°$

j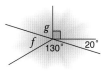
$35°$ i j $75°$ k

9 Toby made an accurate drawing of his family yacht.
He used the scale 1 cm represents 50 cm.
   a The actual yacht is 6 m long and 1·8 m wide.
   How long and wide is the yacht on Toby's drawing?
   b On his drawing the mast is 15 cm high.
   How high, in m, is the actual mast?

10 Write these imperial measurements as approximate metric equivalents.
   a 30 miles   **b** 11 lb   **c** 36 gallons   **d** 12 inches
   e 2·5 oz   **f** 5·25 pints   **g** 6 yards

Shape, Space and Measures

11 Calculate the value of *y*.

a

b

* c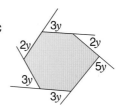

12 Find the circumference and areas of these circles.
Use 🔘π on your calculator. Round your answer to 2 d.p.

a
6 cm

b
2·4 m

* c
8·6 cm

13 Use a pair of compasses to draw a circle with radius 6 cm.
   a Draw and label these parts of your circle.
   radius   diameter   arc   sector   tangent
   b Make another circle and draw and label these parts.
   chord   segment   circumference

14 Draw a regular hexagon by dividing the circumference of a circle into six equal arcs using a protractor.
Explain why a regular hexagon can be constructed this way.

15 Find the red lengths and the angles marked with letters.
Show your reasoning clearly.

a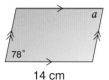
78°   *a*
14 cm

b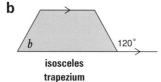
b   120°
isosceles trapezium

c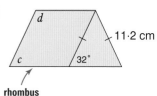
d   11·2 cm
c   32°
rhombus

d
e
141°
26°
kite

e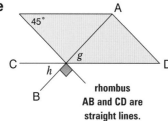
45°   A
C   *g*   D
h
B
rhombus
AB and CD are straight lines.

* f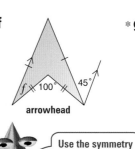
f   100°   45°
arrowhead

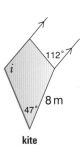
Use the symmetry properties to help.

* g
112°
i
47°   8 m
kite

16 Prove that *x* has the value shown. Show each step clearly and give reasons.

a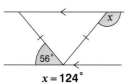
x
56°
x = 124°

b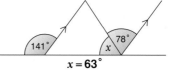
141°   78°
x
x = 63°

c
24°   *x*
68°
x = 32°

17 Calculate the areas and perimeters of these shapes.

a

b

c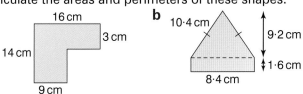

18 What is the bearing of A from B?

a **b** **c**

19 a A circle has a radius of 12 cm.
Calculate the **area** of the circle.
Show your working.

b A different circle has a **circumference** of 140 cm.
What is the **radius** of the circle?
Show your working.

Use π on your calculator.

20 Triangle OAB is rotated through 180° to triangle OXY.

a What is the length of XY?

b What is the size of ∠OYX?

c What is the size of ∠OXY?

d Triangle AOB is isosceles. Explain why.

e What is the length of OX?

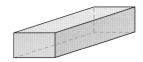

21 Use three copies of this cuboid.
Sketch a different plane of symmetry on each.
Shade your planes.

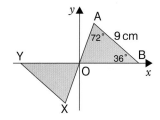

22 Draw a possible path for these.

a the tip of a windscreen wiper

b the head of a boy on a merry-go-round

c a car going halfway round a roundabout

d a horse running parallel to a straight ditch

23 Which of these are true?

a A reflection in two perpendicular lines is equivalent to a single reflection.

b Two rotations about the same centre are equivalent to a single rotation about the same centre.

c Two translations are equivalent to a single translation.

24 This diagram shows the positions of some features at a park.
Measure and write down the bearing of these.

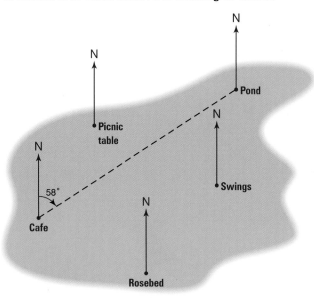

a the pond from the café
b the picnic table from the café
c the swings from the picnic table
d the café from the swings

25 In question **24** what is on a bearing of 220° from the swings?

26 Maria bought six 50 g balls of wool.
She had an old knitting pattern that said she needed 12 oz of wool.
Did she buy enough wool?

27 A ladder is leaning against a wall.
The top of the ladder is 4 m up the wall.
The foot of the ladder is 1 m from the wall.
Make a scale drawing using a scale 2 cm represents 1 m.
Use a ruler and protractor to find
a the actual length of the ladder
b the angle between the ladder and the ground.

28 Which is the odd one out in this list? Explain why.
square   rhombus   rectangle   parallelogram   kite
Is there more than one possible answer?

29 Grace sliced a cone horizontally near the top.
a What shape will the cross-section be?
b What if Grace sliced the cone vertically through the top?

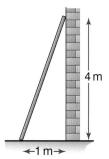

30 Use a copy of this.
a Use compasses to construct the line through P that is
perpendicular to AB.
b Name the point where this line meets BC as X.
c Bisect angle ABC.
d Name the point where this bisector meets PX as Y.
e Measure the length YB to the nearest millimetre.

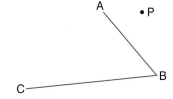

31 a Name the triangles which are congruent to ABC.
  b For each of the congruent triangles name the side equal to
    i AB   **ii** BC.
  c For each of the congruent triangles name the angle equal to
    i A   **ii** C   **iii** B.
  d The transformation that maps ABC onto GHI is a:
    A rotation of 90°   **B** rotation of 180°
    C reflection   **D** translation.

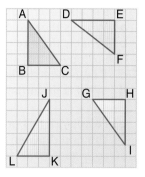

32

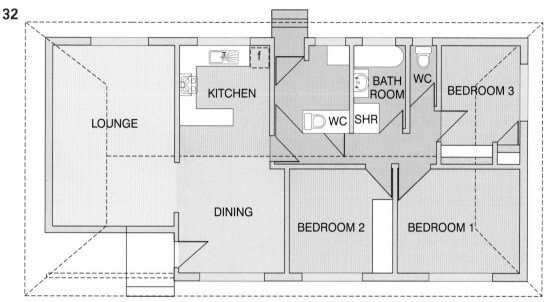

scale 1:100

This is a scale drawing of a new house.
  a What are the inside dimensions of the lounge?
  b What are the inside dimensions of bedroom 1?
  c How long is the window in bedroom 3?
  d The outside dotted line is the edge of the roof. What are the dimensions of the roof?
  e How deep is the wardrobe in bedroom 2?
  f How long is the bath?

33 Triangle ABC is shown on the grid.
  Use a copy of the grid.
  a Reflect ABC in the x-axis.
    Label the image A'B'C'.
  b Reflect A'B'C' in the y-axis.
    Label the image A″B″C″.
  c Write down the coordinates of A″B″ and C″.
  d Which single transformation maps triangle ABC
    onto triangle A″B″C″?
    A Reflection   **B** Rotation
    C Translation   **D** Enlargement

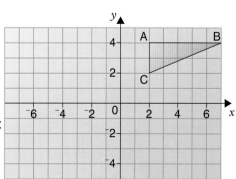

Shape, Space and Measures

T **34** This shape is a regular pentagon.
   a What is the size of angle x?
   b What is the size of angle y?
   Use a copy of this shape.
   c Draw on all the lines of symmetry.
   d What is the order of rotational symmetry?

T **35** Use a copy of this grid.
   On the grid draw an
   enlargement, scale factor 2, of
   the shape.
   Use point O as the centre of
   enlargement.

36 Farmer Jones has these three fields.
   a Find the area of each in
   hectares.
   ***b** Farmer Jones has
   600 m of fencing. What
   is the area of the largest
   field he can enclose?

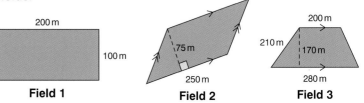

Field 1   **Field 2**   **Field 3**

37 This pattern is made from isosceles
   triangles. It has rotation symmetry of
   order 8.
   Find the size of angle x.
   Show your working.

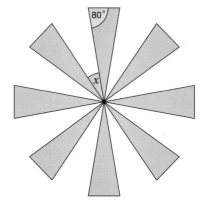

38 The diagram shows a model made with nine cubes.
   Five of the cubes are pink. The other four cubes are blue.
   a The drawings below show the four side-views of the model.
   Which side-view does each drawing show?

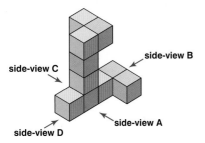

side-view C
side-view B
side-view D
side-view A

Shape, Space and Measures Support

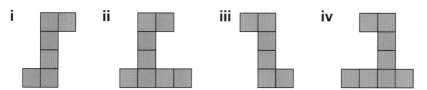

b Complete the top-view of the model by shading the squares which are pink.

top-view

c Imagine that you turn the model upside down.
What will the new top-view of the model look like?

Complete the new top-view of the model by shading the squares which are pink.

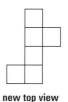

new top view

39 The blue shape is reflected in the x-axis to the green shape.
   a Describe the transformation that maps the blue shape onto the purple shape.
   b Describe a combination of transformations that will map the green shape onto the purple shape.
   c Show that the blue shape can also be transformed to the green shape by a combination of two transformations.

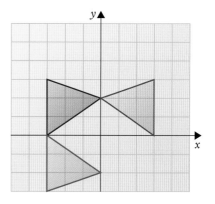

40 The large triangle, A, is an enlargement of the small triangle, B.
What is the value of
   a x **b** the shaded angle?

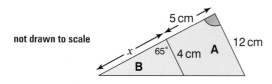

not drawn to scale

5 cm   12 cm x 65°   4 cm   **A**   **B**

41 a Make an accurate scale drawing of the triangular play area shown. Use the scale 1 cm represents 2 m.
   b Find the length of RF on your drawing.
   c Find the actual distance between the climbing frame and the roundabout.
   Show your working and write units with your answers.

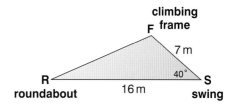

climbing frame
F
7 m
40°
R   S
roundabout   16 m   swing

243

Shape, Space and Measures

42 a Any quadrilateral can be split into 2 triangles.

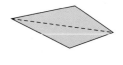

Explain how you know that the angles inside a **quadrilateral** add up to 360°.

b What do the angles inside a **pentagon** add up to?

c What do the angles inside a **heptagon** (7-sided shape) add up to? Show your working.

d Prove, using diagrams, that the sum of the exterior angles of a pentagon is 360°.

43 What is the surface area of this open shoe box in
   a cm² **b** mm²?

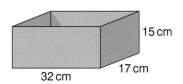

15 cm

17 cm

32 cm

44 Calculate the value of x. Show your working clearly and give reasons.

a

$x + 15°$ $2x$

b

$2x - 10°$ $74°$
$24°$ $3x$ $x + 20°$

c

$4x$
$x + 35°$
$2x - 5°$

d

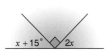

$4x - 10°$ $2x + 16°$

e

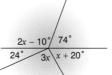

$x - 40°$
$58°$

45 A bakery sells mini pizzas with a diameter of 14 cm.
   Their regular size pizzas have a diameter of 25 cm.
   a What is the area of a mini pizza to the nearest cm²?
   b How many people could share a regular pizza if they each eat about the same amount of pizza as the mini one?

46 a Construct a triangle, using your ruler and compasses, with sides of length 4 cm, 6·4 cm and 7 cm. Measure the angle between the sides of length 4 cm and 6·4 cm.
   b Explain why it is not possible to draw triangle ABC such that AB = 10 cm, AC = 5 cm and BC = 3 cm.

47 Find the mid-point of the line joining these points.
    a (1, 2) and (7, 12)   **b** (⁻2, 5) and (6, 15)

48 Find the surface area and volume of this prism.

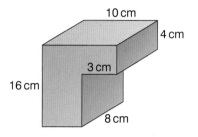

49 This diagram represents the plan view of the model on the right.

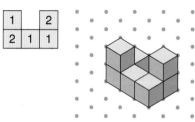

    a How many cubes does the model have?

    b On isometric paper, draw the shapes that these represent?

    i   **ii**   **iii** 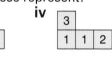   **iv**

50 **a** Calculate the size of an exterior angle of a regular 8-sided polygon.
    b Calculate the size of an interior angle of a regular 8-sided polygon.
    c Can a regular polygon be drawn with each exterior angle equal to 62°?
       Give reasons for your answer.

51 A blue chair and a black chair are placed some distance apart.
Sketch the locus of all the places pupils could stand so they are equal distance from the blue and black chairs.

You need to know

✓ lines and angles page 229

············ **Key vocabulary** ·······································

alternate angles, convention, corresponding angles, definition, derived property, exterior angle, hypotenuse, interior angle, proof, prove, Pythagoras' theorem

 Snap it!

Photos often contain interesting lines and angles.
Find some interesting photos.

You could look at home, in magazines, newspapers, on the Internet or on a CD-ROM. Make a poster or collage of your photo.

Conventions, definitions and derived properties

Discussion

- *Arrows are used to show parallel lines.*

  This is a **convention** that was *agreed* so that everyone easily recognises parallel lines.
  Something different *could* have been chosen to show parallel lines.

  What are some other mathematical conventions? **Discuss.**

  > Think about perpendicular, decimal points, etc.

- *A rhombus is a quadrilateral with all sides equal.*

  This is a **definition**. It tells us the minimum amount of information needed to identify a rhombus.
  What might the definition of a square be? **Discuss.**
  What about a rectangle?

  $a + b + c = 180°$

- *The angles inside a triangle add to 180°.*

  Do we *need* to know this to identify a triangle? **Discuss.**
  A **derived property** is not essential to a definition but it follows as a result of the definition.
  Think of at least one other derived property. **Discuss.**

- What do you think these are, a convention, a definition or a derived property? **Discuss.**

  a We label the vertices of a shape with upper-case letters and the side opposite each vertex with the same lower-case letter.

  b If no direction is given for a rotation we turn anticlockwise.

  c The diagonals of a rectangle are equal.

  d One complete turn is divided into 360°.
  A degree is a unit for measuring angles.

  e The diagonals of a kite cross at right angles.

  f A parallelogram has no lines of symmetry.

  g The inverse of a reflection in the mirror line m, is a reflection in the mirror line m.

  h A reflection is a transformation that maps P to P' such that P and P' are equidistant from the mirror line and PP' is perpendicular to it.

  i A parallelogram is a quadrilateral with two pairs of parallel sides.

  j A circle is the set of points equidistant from one fixed point.

Shape, Space and Measures

Demonstrations and proofs

We can **demonstrate** that the interior angles of a quadrilateral add to 360°. We could cut off the corners of any quadrilateral and arrange the angles at a point to show they add to 360°.

To **prove** this we must make deductions based on the properties of angles and triangles that show it is true for all cases.

First **prove** the interior angles of a triangle add to 180°.

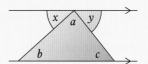

$x = b$ (alternate angles parallel lines are equal)
$y = c$ (alternate angles parallel lines are equal)
$x + a + y = 180°$ (angles on a straight line add to 180°)
so $b + a + c = 180°$ (substituting for x and y)

The interior angles of a triangle add to 180°.

Note This proof relies on the assumption that these are true.
         Alternate angles on parallel lines are equal.
         Angles on a straight line add to 180°.

Now **prove** as a result that interior angles of a quadrilateral add to 360°.

Any quadrilateral can be divided into two triangles.

$d + e + f = 180°$ (angles in a triangle add to 180°)
$g + h + i = 180°$ (angles in a triangle add to 180°)
$d + e + f + g + h + i = 360°$

The interior angles of a quadrilateral add to 360°.

Exercise 1

1 Jon wrote this.

> 2, 5, 10, 17, 26, 37, ...
> This quadratic sequence has the rule $T(n) = n^2 + 1$.

Term	2	5	10	17	26	37
1st difference		3	5	7	9	11
2nd difference			2	2	2	2

> The 2nd difference of a quadratic sequence is always constant and equal to twice the number multiplying n^2.

Is what Jon wrote a proof or a demonstration? Explain.

2 Becky folded the corners of 100 triangles down to meet at a common point on the base. She wrote this.

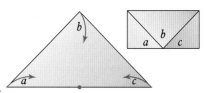

> $a + b + c = 180°$ (angles on a straight line add to 180°)
> Internal angles of a triangle add to 180°.

Is what Becky wrote a proof or a demonstration? Explain.

3 Fraser and Henry were asked to **prove** that AĈD = AB̂C + DÊC.

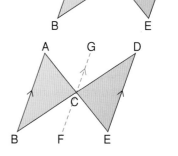

a Fraser drew the diagram on paper.
He cut off angles AB̂C and DÊC and showed that they fitted exactly into AĈD.
Has Fraser proved that AĈD = AB̂C + DÊC?

b Henry began like this.

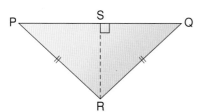

> Add a third line parallel to AB and ED through C.
> Label it FG.
> AB̂C = B̂CF (alternate angles on parallel lines are equal)
> DÊC = EĈF (alternate angles on parallel lines are equal)

Finish what Henry has written to prove that
AĈD = AB̂C + DÊC.
What assumptions do you need to make for the proof?

4 Use deductive reasoning and the properties of angles and parallel lines that you already know to answer these.

a AB is parallel to CD. ACE is a straight line.
BĈD = DĈE
Prove that triangle ABC is isosceles.

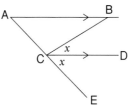

b PR = RQ and QSR is a right angle.
Prove that PS = SQ.

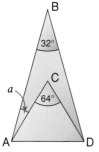

*** c** *Prove* that the exterior angles of any polygon add to 360°.
Hint Prove this for say, a pentagon first and then generalise your proof.
What assumptions do you need to make?

5 Two isosceles triangles have the same base, AD, so that AB = DB and AC = DC.

[SATs Paper 1 Level 8]

a Show, by calculating, that angle a is 16°.
b Other pairs of isosceles triangles can be drawn from the same base, AD.
Angle ACD is twice the size of angle ABD.
Call these angles $2x$ and x.
Prove that angle a is always half of angle x.

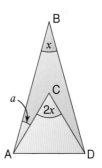

Shape, Space and Measures

Review 1

a $54° + 65° + c = 180°$ (sum of angles of a triangle equal 180°)

   $54° + 65° = 180 - c$

   $c + d = 180°$ (angles on a straight line add to 180°)

   $d = 180° - c$

   so $d = 180° - c = 54° + 65°$

This shows that the exterior angle of the triangle is equal to the sum of the two interior opposite angles.

Is this a proof or a demonstration? Explain.

b Using the above method prove the result is true for any triangle. What assumptions did you make?

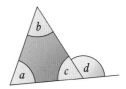

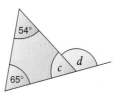

Review 2 $\angle AED = \angle DCB$
Prove that $\angle ADE = \angle EBC$

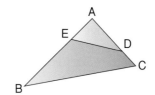

Finding angles

Remember

See page 229 to remind yourself of the angle properties of intersecting and parallel lines, triangles, quadrilaterals and polygons.

We often use **geometrical reasoning** to find an angle.

    1 Write down what you know. Mark any known angles, parallel lines, equal sides etc.

    2 Name any other angles that need naming.

    3 Write down the steps that are needed to find the angle you want.

       Always give reasons.

Worked example

Find y.

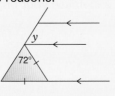

> You need an equation and a reason for each fact you write down.

Answer

Name angle x.

$x = 72°$ (base angles of an isosceles triangle)

$y = x$ (corresponding angles on parallel lines are equal)

$y = \mathbf{72°}$ (because $x = 72°$)

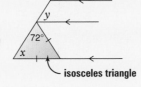

isosceles triangle

Worked example

Find the value of x.

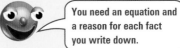

Answer

ABC is an isosceles triangle.

$\angle B = \angle C$

   $2x + 10° = 3x - 4°$

$2x + 10° - \mathbf{2x} = 3x - 4° - \mathbf{2x}$ subtracting $2x$ from both sides

   $10° + \mathbf{4°} = x - 4° + \mathbf{4°}$ adding 4° to both sides

         $x = \mathbf{14°}$

1 ABCD is a rectangle.
Work out the size of b.
Show your working clearly.

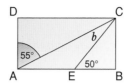

2 Find the sizes of the angles marked with letters.
Show your working clearly and give reasons.

a

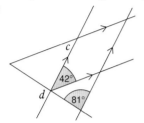

b

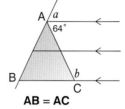

AB = AC

3 Prove that
   i $c = 55°$
   ii $d = 50°$

4 **a** PQRS is a rectangle.

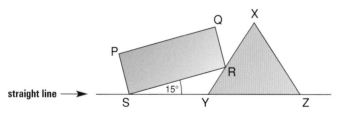

straight line →

XYZ is an equilateral triangle
Find the size of ∠SRY.
Show your working clearly and give reasons.

b ABC is a straight line.
BCD is an equilateral triangle.

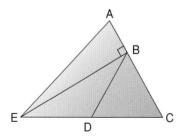

Show that triangle EBD is isosceles.

5 The drawing shows how shapes A and B fit together to make a right-angled triangle. Work out the size of each of the angles in shape B. Write them in the correct place in shape B below.

[SATs Paper 1 Level 6]

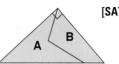

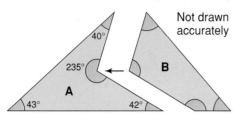

Not drawn accurately

6 This triangle has been drawn on a straight line ST.
  a Write b in terms of a.
  b Write b in terms of d and c.
  c Use your answers to parts **a** and **b** to show that $a = d + c$.

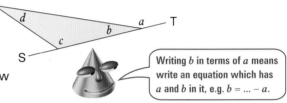

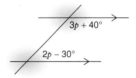

Writing b in terms of a means write an equation which has a and b in it, e.g. $b = \ldots - a$.

7 Calculate the value of p by writing and solving an equation.
  a

$2p \quad p + 30°$
$3p \quad p + 50°$

  b

$p + 20°$
$2p + 5°$

  *c

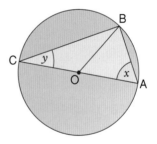

$3p + 40°$
$2p - 30°$

8 The diagram shows 3 points, A, B and C, on a circle, centre O.
AC is a diameter of the circle.
  a Angle BAO is $x°$ and angle BCO is $y°$. Explain why angle ABO must be $x°$ and angle CBO must be $y°$.
  b Use algebra to show that angle ABC **must** be 90°.

[SATs Paper 1 Level 8]

9 Look at the diagram:

[SATs Paper 1 Level 8]

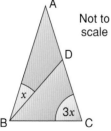

Not to scale

Side AB is the same length as side AC.
Side BD is the same length as side BC.
Calculate the value of x.
Show your working.

*10 Find the size of angle ADC in terms of z.

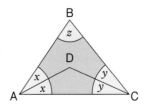

*11 AEB and ADC are straight lines.
Find the size of angle C.
Show your working.

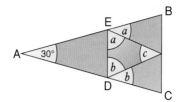

Review 1 Find the size of x. Show your working clearly and give reasons.

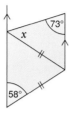

Review 2 Find the value of x and the size of the biggest angle in the quadrilateral.

Review 3 Prove that triangle BDE is isosceles.

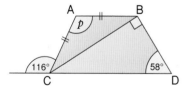

Review 4

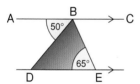

a Find the value of p. Give reasons and show working.
b What can you tell about AB and CD? Explain.

Review 5 O is the centre of the circle.
Find the values of a and b.
Show your working and give reasons.

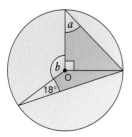

Pythagoras' theorem

Investigation

Pythagoras

On a loose piece of paper, draw a right-angled triangle with sides of length 8 cm, 6 cm and 10 cm.
Measure and draw carefully.

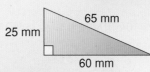

Draw the squares on the sides of the triangle.
Cut up the squares P and Q and fit them onto square R.
You should be able to cut them in such a way that they fit exactly onto R.
What is the relationship between these three squares?

Note You could draw the triangle on 1 cm squared paper and count squares to find the areas of P, Q and R.

What if you began with this triangle?

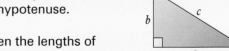

What if you began with any other right-angled triangle? **Investigate**.
What if you began with a scalene triangle? **Investigate**.

The **hypotenuse** is the longest side in a right-angled triangle. It is the side opposite the right angle. In this triangle, c is the hypotenuse.

Pythagoras' theorem gives the relationship between the lengths of the sides in a right-angled triangle.
In words, this theorem is 'the square on the hypotenuse equals the sum of the squares on the other two sides'.

$$c^2 = a^2 + b^2$$

Discussion

● Pythagoras' theorem is a *property of areas* and a *property of lengths*.
  Discuss what this means.

● Pythagoras' theorem applies only to right-angled triangles.
  If $c^2 > a^2 + b^2$,
  is angle C acute or obtuse? **Discuss**.
  What if $c^2 < a^2 + b^2$? **Discuss**.

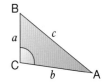

Exercise 3

1 Name the hypotenuse in each of these triangles.

  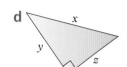 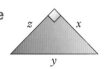

a   b   c   d   e

2 For each of the triangles in question **1**, write down the relationship between the lengths of the sides.

Review Using Pythagoras' theorem, write down the relationship between the lengths of the sides of these triangles.

a   b

We can use Pythagoras' theorem to find the length of the third side of a right-angled triangle, if we know the lengths of the other two sides.

Worked Example

Find the value of x in each of these triangles.

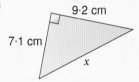 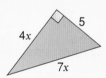

a   b   c

Answer

a $x^2 = 7 \cdot 1^2 + 9 \cdot 2^2$ **Pythagoras' theorem**

    $= 50 \cdot 41 + 84 \cdot 64$

    $= 135 \cdot 05$

   $x = \sqrt{135 \cdot 05}$

    $= \textbf{11·6 cm (1 d.p.)}$

Possible keying sequences are:

Key [7.1] [x^2] [+] [9.2] [x^2] [=] [√] [Ans] [=]

or **Key** [√] [(] [7.1] [x^2] [+] [9.2] [x^2] [)] [=]

b $15^2 = x^2 + 9^2$ **Pythagoras' theorem**

Rewrite with x^2 as the subject: $x^2 + 9^2 = 15^2$

                     $x^2 = 15^2 - 9^2$ subtracting 9^2 from both sides

                       $= 225 - 81$

                       $= 144$

                $x = \sqrt{144}$

                       $= \textbf{12}$

Possible keying sequences are:

Key [15] [x^2] [−] [9] [x^2] [=] [√] [Ans] [=]

or **Key** [√] [(] [15] [x^2] [−] [9] [x^2] [)] [=]

c

$$(7x)^2 = 5^2 + (4x)^2$$ Pythagoras' theorem

$$49x^2 = 25 + 16x^2$$

$$49x^2 - 16x^2 = 25$$ subtracting $16x^2$ from both sides

$$33x^2 = 25$$

$$x^2 = \frac{25}{33}$$ dividing both sides by 33

$$x = \sqrt{\frac{25}{33}}$$

$$= \mathbf{0 \cdot 87 \ (2 \ d.p.)}$$

Exercise 4   **Round your answers to 1 d.p. when rounding is necessary.**

1 Find the value of a.

a

6, 5, a

b

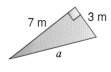

7 m, 3 m, a

c

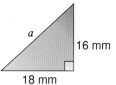

a, 16 mm, 18 mm

d

28, a, 24

e

a, 3·4 cm, 2·8 cm

f

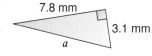

7.8 mm, 3.1 mm, a

g

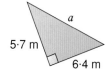

a, 5·7 m, 6·4 m

2 Find the length of the unknown side.

a

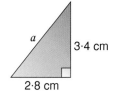

5 m, 6 m, x

b

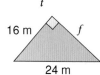

16 cm, 12 cm, t

c

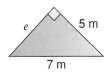

26 mm, a, 19 mm

d

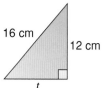

e, 5 m, 7 m

e

14 mm, x, 8 mm

f

16 m, f, 24 m

3 Find the value of p.

a

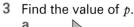
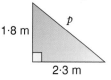

1·8 m, p, 2·3 m

b

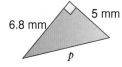

12.7 cm, 8·4 cm, p

c

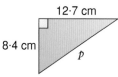

p, 5 cm, 6·8 cm

d

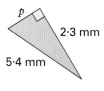

p, 2·3 mm, 5·4 mm

e

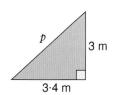

p, 3 m, 3·4 m

f

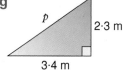

6.8 mm, 5 mm, p

g

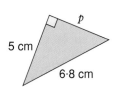

p, 2·3 m, 3·4 m

h

2·9 cm, p, 1·4 cm

4 a Calculate the length of the unknown side of this right-angled triangle.
Show your working.

[SATs Paper 2 Level 7]

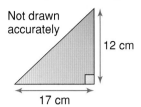

Not drawn accurately

12 cm

17 cm

b Calculate the length of the unknown side of the right-angled triangle below.
Show your working.

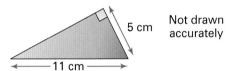

5 cm   Not drawn accurately

11 cm

5 The two shorter sides of a right-angled triangle are 5 cm and 4 cm. How long is the hypotenuse?

6 Anne cycles 8 km from A to B, then 5 km from B to C. She returns to A. How far is her return journey from C to A?

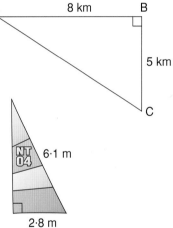

A   8 km   B

5 km

C

7 A sail, the shape shown, is made for a boat. A binding is sewn right around the edge of this sail. What total length of binding is needed?
Round your answer sensibly.

NT 04   6·1 m

2·8 m

8 A helicopter flew 24 km to the West, then 15 km to the North. How far is the helicopter then from its starting point?

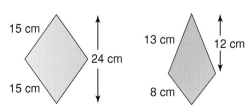

9 The two equal sides of a right-angled isosceles triangle are 25 mm. What is the length of the third side?

10 Find the length of the diagonals of the square, rectangle, rhombus and kite shown below.

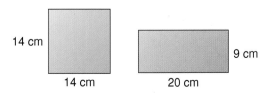

14 cm

14 cm

9 cm

20 cm

15 cm

15 cm

24 cm

13 cm

8 cm

12 cm

11 What is the length of the longest straight line that can be drawn on a piece of A4 paper which measures 298 mm by 210 mm?

12 Find the heights of the equilateral and isosceles triangles shown below.

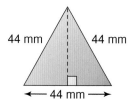

44 mm   44 mm   25 mm

←— 44 mm —→   ←— 30 mm —→

13 Find the perimeter of this shape.
Give your answer to the nearest centimetre.

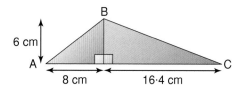

6 cm

A   8 cm   16·4 cm   C

B

14 Find the length of *a* to 1 d.p.

a   **b**   **c** 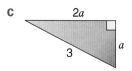   **d**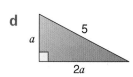

5   *a*   *a*   2*a*   5

a   7   3   *a*   *a*   2*a*

e 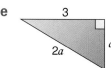   **f** 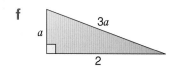   **g**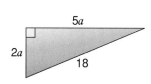

3   3*a*   5*a*

2*a*   *a*   *a*   2   2*a*   18

15 The diagonals of a square are 8 cm long. What is the length of a side of this square?

16 The perpendicular height of an equilateral triangle is 48 mm. Find the length of the sides of this triangle.

Review 1 Find the length of the side marked as *x*.

a 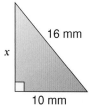   **b**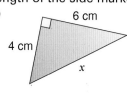

16 mm   6 cm

x   4 cm

10 mm   *x*

Review 2 How long are the diagonals of a square which has sides of length 140 mm?

Review 3 A new section of road cuts off one of the dangerous right-angled bends on a country lane. To the nearest 10 m, how much shorter is the new section of road?

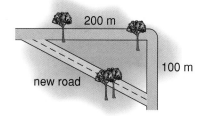

200 m

100 m

new road

Review 4 Find *a*.

a 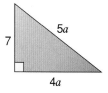   **b**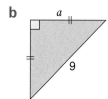

5*a*   *a*

7   9

4*a*

Investigations

1 Square roots and spirals

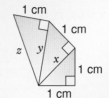

Which lengths in these diagrams are equal to $\sqrt{2}$? Which are equal to $\sqrt{3}$?
How could lengths of $\sqrt{5}$, $\sqrt{6}$, $\sqrt{7}$, ... be drawn?
Investigate to find the spiral formed by continuing the sequence of diagrams.

2 Hexagons
A regular hexagon has a perimeter of 48 cm.
How could you use this to find the area of the hexagon?

∗**What if** the perimeter of the hexagon is P cm? **Investigate**.

Pythagorean triples

Investigation

Pythagorean triples

You will need a spreadsheet or calculator.

● Using Pythagoras' theorem, we find the value of x is 5.
The lengths of the sides of this right-angled triangle are all whole numbers.
The three whole numbers 3, 4, 5 are called a **Pythagorean triple**.

There are many other sets of three whole numbers which can be the lengths of the sides in right-angled triangles.
Investigate to find more of these Pythagorean triples. You could use a spreadsheet.
Hint 5, 7 and 8 are the *shortest* sides in some of the other Pythagorean triples.

● 3, 4, 5 is a Pythagorean triple.
Is 2×3, 2×4, 2×5 or 6, 8, 10 also a Pythagorean triple?
What about 9, 12, 15?
Are the multiples of all Pythagorean triples also Pythagorean triples? **Investigate**.

●

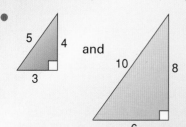

and

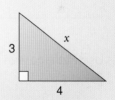

are similar triangles.
One is an enlargement of the other.

See page 276 for similar triangles.

Do a Pythagorean triple and a multiple of a Pythagorean triple always form similar triangles?
Investigate.

Some common **Pythagorean triples** are
    3, 4, 5;   5, 12, 13;   7, 24, 25;   8, 15, 17.

Example

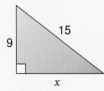

$x = 12$ 3 × (3, 4, 5) is 9, 12, 15

Worked Example
Find x and y.

Answer
    $x = 50$ 10 × 3, 4, 5 triangle.

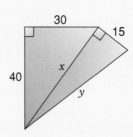

$y^2 = 50^2 + 15^2$ **Pythagoras' theorem**
$y = \sqrt{50^2 + 15^2}$
$= \textbf{52·2 (1 d.p.)}$

| Exercise 5 | | **Try to use Pythagorean triples and their multiples wherever possible.** |

1 Copy and complete the table below for the triangle shown.

a	3		6		5	10	50	15	15		2·5
b	4	4		16	12				36	2	6
c		5	10	20		26	130	25		2·5	

2 Use Pythagorean triples to find the value of e.

a 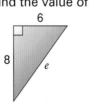   b 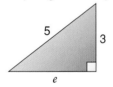   c 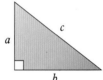   d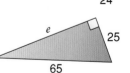

e 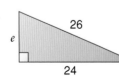   f 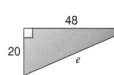   g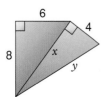

3 Find the values of x and y in each of the following.

a 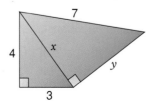   b   c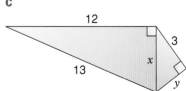

4 The three sides of a triangle are given.
Do *not* construct the triangles.
Decide if angle A is a right-angle, an obtuse angle or an acute angle.
  a AB = 5 cm, AC = 12 cm, BC = 13 cm
  b AB = 5 cm, AC = 8 cm, BC = 11 cm
  c AB = 6 mm, AC = 7 mm, BC = 8 mm
  d AB = 16 m, AC = 29 m, BC = 35 m
  e AB = 14 mm, AC = 17 mm, BC = 20 mm

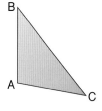

5 Find whole numbers for x and y.
Is there more than one answer?

a

b

c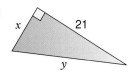

6 Find a.

a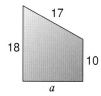

b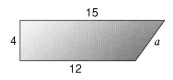

7 A windowsill in an office building is 12 m above the ground. The top of a 13 m ladder is resting against the windowsill. How far out from the wall of the office building, is the foot of the ladder?

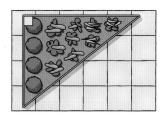

8 The stay on a gate is 1·5 m long.
How far apart are the horizontal rails?

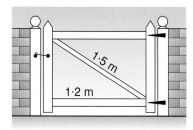

9 Jane put a wooden edging around this triangular garden.
She used a 10 m length of wood for the longest side and a 6 m length for the shortest side. What total length of wood did Jane use?

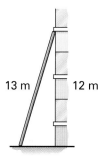

10 Kyle walks 200 m East, then 480 m South. How far is Kyle then from his starting point?

Review 1 Find the value of p in each of the following.

a

b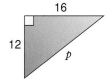

c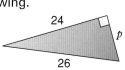

Shape, Space and Measures

Review 2 Find the values of a and b.

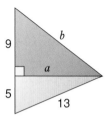

Review 3 A suitcase is 80 cm long and 60 cm wide.
Jane has an umbrella of length 1·05 m.
Will it fit in the bottom of her suitcase? Explain.

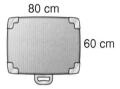

80 cm

60 cm

Investigation

Predicting Pythagorean triples

Consider the Pythagorean triple 3, 4, 5.
Notice that $3^2 = 9$ and $4 + 5 = 9$.
Can you use this relationship to predict the other numbers in a Pythagorean triple that has 7 as its smallest number?

What if the smallest number was 9?

What if the smallest number was 6?

What if ...

Investigate.

Given all three sides of a triangle, we can use **the converse of Pythagoras' theorem** to find whether or not the triangle is a right-angled triangle. The converse of Pythagoras' theorem is 'if the square on one side of a triangle is equal to the sum of the squares on the other two sides, then the triangle is a right-angled triangle'.

Example To test if a triangle with sides 25 mm, 65 mm and 60 mm is right-angled, test if Pythagoras' theorem is true.

The longest side is 65 mm. $65^2 = 4225$
The sum of the squares on the other sides $= 25^2 + 60^2$
$$= 4225$$

$$65^2 = 25^2 + 60^2$$
So the triangle is right-angled.

Example If the lengths of the sides are 18 cm, 25 cm and 36 cm then:

the longest side is 36. $36^2 = 1296$
the sum of the squares on the other two sides $= 18^2 + 25^2$
$$= 949$$

Since $36^2 \neq 18^2 + 25^2$, the triangle is not right-angled.
Since $36^2 > 18^2 + 25^2$, the angle opposite the longest side is an obtuse angle.

Exercise 6

1 a Decide whether or not the triangles, with sides of the following lengths, are right-angled triangles.

    i 2, 3, 4   **ii** 28, 35, 21   **iii** 40, 75, 85

    iv 22, 32, 45   **v** 7, 17, 15   **vi** 50, 14, 48

b For each non-right angled triangle, decide if the angle opposite the longest side is obtuse or acute.

2 a Explain why angle t must be a right angle.

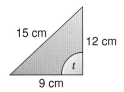

b Find the volume of this triangular prism.

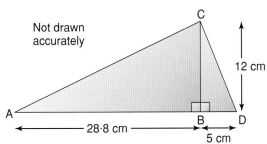

3 Two right-angled triangles are joined together to make a larger triangle ACD.   **[SATs Paper 2 Level 8]**

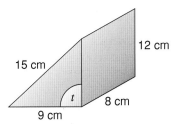

a Show that the perimeter of triangle ACD is 78 cm.

b Show that triangle ACD is also a right-angled triangle.

Review 1 Is ∠BAC a right angle? Explain.

a

b

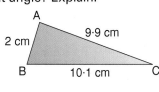

Review 2 A builder laying the foundations of a garage wants to make sure the walls of the garage are at right angles.

The length of the garage has to be 8·4 m and the width 4·3 m.

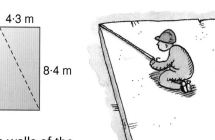

To check for a right angle the builder measures the diagonal.

a Explain how measuring the diagonal will check that the walls of the garage are at right angles.

b How long should the diagonal be so that the walls are at right angles?

Discussion

Jon laid the boxing for the foundation of a house. To check that the walls of the house would be at right angles, Jon measured the diagonals. Would Jon need to measure anything else or do any calculations? **Discuss**.

Where, on a building site, might the Pythagoras' theorem or its converse be used? **Discuss**.

Where else, in the workplace, might the Pythagoras' theorem be applied? **Discuss**.

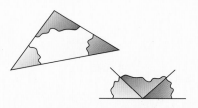

Summary of key points

A The sides of a triangle are labelled with the lower-case letter of the opposite angle.
This is called a **convention**.

A **definition** is the minimum amount of information needed to specify a geometrical term.
Example A polygon is a closed shape with straight sides.

A **derived property** follows as a result of a definition. It is not essential to a definition.
Example The angles of a triangle add to 180°.

B We can **demonstrate** some geometrical facts.
Example We can demonstrate that the sum of the angles in a triangle is 180° by cutting out the angles and placing them on a straight line.

We can **prove** geometrical facts by using the properties of angles and shapes to show they are true for all cases.

Example   We can prove that the sum of the interior angles of a hexagon equals 720°.

A hexagon can be split into four triangles as shown.

The sum of the angles in each triangle is 180°.

The sum of the interior angles = the sum of the angles in all four triangles

$$= 4 \times 180°$$
$$= 720°$$

  We often use **geometrical reasoning** to find an unknown angle or prove something.

Write down the steps clearly, one by one, and give reasons.

Example   Prove that $x = 63°$.

$$a = 180° - 126° \qquad \text{angles on a straight line add to 180°}$$
$$a = 54°$$

$$x = b \qquad \text{base angles of isosceles } \triangle$$

$$x + b = 180° - 54° \qquad \text{angles of a triangle add to 180°}$$
$$2x = 126°$$
$$x = 63°$$

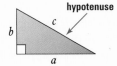

D   In a right-angled triangle the **hypotenuse** is the side opposite the right angle. It is the longest side.

Pythagoras' theorem says 'The square on the hypotenuse equals the sum of the squares on the other two sides'.

$$c^2 = a^2 + b^2$$

We can use Pythagoras' theorem to find the length of the third side in a right-angled triangle if we know the length of the other two sides.

Examples

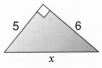

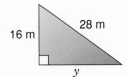

$$x^2 = 5^2 + 6^2$$
$$= 61$$
$$x = \sqrt{61}$$
$$= 7.8 \text{ (1 d.p.)}$$

$$y^2 + 16^2 = 28^2$$
$$y^2 = 784 - 256$$
$$y = \sqrt{528}$$
$$= 23.0 \text{ m (1 d.p.)}$$

E **Pythagorean triples** are sets of three **whole** numbers which can be the lengths of the sides of a right-angled triangle.

The common triples are:

3, 4, 5	$5^2 = 3^2 + 4^2$
5, 12, 13	$13^2 = 5^2 + 12^2$
7, 24, 25	$25^2 = 7^2 + 24^2$
8, 15, 17	$17^2 = 8^2 + 15^2$

Multiples of these triples are also Pythagorean triples.

Example $9^2 + 12^2 = 15^2$
  3×3 3×4 3×5

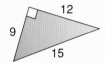

 F We can use the **converse of Pythagoras' theorem** to find out whether a triangle is right-angled.

Example $4^2 + 5^2 \neq 11^2$

  longest side

This is not a right-angled triangle.

$11^2 > 4^2 + 5^2$ so the angle opposite the longest side is obtuse.

Test yourself   **Except for question 7.**

1 State whether each of these is a **convention**, a **definition** or a **derived property**. **A**
   a The vertices of a shape are labelled with upper-case letters.
   b A half-turn is 180°.
   c The equal sides of a shape are indicated with dashes.
   d An isosceles trapezium has one pair of parallel sides and a pair of equal sides.
   e The inverse of a rotation of 90° is a clockwise rotation of 90°.
   f A rhombus is a parallelogram with four equal sides.

2 Simon drew squares on each of three sides of this right-angled triangle as shown. **B**
   He then cut the squares and showed that he could fit the two smaller squares exactly onto the largest square.
   Thus he said that 'The square on the hypotenuse equals the sum of the squares on the other two sides'.
   Is what Simon found out a demonstration or a proof? Explain.

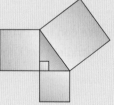

3 Use deductive reasoning and the properties of angles and parallel lines to find a and b. **B**
   MN = MO. QP is parallel to NO.

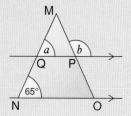

4 Find the sizes of the angles marked with letters.
Show your working clearly and give reasons.

a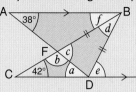

b

AB is parallel to CD.
BF = BD

△ABC is an equilateral triangle.
△BCD is an isosceles triangle.

5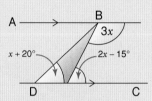

AB is parallel to DC.
Find the value of x. Show your working clearly and give reasons.

6 This is a sketch of the tiles in a child's puzzle.
PQ and RS are parallel. Find the angles marked with letters. Show your working and
give reasons.

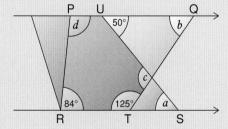

7 Find the length of the unknown side.
Round your answers to 1 d.p. when rounding is necessary.

a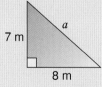

b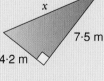

c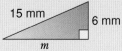

d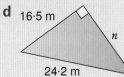

8 ABC and ACD are both right-angled triangles.   [SATs 2000 Paper 2 Level 7] D E

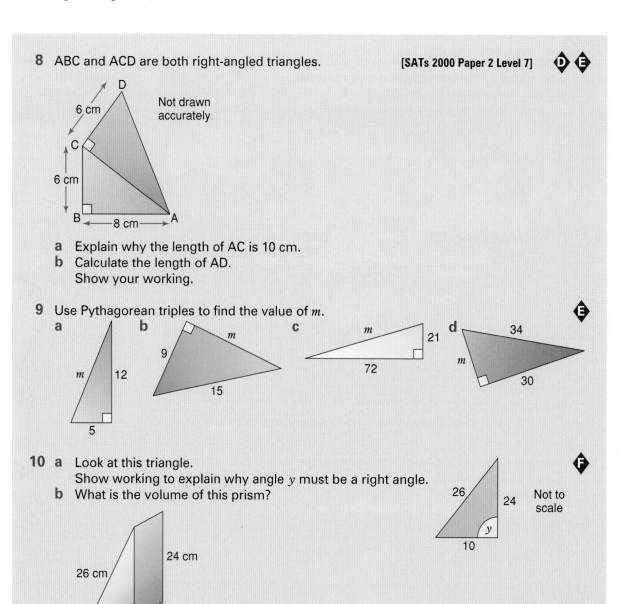

a Explain why the length of AC is 10 cm.
b Calculate the length of AD.
   Show your working.

9 Use Pythagorean triples to find the value of m.   E

10 a Look at this triangle.
      Show working to explain why angle y must be a right angle.
   b What is the volume of this prism?   F

   You must show each step in your working.

11 Shape, Construction and Loci

You need to know

✓ 2-D shapes   page 230

                   – polygons

✓ constructions   page 231

✓ locus   page 233

✓ 3-D shapes   page 233

Key vocabulary

centre (of circle), chord, circumference, congruent, cross-section, diameter, hypotenuse, radius, region, segment, similar, tangent (to a curve)

 Mission impossible

- Maurits Escher drew this picture.
  Look closely at it to see if you can see anything 'impossible'.
  Find other M. C. Escher 'impossible' drawings.

- This rectangle is an 'impossible rectangle'.

  Try to draw an 'impossible triangle'.

Constructing triangles

To be able to **construct a triangle** accurately, we need three pieces of information about sides and angles.
Sometimes these three pieces of information give a unique triangle, sometime it is possible to draw more than one triangle and sometimes it is not possible to draw one at all.

Investigation

Unique or Not Unique?

Construct triangle ABC with $\angle A = 57°$, $\angle B = 72°$, $\angle C = 51°$.
Is it possible to construct a different triangle with the same angles?
How many different triangles could you draw with these same angles?
Investigate.

What if you are given three sides (SSS) for example AB = 8 cm, BC = 5 cm, AC = 6 cm? Is the triangle unique?

What if you are given two angles and a side (ASA)?

Try these.　　AB = 6 cm, $\angle A = 30°$, $\angle B = 60°$
　　　　　　　AB = 7 cm, $\angle A = 30°$, $\angle B = 60°$
　　　　　　　AB = 6 cm, $\angle A = 60°$, $\angle B = 30°$
　　　　　　　AB = 8 cm, $\angle C = 60°$, $\angle B = 30°$
　　　　　　　AC = 6 cm, $\angle B = 55°$, $\angle C = 35°$

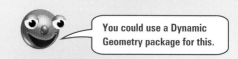

You could use a Dynamic Geometry package for this.

Can you draw more than one triangle for any of them?
Try some measurements of your own.

What if you are given two sides and an included angle (SAS)?

Try these.　　AB = 8 cm, BC = 5 cm, $\angle B = 35°$
　　　　　　　AB = 6 cm, AC = 7 cm, $\angle A = 40°$

Can you draw more than one triangle for either of them?
Try some measurements of your own.

What if you are given two sides and an angle not included (SSA)?

Try these.　　AB = 5 cm, AC = 3 cm, $\angle B = 30°$
　　　　　　　AB = 5 cm, AC = 4.4 cm, $\angle B = 65°$
　　　　　　　AB = 8 cm, AC = 2 cm, $\angle B = 45°$
　　　　　　　AB = 5 cm, AC = 6.5 cm, $\angle B = 65°$

How many different triangles can you draw for each set of information?

When **constructing triangles**, if you are given

- SSS, SAS, ASA or RHS (right angle, hypotenuse and another side) a unique triangle can be constructed,
- AAA an infinite number of triangles can be constructed, all similar
- SSA it is possible the triangle will not be unique.

For SSA there are three cases to consider.

1 The arc with radius, the second side, cuts side 3 in two places.

Example   AB = 4 cm, AC = 3 cm, ∠B = 30°

Two different triangles are possible.

2 The arc with radius, the second side, *touches* side 3, giving a unique right-angled triangle.

Example   AB = 5 cm, AC = 3 cm, ∠B = 37°

3 The arc does not reach side 3 so no triangle is possible.

Example   AB = 5 cm, AC = 2 cm, ∠B = 50°

Note Sometimes SSA does produce a unique triangle.

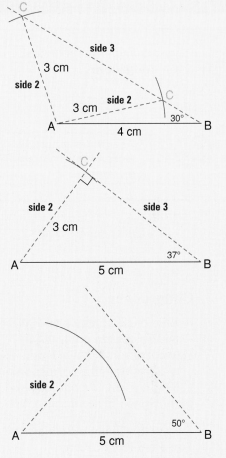

Exercise 1

1 For which of these sets of information will you be able to draw a unique triangle, PQR?
   a PQ = 7 cm, QR = 12 cm, PR = 8 cm   **b** ∠P = 20°, ∠Q = 40°, ∠R = 120°
   c PR = 6 cm, ∠P = 35°, ∠Q = 55°   **d** PQ = 5 cm, ∠P = 38°, ∠Q = 50°
   e PQ = 8 cm, QR = 5 cm, ∠Q = 30°   **f** PQ = 8 cm, ∠Q = 25°, PR = 4 cm
   g PQ = 5 cm, ∠Q = 65°, PR = 4.4 cm

2 Draw triangle ABC if AC = 5 cm, BC = 4 cm and   **a** ∠A = 40°   **b** ∠B = 40°   **c** ∠C = 40°
   Can all three triangles be drawn? Are they all unique?

3 Draw triangle ABC with AC = 10 cm, ∠C = 115°.
   a Can you draw a triangle if   **i** AB = 8 cm   **ii** AB = 12 cm?
   b What can you say about the length of AB if a triangle is possible?
  ∗ c Can there ever be more than one triangle for a value of AB? Explain.

*4   **a**   Write a set of information that gives a unique triangle.
    b   Write a set of information that gives two different triangles.

Review   In triangle ABC, BC = 5 cm and AC = 6 cm. In addition *either* ∠A *or* ∠B is given. In which case will the triangle be unique?

Congruent triangles

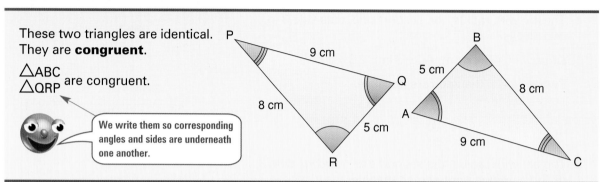

These two triangles are identical. They are **congruent**.

△ABC
△QRP   are congruent.

We write them so corresponding angles and sides are underneath one another.

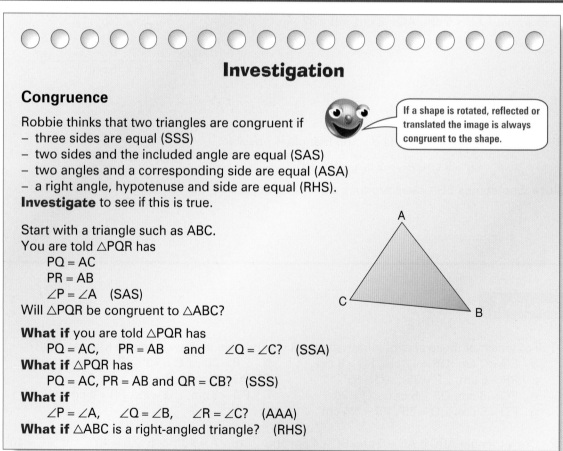

Investigation

Congruence

Robbie thinks that two triangles are congruent if
– three sides are equal (SSS)
– two sides and the included angle are equal (SAS)
– two angles and a corresponding side are equal (ASA)
– a right angle, hypotenuse and side are equal (RHS).
Investigate to see if this is true.

If a shape is rotated, reflected or translated the image is always congruent to the shape.

Start with a triangle such as ABC.
You are told △PQR has
    PQ = AC
    PR = AB
    ∠P = ∠A   (SAS)
Will △PQR be congruent to △ABC?

What if you are told △PQR has
    PQ = AC,   PR = AB   and   ∠Q = ∠C?   (SSA)
What if △PQR has
    PQ = AC, PR = AB and QR = CB?   (SSS)
What if
    ∠P = ∠A,   ∠Q = ∠B,   ∠R = ∠C?   (AAA)
What if △ABC is a right-angled triangle?   (RHS)

Triangles are congruent if
– two sides and the angle between in each triangle are the same (SAS)
– three corresponding sides in each triangle are the same (SSS)
– two angles and a corresponding side in each triangle are the same (ASA)
– a right angle, hypotenuse and side in each triangle are the same (RHS).

Link to constructing triangles, page 270.

Worked Example
PQRS is a square.
A and B are points on QR and SR such that QA = SB.
Use congruent triangles to prove that ∠PAQ = ∠PBS.

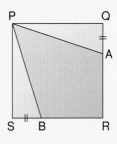

Answer
PQ = PS (sides of a square)
QA = SB (given)
∠Q = ∠S (right angles)
∴   Triangles PQA are congruent. (SAS)
        PSB
∴   ∠PAQ = ∠PBS.

> Remember ∴ means therefore

Exercise 2

1 The diagrams are *not* drawn to scale.

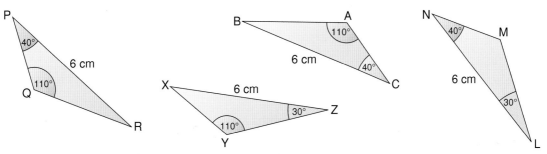

These four triangles are congruent.

  a Copy and complete: △s PQR , △s PQR , △s PQR are congruent.
                                C__   _Y_   __L

  b Use your answers to **a** to write down the angles which are equal to angle R.
  c Use your answers to **a** to find the sides equal to QR.

2

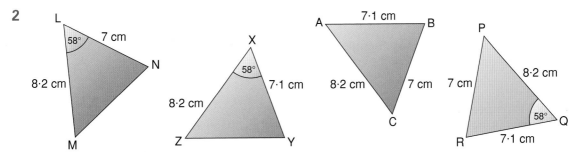

Which of the blue triangles are congruent to the red triangle? Give reasons for your answer.

3 Three of these triangles are congruent. Which three?

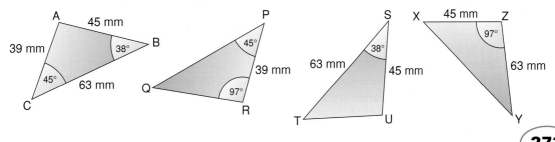

4 Which of the following pairs of triangles, A and B, are congruent? For those that are, state whether the reason is SSS or SAA or SAS or RHS.

a 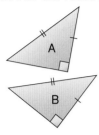 **b** 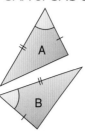 **c** 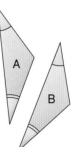 **d**

e 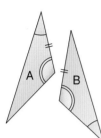 **f** 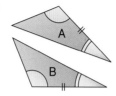 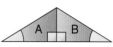 **g**

h   **i**   **j**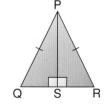

5 The diagram shows five triangles. All lengths are in centimetres.   [SATs Paper 1 Level 8]

a Write the letters of two triangles that are **congruent** to each other.
Explain how you know they are congruent.

6 PQ = PR.
PS is perpendicular to QR.
Show, using congruence, that the two base angles of an isosceles triangle are equal.

7 KLMN is a rhombus.
Show, using congruence, that LN and KM bisect at right angles.

*8 a Both △ABC and △PQR have angles of 80° and 60° and a side of 5 cm.
These triangles are *not* congruent.
Draw possible triangles ABC and PQR.

b Both △LMN and △XYZ have sides of 5 cm and 8 cm and an angle of 50°. These triangles are *not* congruent.
Draw possible triangles LMN and XYZ.

*9 AC is a horizontal line; BD is a vertical line.
Use congruent triangles to prove that AB = BD.

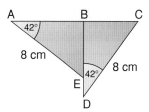

*10 PQRS is a square. △AQR is an equilateral triangle.
 a Prove that triangles APQ and ASR are congruent.
 b Hence, or otherwise, show that △PAS is isosceles.

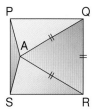

Review 1 Which of these triangles are congruent?

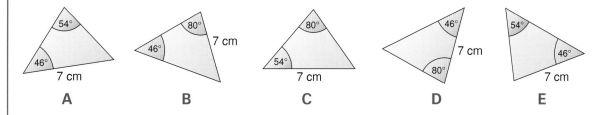

A **B** **C** **D** **E**

Review 2 Which of these triangles are congruent?

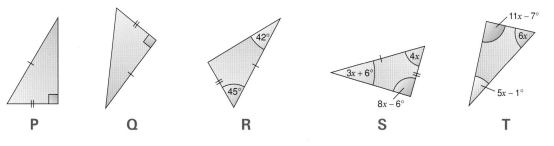

P **Q** **R** **S** **T**

Review 3 Give a reason why these pairs of triangles are congruent.
a **b** **c**

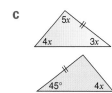

* **Review 4** KLMN is an isosceles trapezium.
Use congruence to prove the diagonals are equal.

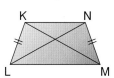

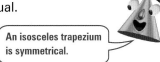

An isosceles trapezium
is symmetrical.

Similar shapes

The term '**similar**' has a special meaning in mathematics.

Shapes that are identical in every way are called congruent shapes.
These shapes are congruent.

Shapes that are the same shape but different sizes are called **similar shapes**.
These shapes are similar.
One is an **enlargement** of the other.

Discussion

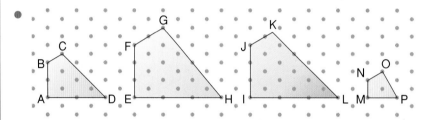

If a shape is enlarged, do we always get similar shapes?
Which of the shapes above are similar shapes? **Discuss**.

What can you say about the angles B and J?
Which angles are equal to angle C? Which angles are equal to angle D?
What other equal angles can you find? **Discuss**.

What can you say about the angles of two similar shapes? **Discuss**.

- The shapes ABCDE and FGHIJ are similar.

$$\frac{\text{Length BC}}{\text{Length GH}} = \frac{8}{4} = 2.$$

Find the ratio of other corresponding lengths of ABCDE and FGHIJ.
What can you say about the ratio of the lengths of the sides in similar shapes. **Discuss**.

Is the shape KLMNO similar to shape ABCDE?
Find the ratio of the lengths of corresponding sides.

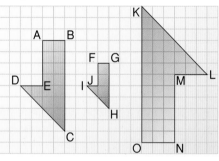

In **similar shapes**
- one is an enlargement of the other
- angles in corresponding positions are equal
- corresponding sides have the same ratio

Example   ABC and PQR are similar.
The corresponding angles are A and Q, C and R, B and P.
The corresponding sides are AB and PQ, BC and PR, AC and QR.

Link to enlargement.

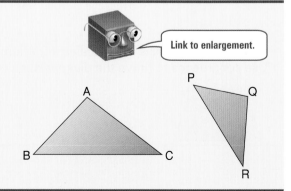

Exercise 3

1 a All of these shapes are regular.
Which ones are similar?

b Explain why regular polygons with the same number of sides are always similar.
c Explain why all circles are similar.

2 One of these diagrams is an enlargement of the other.
Are the diagrams similar?

3 Write the letters of two triangles that are mathematically **similar** to   [SATs Paper 1 Level 8]
each other but **not** congruent.

4 Which word, **always** or **sometimes**, is missing in these statements?   [SATs Paper 1 Level 8]
a A guitar in a photo and the guitar in the negative of the photo are _____ similar
shapes.
b A building and its shadow are _____ similar shapes.
c A slide and its image on a screen are _____ similar shapes.
d The floor plan of a house and the floor itself are _____ similar shapes.
e A photo of the front view of a racing car and the front view of the racing car itself are
_____ similar shapes.
f An oil painting of a daisy and the daisy itself are _____ similar shapes.

5 In this diagram there are two similar triangles.
One is △PTS.
a Name the other similar triangle.
b Which angle corresponds to ∠PRQ?
c Which side corresponds to PS?

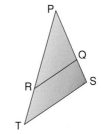

*6 In this diagram there are two similar triangles.
a Name these similar triangles.
b Which angle corresponds to angle B?
c Which angle corresponds to ∠DCE?
d Which side corresponds to AB?
e Which side corresponds to EC?

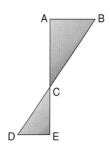

*7 Are the following pairs of triangles similar?
The triangles are NOT drawn to scale.

a

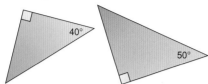

40°

50°

b

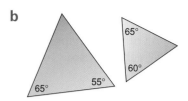

65°

65° 55° 60°

Review 1

a Portia looked at herself in a mirror that enlarged.
Will she be 'similar' to her image?

b What if she looked at herself in a mirror that 'widened' her?

Review 2   These two quadrilaterals are similar.

a Which angle corresponds to angle B?
b Which angle corresponds to angle F?
c Which side corresponds to DC?
d Which side corresponds to HG?

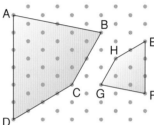
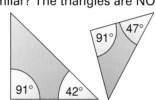

Review 3   Are the following pairs of triangles similar? The triangles are NOT drawn to scale.

a

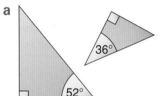

36°

52°

b

91° 47°

91° 42°

Finding unknown lengths

Worked Example

When Jayne planted a tree 5 m from a window
the tree just blocked from view a building 50 m
away.
If the building was 20 m tall, how tall was the tree?

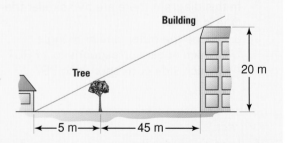

Building

Tree

20 m

←5 m→←—— 45 m ——→

Answer

Label the diagram as shown.
Let x be the height of the tree.
Since $\triangle$ABE is similar to $\triangle$ACD

then $\frac{BE}{CD} = \frac{AE}{AD}$

$\frac{x}{20} = \frac{5}{50}$

$x = 20 \times \frac{5}{50}$

$x = 2$

The tree was **2 m** tall.

or **Using scale factor**

$5 \rightarrow 50$ gives a scale factor of 10.

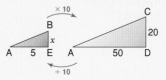

$x = 20 \div 10 = 2$

or **Using proportion**

Worked Example

A light, 3·4 metres above the floor, produces a circular patch of
light on the floor. The radius of this patch of light is 1·8 metres.

A table, which is 1·2 m high, is placed directly under the light. What
is the radius of the patch of light on the table?

Answer

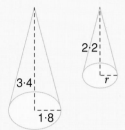

The shaded triangles are similar, so
$$\frac{r}{1\cdot8} = \frac{2\cdot2}{3\cdot4}$$
$$r = 1\cdot8 \times \frac{2\cdot2}{3\cdot4}$$
$$= 1\cdot2 \text{ to 1 d.p.}$$

The scale factor is $\frac{2\cdot2}{3\cdot4}$.

The radius of the patch of
light on the table is **1·2 m**,
to the nearest tenth of a metre.

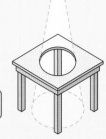

Worked Example

These two triangles are similar.
a Find the perimeter and area of each.
b What is the ratio of the two perimeters?
c What is the ratio of the two areas?

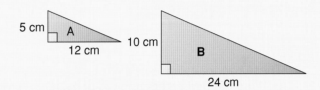

Answer
a Use Pythagoras' theorem to find the hypotenuse of the smaller triangle.
5, 12, 13 is a Pythagorean triple so the hypotenuse is 13.

$$\frac{\text{hypotenuse of B}}{\text{hypotenuse of A}} = \frac{24}{12} \quad \text{similar triangles}$$

hypotenuse of B = 2 × hypotenuse of A
$$= 2 \times 13$$
$$= \textbf{26 cm}$$

Perimeter of A = 5 + 12 + 13   Perimeter of B = 10 + 24 + 26
$$= \textbf{30 cm} \qquad\qquad\qquad = \textbf{60 cm}$$
Area of A = $\frac{1}{2} \times 5 \times 12$ Area of B = $\frac{1}{2} \times 10 \times 24$
$$= \textbf{30 cm}^2 \qquad\qquad\qquad = \textbf{120 cm}^2$$

b Ratio   *perimeter of A : perimeter of B* = **1 : 2**
c Ratio   *area of A : area of B* = **1 : 4**

Discussion

If two triangles are similar and the ratio of corresponding sides
is 1 : *r*, what is the ratio of the perimeter of the triangles?
What about the ratio of the areas?

Link to
enlargement.

Exercise 4

1 Beth, who is 1·53 m tall, gets her friend Jill
to help her find the height of a building.
Jill measures Beth's shadow as 2·42 m.
Beth measures the shadow of the building
as 19·24 m. Beth then uses similar
triangles to find the height of the building.
What answer should she get?

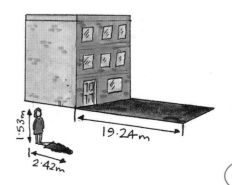

2 Each pair are similar. Find the length x.

a

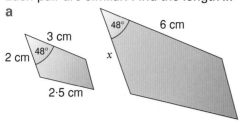

b

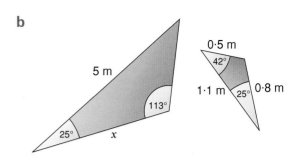

3 Find the value of l.

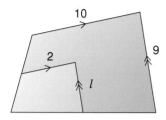

4 These two triangles are similar.
  a Find the perimeter and area of each.
  b Find the ratio of the perimeters.
  c Find the ratio of the areas.

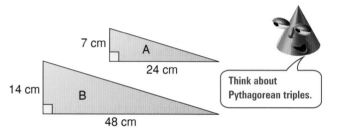

Think about Pythagorean triples.

5 A picture has a board behind it.

[SATs Paper 1 Level 8]

The drawings show the dimensions of the rectangular picture and the rectangular board.

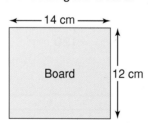

  a Show that the two rectangles are **not** mathematically similar.
  b Suppose you wanted to cut the board to make it mathematically similar to the picture.
    Keep the width of the board as 14 cm.
    What should the new height of the board be?
    Show your working.

6 A tall man, who is sitting up straight in his seat, is completely blocking John's view at the cinema. John is 10 m from the 4 m high screen. The man is 1 m in front of John.
How far would the man need to lower his head and shoulders if John is to be able to see all of the screen?

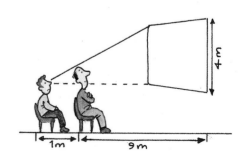

7 Hamish was doing an experiment on light. This is one of the diagrams he drew.

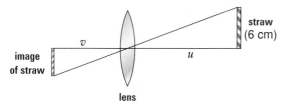

When u was 18 cm the image of the straw was 5 cm high. What was the distance v?

8 These two shapes are similar.

  a What is the ratio of their perimeters?

  b What is the ratio of their areas?

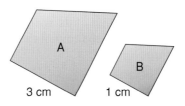

*9 A cone is cut parallel to its base to form a smaller cone. Find the radius of this smaller cone.

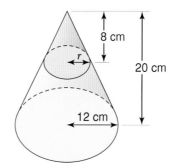

*10 **a** Explain why these two cuboids are similar.

  b What is the ratio of their surface areas?

  c What is the ratio of their volumes?

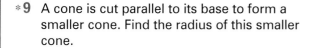

  d P and Q are similar cuboids.
What is the ratio of their surface areas?
What is the ratio of their volumes?

  e Cuboid P has surface area 66 cm² and volume 36 cm³.
What is the surface area and volume of cuboid Q?

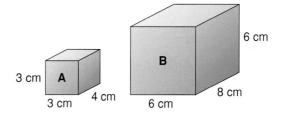

*11 A light is shone onto a screen, through a hole in a piece of cardboard.
The hole is 4 cm wide. The spot of light on the screen has a diameter of 10 cm.
If the cardboard is 5 cm from the light, show that the screen is 12·5 cm from the light.

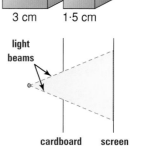

*12 The two triangles in this diagram are similar.
Find the value of x.

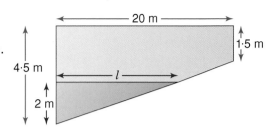

*13 A swimming pool is being filled.
Find the length, *l*, of the surface of the water
when the pool has been filled to a depth of 2 m.

Review 1   Sarah found the height of a tree by placing
a 30 cm ruler upright in the shadow of the tree. She
placed the ruler so that the end of its shadow was at the
same place as the end of the shadow of the tree.
How high was this tree?

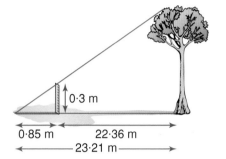

Review 2

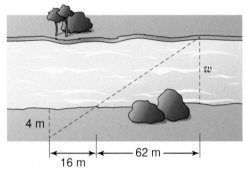

Zeke calculated the width, *w*, of a river by taking the measurements shown, then using similar
triangles.
What answer should Zeke get for the width?

Review 3   These two rectangles are similar.

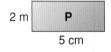

2 m **P**   5 cm

Q   3 cm

a   What is the length of Q?
b   Find the perimeter and area of each.
c   Find the ratio of the perimeters.
d   Find the ratio of the areas.

*Review 4   A cylinder of radius 10 cm just fits inside a
hollow cone of height 24 cm, as shown in the diagram.
a   If the radius of the cone is 16 cm, how tall is the
cylinder?
b   What is the ratio of the areas of the bases of the
cone and the cylinder?

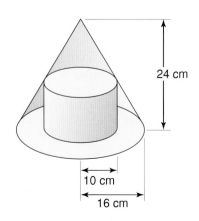

Practical

Use similar shapes to find the height of a tree or building or the width of a road or stream.

Investigation

Similar triangles
These triangles are similar since angles are the same in corresponding positions.

These triangles are similar since the corresponding sides are in the same ratio.

Suppose we are given just one pair of equal angles, as shown in these triangles. What is the least amount of information you need to be given about the sides to be sure the triangles are similar? **Investigate**.

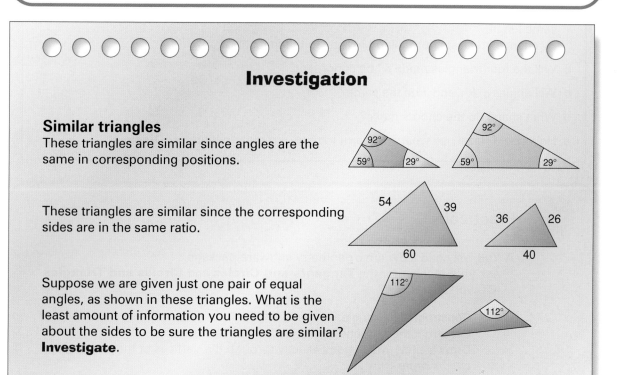

Circles, construction and congruence

Remember
When a line just touches the circle at P, it is called a **tangent** to the circle at that point.

When a line intersects the circle at two points M and N, the line segment MN is called a **chord** of the circle. The chord divides the area enclosed by the circle into two **segments**.

When a line passes through the centre, the line segment CD becomes the **diameter**. It divides the area enclosed by the circle into two **semicircles**.

The diameter is twice the radius.

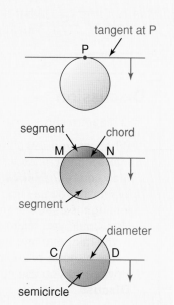

Discussion

Hazel drew a chord on a circle.
She drew the two radii to make a triangle.

a What is the special name of this triangle?

She then drew three more chords **exactly** the same length.
By chance, the last chord joined exactly to the first chord.

b Will the four triangles made all be congruent?

c Will angles a, b, c and d be the same?

d What shape do the chords make?

e How could Hazel use what she learnt to draw a regular pentagon? **Discuss**.

Practical

A You will need a dynamic geometry software package.
Ask your teacher for the **Tangents and Circles** and **Circles and Triangles** ICT worksheets.

B 1 a Construct any triangle.
Construct the perpendicular bisectors of each side.
Draw a circle which goes exactly through the vertices of the triangle, the *circumcircle*.
What do you notice about the centre of the circumcircle?
b Repeat part **a** for another triangle.

2 a Construct any triangle.
Construct the angle bisectors of each angle.
Draw a circle inside the triangle so that the circumference *just* touches the sides of the triangle. This is called an *inscribed* circle.
What do you notice about the centre of the inscribed circle?
b Repeat part **a** for another triangle.

Discussion

This diagram shows two intersecting circles with equal radii.

a What can you say about the common chord, AC, of two intersecting circles with equal radii, and the line joining their centres? **Discuss**.

b What can you say about the shape formed by the radii joining the centre to the points of intersection? **Discuss**.

c How could you relate these properties to the construction of
 i the midpoint and perpendicular bisector of a line segment
 ii the bisector of an angle?

*d How could you use congruence to prove that the standard constructions are exact?
Discuss.

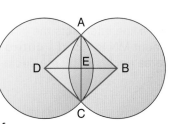

Exercise 5

1 PQ is a tangent. Find the size of a.

a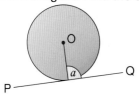

b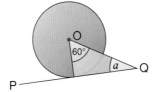

c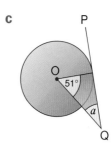

2 **a** Construct an accurate regular hexagon.
   b Explain how to draw any regular polygon accurately.

3

circle A   circle B

A	constructing the bisectors of each angle of the triangle
B	constructing the perpendicular bisectors of the sides of the triangle
C	constructing the perpendicular from each vertex of the triangle to the opposite side

What goes in the gap? Choose from the box.
   a Jasmine drew the circumcircle for a triangle. She found where the centre of the circle was by _____.
   b Padmira drew the inscribed circle for a triangle. She found where the centre of the circle was by _____.

*4 Joe constructed the perpendicular from point A to a line segment as follows.

Step 1	**Step 2**	**Step 3**

With compass point on A, construct arcs to cross the line segment at B and C.

Keep the same length on the compasses.
With compass point first on B then on C, make arcs that cross at D.

Join AD.

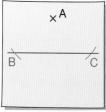 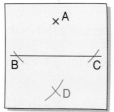 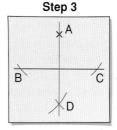

   a We know that the common chord of two intersecting circles of equal radii and the line joining the centres of the circles bisect at right angles.
   Explain how this property relates to Joe's construction.
   b Prove, using congruence, that Joe's construction is exact.

Review Draw any circle. On your circle draw two chords PQ and QR. Construct the perpendicular bisector of each chord. The perpendicular bisectors meet at C.
Prove, using congruence, that C is the centre of the circle.

3-D shapes

Example When a **3-D shape** is sliced we can get several different shaped planes and solids.

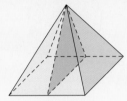

square
cross-section

triangular
cross-section

Exercise 6

1 Use several copies of this cube.
   Show how you could slice a cube so that the cross-section is
   a a triangle   **b** a hexagon.

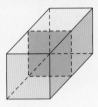

2 Ros sliced a symmetrical piece off the corner of the cube as shown.
   a What shape is the face?
   b Is it possible to cut a symmetrical slice off the corner of a cube that
         will give a different shaped face? If so, sketch it.
   c Describe the shape left if a symmetrical slice, the same size, is cut off
         each corner of the cube? Is there more than one possible answer?
   *d Repeat part **c** for a regular tetrahedron.
   *e Repeat part **c** for a regular octahedron.

3 **a** Paulo imagined a cube. In his mind he put a dot in the centre of
         each of the six faces.
         He imagined the dots on adjacent sides joined inside the cube with
         a piece of string as shown.
         What shape did the string make?
   b Paulo imagined an octahedron. He again put dots at the centre of
         each of the eight faces and joined dots on adjacent faces with string.
         What shape did the string make?

4 If we join any three vertices of a cuboid we get a triangle.
   If the width, length and height of the cuboid are all different, how many different triangles
   can you make by joining three vertices?
   Sketch each one.

5 **a** How many different shaped cross-sections can you make by cutting a square-based
         right pyramid horizontally or vertically?
         Sketch each one.
   b Which ones are cut along a plane of symmetry?

Review A cone is cut
a horizontally
b vertically.
Draw the shape of the cross-section in each case.
c If the cone is cut at an angle, what is the cross-section?

Locus and construction

Remember

A **locus** is a set of points that satisfy a rule or set of rules.
The locus of a moving object is the path that it follows.

Example A fly walks so that it is always the same
distance from two spider webs.
The locus is the perpendicular bisector
of the line joining the two webs.

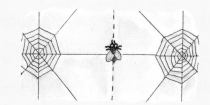

We can construct this locus.

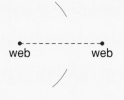

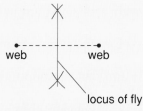

For full instructions
see page 231.

**With compass point on
one web, make two arcs.**

**Repeat with compass
point on the other web.
Join the points where
the arcs meet.**

Discussion

● There are two fences at the dog trials.

A dog must run so that it is exactly the same distance from both
fences.
Which of these is the locus of the dog? **Discuss.**

**circle
perpendicular bisector of the line between the fences
bisector of the angle between the fences**

How could you construct this locus? **Discuss.**

● Juanita made a spinner by glueing a piece of rectangular card onto a stick.
She spun the stick between her palms as fast as she could.
If you were watching, which of these shapes would you appear to see?
Discuss.

sphere   cone   cylinder   cuboid

Shape, Space and Measures

LOCI

Exercise 7

1 Match these loci and descriptions.
   a A piece of machinery moves so that it is always the same distance from a fixed point.
   b Russ runs so that he is always the same distance from two gates on opposite sides of a rectangular park.
   c Russ runs so that he is always the same distance from two dry stone walls that meet at an angle.

A	perpendicular bisector of the line joining two points
B	a circle
C	bisector of the angle between the lines

2 The diagram below shows two points A and B that are 6 cm apart. **[SATs Paper 1 Level 7]**
   Around each point are six circles of radius 1 cm, 2 cm, 3 cm, 4 cm, 5 cm and 6 cm. Each circle has either A or B as its centre.

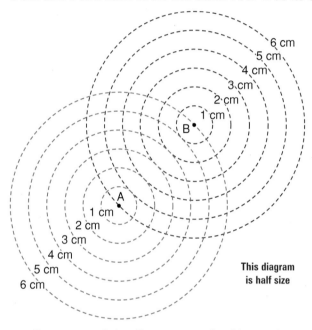

This diagram is half size

T a On a copy of the diagram, mark with a cross any points that are 4 cm away from A **and** 4 cm away from B.
   b Now draw the locus of **all** points that are the **same distance** from A as they are from B.
   c Draw two points C and D, 10 cm apart.
   Use a straight edge and compasses to draw the locus of all points that are the **same distance** from C as from D.
   Leave in your construction lines.

T 3 Use a copy of this.
   This is a scale drawing of a field.
   The scale is 1 mm = 2 m.
   AB, BC, CD and DA are hedgerows.
   Construct the following loci, using compasses and a straight edge. Leave your construction marks.
   a The locus of a ball kicked by a boy at D so that it is always the same distance from AD and DC.
   b The locus of a dog tethered at S on a 20 m rope. Can the dog get to the ball?
   c The locus of a man walking through the field so that he is always the same distance from A and B. Could the ball hit the man?

A   100 m   B
60 m   54 m
•S
D   140 m   C

4 The diagram shows the locus of all points that are the **same distance** from A as from B.
The locus is one straight line.

[SATs Paper 1 Level 7]

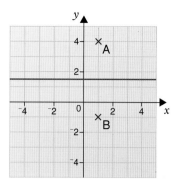

a The locus of all points that are the **same distance** from (2, 2) and (⁻4, 2) is also one straight line. Draw this straight line.

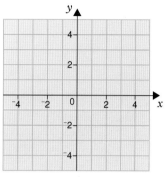

b The locus of all points that are the **same distance** from the x-axis as they are from the y-axis is **two** straight lines. Draw both straight lines.

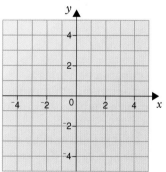

5 A spider is hanging on a single strand of web.
A fly flies so that it is always 10 cm from the spider.
The spider does not move.
What is the locus of the fly?

6 In a play, a boy is pretending to turn a frog into a handsome prince.
He has a wand and he moves the tip of his finger so that it is always 7 cm from the wand.
What is the locus of his finger tip?

7 An oak tree and a yew tree stand either side of a park.
The park ranger wants to plant a row of trees.
Where should he plant them so that their centres are
  a equidistant from the centres of the oak and yew tree
  * **b** twice as far from the centre of the oak tree as from the centre of the yew tree?

8 A plot has two large trees 10 m apart.
An architect is designing a square house for the plot.
She decides she wants the trees right next to two adjacent sides of the house.
What is the locus of all the different positions she could put A, the corner of the house?

Shape, Space and Measures

*9 A child's toy has a rod fixed at one end and is pivoted at the other. It is attached halfway along another rod at the pivot. This rod has a wheel on one end and a 'head' at the other. The wheel moves along the floor. What is the locus of the head?

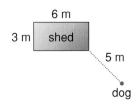

T **Review 1** Use a copy of this.

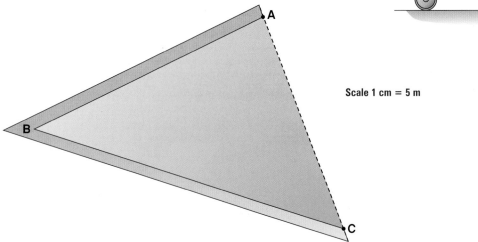

Scale 1 cm = 5 m

This is a triangular piece of land owned by a market gardener. BA and BC are boundary paths and AC is a fence.

a How long is the fence?

Remember to show all your construction lines.

b The gardener wants to put another path through his property, equidistant from AB and BC. Construct the locus of points equidistant from AB and BC using a compass and ruler. This is the new path. Call it BD.

c He wants to build a shed on this path exactly halfway along. Construct the locus of points the same distance from B and D using compass and ruler. Mark the point where he will build his shed H.

d He wants to water an area 20 m from C inside his property. Show this area. Will the water spray reach the shed?

Review 2 A dog is tied by a 5 m rope to the corner of a shed 3 m by 6 m. Sketch a diagram and shade the area which the dog can reach.

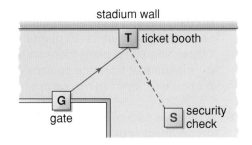

*Review 3 People are queueing to go into a stadium. They stand behind a gate at G then go to show their ticket at the booth T. They then go to S, for a security check. The officials want to put T against the outside wall of the stadium. Where should the ticket booth be placed so that people have the shortest possible distance to walk from the gate to the security check?

Investigation

A Special curves

a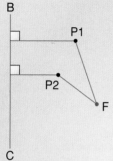

Investigate the locus of a point P which is the same perpendicular distance from the line BC as it is from the point F. Two possible positions of P are shown.

b **What if** BC was a sloping line?

c **Investigate** the locus of a point R which moves so that the sum of the distances RD and RE is 5 cm.

d **What if** the sum of these distances was 6 cm?
What if the sum of these distances was 8 cm?
What if ...

e **Investigate** the locus of a point Q which moves so that the difference of the distances from two fixed points is constant.

 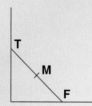

B Ladders and locus

These drawings show three of the positions of a ladder which slides from a nearly vertical position to a nearly horizontal position. As the top of the ladder (T) slides down the wall, the foot of the ladder (F) slides along the ground.

Investigate the path of the point M, the mid-point of the ladder. To help in your investigation you could use a ruler to represent the ladder, and a desk pushed against a wall of your classroom.

What if M was closer to F than to T?
What if M was closer to T than to F?
What if the wall was sloping?
What if the ground was sloping?

Shape, Space and Measures

Worked Example

Sketch the region in which a point P would be if it is
a always less than 15 mm from the point A
b always closer to the line BD than the line BC, and
within the acute angle DBC.

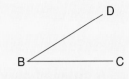

Answer

a

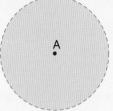

Locus is shown shaded.
It is the interior of the circle,
centre A, radius 15 mm.

b

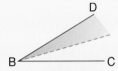

Locus is shown shaded.
It is the area between the line BD
and the bisector of the angle DBC.

Note A boundary which is not included in the region is dotted.

Exercise 8

1 Draw two points A and B.
   Sketch the region in which a point P would be if it is always
   closer to A than to B.

   •A

   •B

2 Sketch the region in which a point Q would be if it is always less than 2 cm from a fixed
   point C.

3 Sketch the region in which a point R would be if it is always
   further from the line AB than from the line CD.

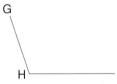

4 A point T is within the obtuse angle GHI and is closer to the line GH
   than to the line IH. Sketch the region in which T would be.

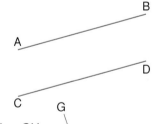

5 Two roads, AB and CB, between villages A, B and C are shown.
   Trace this diagram.

   Samantha walks her dog each day. The region in which she walks
   the dog is further from BC than from AB and closer to A than B.
   Sketch the region in which Samantha walks her dog.

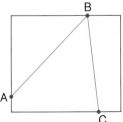

6 Two villages A and B are 20 km apart; B being due south of A. The
   fire brigade from A services an area of radius 10 km around A while
   the brigade from B services an area of radius 12 km around B.
   Using the scale 1 cm represents 4 km, shade the area that is serviced by both brigades.

7 Use a copy of this. **Scale: 1 cm represents 10 km**
This diagram shows three towns.
The towns need a mobile phone tower.
It must be nearer to Crocksford than Ramsby and less
than 25 km from Raydown.
Construct on the diagram the region where the new
tower can be placed.

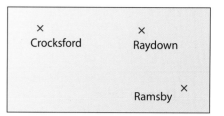

8

Scale: 1 cm represents 5 m

AB represents a brick wall; X and Y represent two rotating sprinklers. Use a copy of this
diagram.
The sprinklers are able to water a radius of 25 m. The wall is so high that no water, from
the sprinklers, reaches the ground for a distance of 2 m on the other side of the wall.
On your diagram, shade the region that is watered by both sprinklers.

9
•
A
•
B

Scale: 2 mm represents 1 km

L ————————————— M

A and B are radio stations which can be received for a 20 km radius. LM is a hill which
creates a reception 'shadow' so that for 5 km on the far side no reception is received and
for a further 5 km the reception is poor.
Use a copy of the diagram. On your diagram, shade the area that receives poor reception
from both radio stations.

Review 1 *Use the scale 1 cm represents 2 m for this question.*
Two rotating sprinklers are placed 10 m apart on a lawn. One can water an area of radius 5 m,
the other can water an area up to 7 m away. Draw a diagram to show the area that is watered
by both sprinklers.

Review 2

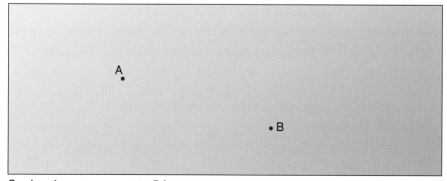

Scale: 1 cm represents 5 km

This diagram represents two towns at A and B. Trace this diagram.
The hospital at A will admit patients who live closer to A than to B. The fire brigades from
A and B will travel a maximum of 15 km.
On your diagram, shade the area in which the ambulance from A and the fire brigade from
either A or B would attend an accident.

293

Investigation

Doggy tales

This picture shows a dog tethered to a chain that is attached to two stakes. The end of the dog's lead is attached to a ring that can slide along the chain.

Investigate the possible loci of the dog's head.

Practical

You will need LOGO

Ask your teacher for the **STARS** ICT worksheet.

Summary of key points

  We can **construct unique triangles** accurately if we are given

- three sides (SSS)
- two sides and the included angle (SAS)
- two angles and the side between (ASA)
- right angle, hypotenuse and another side (RHS).

An infinite number of similar triangles can be constructed if we are given AAA.

If we are given SSA the triangle may or may not be unique. See page 271 for the three possible cases.

  Two triangles are **congruent** if they are identical.

They will be congruent if

- two sides and the angle between in each triangle are the same (SAS)
- three sides in each triangle are the same (SSS)
- two angles and a corresponding side in each triangle are the same (AAS)
- a right angle, hypotenuse and side in each triangle are the same (RHS).

See page 273 for an example.

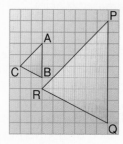

C Shapes that are the same shape but are different sizes are called **similar shapes**.

When shapes are similar

- one is an enlargement of the other
- angles in corresponding positions are equal
- corresponding sides have the same ratio.

Example △PQR is similar to △ABC.

$$\frac{PQ}{AB} = \frac{PR}{AC} = \frac{QR}{BC} = \frac{3}{1}$$

$$\hat{P} = \hat{A} \qquad \hat{Q} = \hat{B} \qquad \hat{R} = \hat{C}$$

△PQR is an enlargement of △ABC.

Remember P̂ means angle P.

D We can use similar triangles to find **unknown lengths**.
See page 278 for an example.

E The angle between the **tangent and the radius** at that point is always 90°.

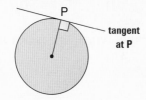

F We can make different **planes** and shapes from **3-D shapes**.

circular
cross-section

rectangular
cross-section

Example When a cylinder is sliced in different ways we can get different cross-sections.

G A **locus** is a set of points that satisfy a rule or set of rules.
We can **construct a locus**.

A locus can be a region.

Example The locus of the points which are closer to MN than to PQ is shown shaded.

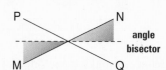

angle bisector

Test yourself

1 For which of these sets of information can you draw one unique triangle, ABC?

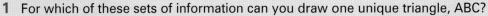

a AB = 6 cm, BC = 7 cm, CA = 10 cm   b AB = 6 cm, BC = 4 cm, B̂ = 30°
c Â = 45°, B̂ = 55°, AB = 5 cm   d AB = 7·5 cm, B̂ = 36°, AC = 3·5 cm

A

2 Two of the triangles are congruent.
Which two? Give the reason.

a

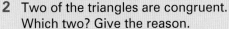

b

B

3 ABCD is a parallelogram.
Use congruence to prove that the opposite sides of a
parallelogram are equal.

B

4 a One of the rectangles is not similar to the other three. Which one is not similar?

C

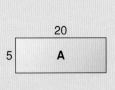

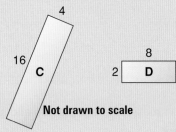
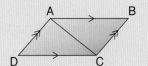

Not drawn to scale

b Two of these triangles are similar. Which ones?

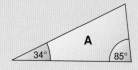
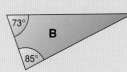

34°   A   85°

73°   B
85°

34°
C
61°

Not drawn to scale

5

D

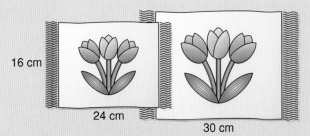

16 cm

24 cm

30 cm

Hilary is making rectangular tablemats in two sizes. The small size measures 24 cm by
16 cm. The large size is 30 cm long.
If Hilary's tablemats are similar shapes, find the width of a large mat.

6

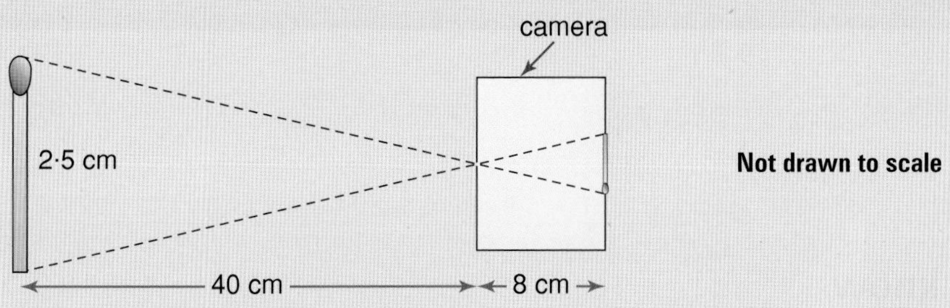

camera

2·5 cm

Not drawn to scale

40 cm ←→ 8 cm

This diagram represents a match and its image on the screen of a pinhole camera. The match, which is 2·5 cm tall, is placed 40 cm from the pinhole. If the pinhole is 8 cm from the screen, how high is the image of the match?

7 AB is a tangent. Find the size of x.

a 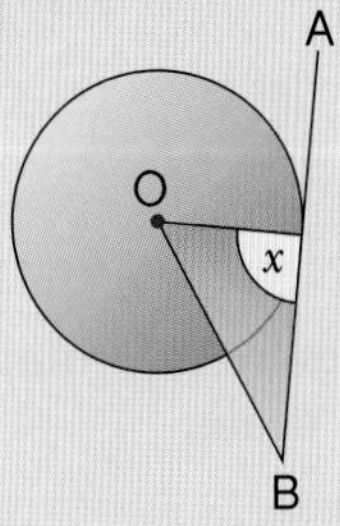 A, O, x, B

b 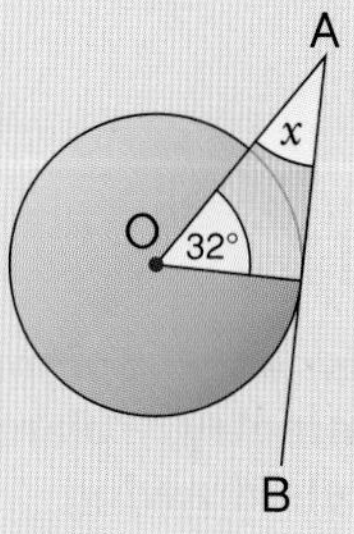 A, x, O, 32°, B

8 Construct accurately a circumcircle round a copy of this triangle. Explain how you did it.

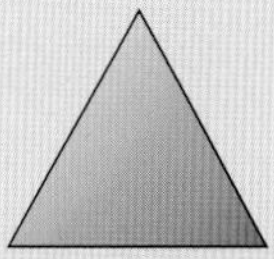

9 If we join any three vertices of a cube we get a triangle. How many different shaped triangles can you make like this? Draw sketches of them.

10 Sketch and describe each of these loci.
  a The locus of a point P which moves so that it is always 28 mm from the point A.
  b The locus of a point R which is always the same distance from two parallel lines CD and EF.
  c The locus of a point Q which is within the quadrilateral GHIJ and equidistant from the lines GH and GJ.

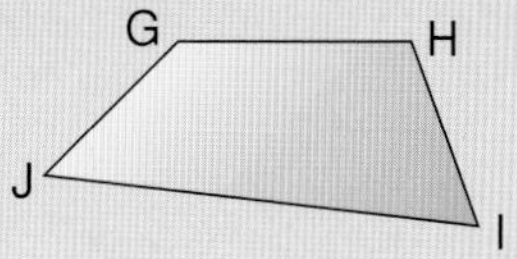

11 A dog is on a leash which is 2 m long. The leash is attached to a ring on a 5 m long piece of wire which is firmly attached to the ground. The ring can move along the wire from A to B but cannot slip off the wire. Shade the area the dog can reach.

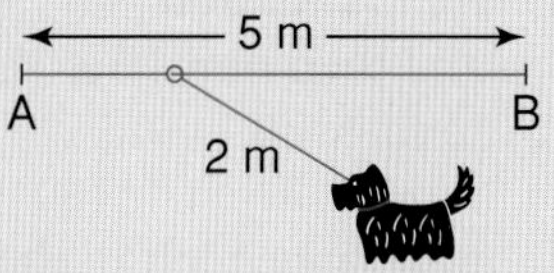

5 m, A, B, 2 m

12 The Martin family are wanting to buy a new house. They want to buy a house which is within 3 km of the school, S, and closer to the airport, A, than to the centre of town, C. The school, airport and town centre are all 4 km apart from each other.
Use a copy of the diagram and show the places they could look for a house to buy.

×C, S×, ×A

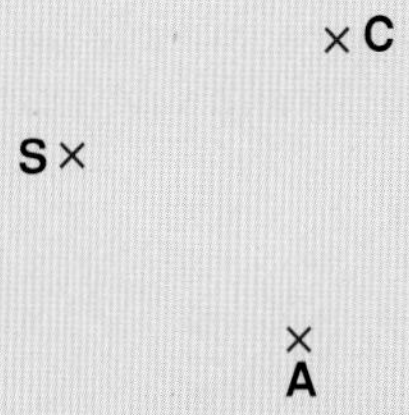

Scale: 1 cm represents 1 km

297

You need to know

✓ coordinates — page 233

✓ symmetry — page 234

✓ congruence — page 231

✓ transformations — page 234

coordinates   page 233

symmetry   page 234

congruence   page 231

transformations   page 234

·· Key vocabulary ·······

centre of enlargement, commutative, enlarge, enlargement, map, plan, plane symmetry, plane of symmetry, similar, scale, scale factor, scale drawing

Room for improvement

This is a picture of Cassandra's room.

She decides to change her bedroom around.
She

> rotates the bookcase 90°
> translates the desk left
> rotates the bed and table 180°
> and then translates them right
> to the wall
> rotates the chair 180° and then
> translates it left to the wall.

This is the floor plan.

Draw a possible floorplan for
Cassandra's new arrangement.

Combinations of transformations

Remember
When we **reflect, rotate or translate** a shape,
the image is always congruent to the original shape.

See page 234 in the support chapter
for examples of transformations.

Worked Example
a Rotate triangle A 90° about (0, 1) to triangle B.
b Describe the transformation that maps triangle B onto triangle C.
c Which of the following rotations maps triangle A onto triangle C?

   A rotation 90° about (0, 0)
   B rotation 180° about (⁻1, 0)
   C rotation 90° about (⁻1, 0)
   D rotation 90° clockwise about (⁻1, 0)

Remember to turn
anticlockwise if no
direction is given.

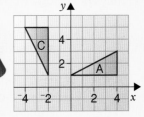

Answer
a A is rotated to B as shown.
b The transformation which maps B onto C
   is a translation 2 units left.
c By trying each of the rotations given, we
   find that C is the correct answer.

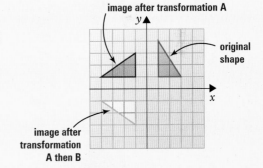

Worked Example
Investigate if this pair of transformations, when carried
out one after the other, is commutative.

   A is a rotation of 90° centre (0, 0).
   B is a reflection in the x-axis.

'Commutative' means the same
result is achieved no matter in
what order they are done.

Answer
Choose a shape to transform.
A right-angled triangle that is **not** isosceles is a good
choice, but you can choose any shape.
Draw the shape in any position.
Transform it by A and then B.

image after transformation A

original
shape

image after
transformation
A then B

Now draw the original shape again in its original position.
Transform it by B and then A.

The shape ends up in a different position.

Transformations A and B are *not* commutative.

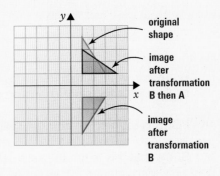

original
shape

image
after
transformation
B then A

image
after
transformation
B

Shape, Space and Measures

Exercise 1

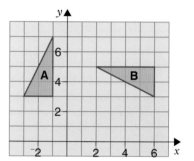

T 1 This diagram shows a piece of land divided into 3 sections.
A house, shown shaded, has been built on one of the sections.
The builder wants to build two more houses.
One house is to be a reflection of the existing house in the
fence line AB. The other is a reflection of the existing house in
fence line BC.
  a Use a copy of the diagram. Show the positions of the two
     new houses.
  b Which of the following would map the new houses onto each other?
         A a reflection   **B** a rotation   **C** a translation   **D** an enlargement

2 Which of these maps triangle A onto triangle B?
         A rotation of 90° about (1, 2) followed by translation
           5 units right and 5 units up
         B rotation of 90° clockwise about (0, 0) followed by
           rotation of 180° about (5, 3)
         C rotation of 90° clockwise about (1, 3) followed by
           translation 4 units right and 3 units down
         D rotation of 90° about (1, 5) followed by translation
           3 units right and 2 units up.

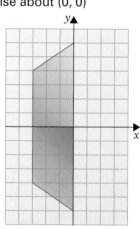

3 a Which of these transformations would map the pink shape onto the green shape?
         A rotation of 180° about (0, 0) followed by reflection in the x-axis
         B reflection in the y-axis followed by reflection in the x-axis
         C translation 3 units right followed by reflection in the x-axis
         D reflection in the line $y = x$ followed by rotation of 90° clockwise about (0, 0)

  i   **ii**   **iii**

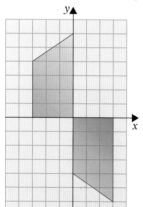   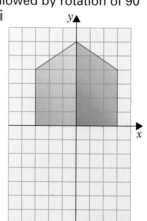

  b In **a ii**, what single transformation would map the pink shape onto the green shape?
  c In **a iii**, what single transformation would map the pink shape onto the green shape?

4 PQRS is rotated 90° about A then reflected in the line
  $y = x$.
  Show that this is not the same as reflection in the line
  $y = x$ then rotation 90° about A.

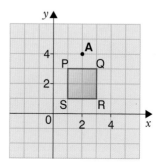

5 Some transformations are given.

L is a reflection in the x-axis.
M is a reflection in the y-axis.
N is a rotation 90°, centre the origin.
P is a rotation 180°, centre (0, 0).
Q is a rotation 270°, centre (0, 0).
I is the identity transformation.

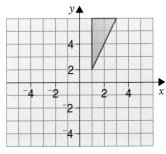

Remember: commutative means you get the same result whatever order you do the transformations.

Note The identity transformation maps a shape onto itself.
The shape does not change position or size.

a Choose a shape and draw it on a set of axes.
b Investigate to see if each of these pairs of transformations are commutative.

i L and M	**ii** L and N	**iii** L and P	**iv** L and Q	**v** L and I
vi M and N	**vii** M and P	**viii** M and Q	**ix** M and I	**x** N and P
xi N and Q	**xii** N and I	**xiii** P and Q	**xiv** P and I	**xv** Q and I

c For the pairs of transformations you found to be commutative, test to see if they are commutative for other shapes.
d Write a short summary of which pairs of transformations you found to be commutative.

T

6 Use a copy of this.
a Q maps onto P after rotation through 180°, centre A (⁻1, 2).
P maps onto R after rotation through 180°, centre B (3, 3).
What single transformation will map Q onto R?
b Plot A and B on the grid and join them.
Join two corresponding points on Q and R.
What two things do you notice about the two lines you have just drawn?
Look at the slope of the lines and their lengths.
c Choose a shape and draw it on a set of axes.
Transform your shape by S then T.
S is a half-turn rotation centre M (0, 0).
T is a half-turn rotation centre N (0, 1).
What single transformation is equivalent to S then T?
Plot and join MN.
Join two corresponding points on your shape and its image after transformation by S then T.
What do you notice about the two lines you have just drawn?
d Check to see if what you noticed is true for two half-turn rotations about other centres of rotation.

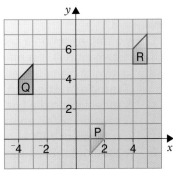

*7 Copy the shape given onto a set of axes.

C is a reflection in the x-axis.
D is a reflection in the y-axis.
E is a reflection in the line $y = x$.
F is a reflection in the line $y = {}^-x$.
G is a reflection in the line $y = x + 1$.
H is a reflection in the line $y = {}^-x - 1$.

See the ICT sheet 'Combinations of Transformations 2' in the Practical page 303. Investigate the concepts in questions 6 and 7 further.

a Which of the following combinations of transformations is equivalent to a single rotation about some point?

i C then D	**ii** D then E
iii E then F	**iv** C then D then E
v F then G then H	**vi** C then D then E then F
vii E then F then G then H	

b Investigate to test if this statement is true.
Only an even number of reflections can be equivalent to a rotation.

Review 1   ABCD is the image of QPSR under
    A   reflection in the y-axis followed by rotation
    B   reflection in the y-axis followed by translation
    C   rotation followed by reflection in the y-axis
    D   reflection followed by reflection.

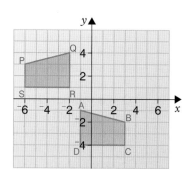

T

Review 2   The flag in the diagram is rotated
clockwise 90° about (2, ⁻1) then reflected in the y-axis.
  a   Use a copy of the diagram and show the image after
    the two transformations.
  b   Are the transformations commutative? Why or why
    not?

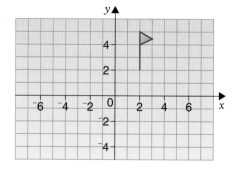

Review 3
  a   Draw a trapezium on a set of axes.
  b   Transformation **A** is a reflection in the line $x = 2$. Use the trapezium to investigate which of
    the following transformations would make this sentence true.
        Transformations A and ___ are commutative.
  B   A reflection in the line $y = {}^-1$
  C   A rotation of 90° about (0, 0)
  D   A rotation of 180° about (0, 2)
  E   A rotation of 270° about (0, 0)
  F   A reflection in the x-axis
  G   A reflection in the y-axis

Investigation

Combining transformations

A   Draw a triangle ABC on a set of axes.
    Choose a point, P, and rotate the shape through 90° clockwise about P to A'B'C'.
    Is it possible to map ABC onto A'B'C' by first rotating ABC about a point *other* than P
    and then translating it?

    Can a rotation about one point always be replaced by a rotation about another point
    followed by a translation? **Investigate**.

    Can a translation followed by a rotation always be replaced by a rotation about
    another point? **Investigate**.

T

B Use a copy of this.

These lines are at 60° to each other.
Reflect A in PQ. Call the image B.
Reflect B in SR. Call the image C.
Reflect C in TH. Call the image D.
Continue reflecting the image in successive lines to fill
the 6 spaces.
Would the last image, F, reflect in the line TU to give A?
How does this relate to rotation symmetry?
Investigate.

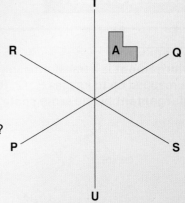

What if you began with lines at 45°
degree angles?

What if you began with lines at these angles?

What if ...?

C **Investigate** how a kaleidoscope works.
Relate what you discovered in part B to kaleidoscopes and other phenomena in the
natural world.

D Remember
A plane of symmetry divides a 3-D shape into two halves. One half is the reflection of
the other in the plane of symmetry.

The purple dotted line is an **axis of rotation symmetry** for this
cube.
If we rotate the cube around this axis, it will look the same in four
different positions.

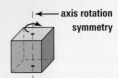

Investigate other axes of rotation symmetry for a cube.
What if the shape is a cuboid?

Practical

You will need a dynamic geometry software package.

Ask your teacher for ICT worksheets '**Combinations of Transformations
1 and 2**'.

Enlargement

Remember

Enlargement is a transformation of a plane (2-D) shape in which points such as P, Q, R and S are mapped on to images P', Q', R' and S', by multiplying the distances from a fixed **centre of enlargement** by the same **scale factor**.

Example

scale factor = 2.

$OP' = 2 \times OP$, $OQ' = 2 \times OQ$, $OR' = 2 \times OR$, $OS' = 2 \times OS$

If the **scale factor is greater than 1**, the image is larger than the original.
If the **scale factor is less than 1**, the image is smaller than the original.

Example

The bright blue shape has been enlarged by a scale factor of $\frac{1}{2}$, centre (⁻2, 1) to get the pink shape.

Each length on the image is **half** as long as the corresponding length on the original.

Discussion

Sita drew this enlargement.
Find the length of AB and A'B'.
What is the ratio   *A'B' : AB*?
What is the ratio   *C'D' : CD*?
What is the scale factor for this enlargement?
What do you notice? **Discuss**.

Are the shaded angles the same size? **Discuss**.
Are the shapes ABCDE and A'B'C'D'E' similar?
Discuss.

> This is linked to ratio.

> Remember similar shapes have ratios of corresponding sides constant and corresponding angles equal.

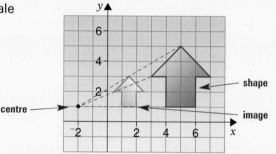

The inverse enlargement maps A'B'C'D'E' onto ABCDE. What is the scale factor and centre for the inverse enlargement? **Discuss**.

Worked Example

Enlarge the shape ABCD by a scale factor of $\frac{1}{3}$, with P as the centre of enlargement.

Answer

Join P to each of A, B, C and D.

Find A′, B′, C′ and D′ using

$PA' = \frac{1}{3}PA,$ $PB' = \frac{1}{3}PB,$ $PC' = \frac{1}{3}PC,$ $PD' = \frac{1}{3}PD.$

Note You can do this by measuring carefully or counting squares.

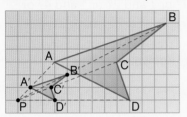

You could use vectors to describe the distances of points from the centre of enlargement.

B is $\binom{12}{6}$ from P.

So B′ is $\binom{4}{2}$ from P.

Practical

A **You will need** some sheets of 1 cm squared paper.

1 Draw this 'face' in the middle of a sheet of squared paper.

2 Use C_1, then C_2, then C_3 as the centre of enlargement. For each centre, enlarge the face by a scale factor of $\frac{1}{2}$.

3 Repeat **2** for a scale factor of $\frac{1}{4}$.

4 Draw a shape of your own on squared paper.

Enlarge it by a scale factor of $\frac{1}{3}$.

Choose a centre of enlargement which is **inside** the shape you drew.

5 Fill in a table like this one for each centre of enlargement.

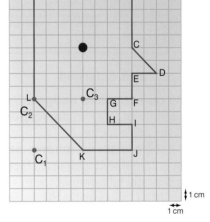

Length	Length on original (cm)	Scale factor		
		$\frac{1}{2}$	$\frac{1}{4}$	$\frac{1}{3}$
AB				
BC				
CD				
GF				
KJ				
LK				

Write down everything you can about enlarging by a scale factor less than one. Think about similar shapes, ratios of corresponding lengths, scale factor for inverse enlargements. Compare the areas of the original shape and the images. Make some conclusions.

Compare the areas of the original shape and the images. Make some conclusions.

You will learn more about this on page 310.

Shape, Space and Measures

B **You will need** some copies of this shape on squared paper.

Use centre of enlargement O.

Enlarge shape A by scale factor 2. Label the image B.
Now enlarge shape B by scale factor 3.
Label the image C.
What single enlargement would map shape A onto shape C?

What if the centre of enlargement was P?
What if the scale factors were 2 and 4? or 3 and 4? or $\frac{1}{2}$ and 2? or $\frac{1}{2}$ and 4? or $\frac{1}{2}$ and $\frac{1}{4}$?

Make a general statement about two successive enlargements with scale factors k_1 and k_2. What single enlargement are they equivalent to?

C **You will need** a photocopier.

Draw a simple shape on a sheet of paper.
Use the 'enlargement' facility on your photocopier to enlarge your shape to 65%.
Compare the shape and its enlargement.
Measure lengths and angles.

When a shape is **enlarged**

- the **object** and its **image** are **similar**
- the ratio of any two **corresponding line segments** is equal to the scale factor, k

  See similar shapes on page 276.

- the scale factor for **inverse enlargement** is $\frac{1}{k}$.
- an enlargement with scale factor k_1 followed by an enlargement with scale k_2 is equivalent to a single enlargement with scale factor k_1, k_2.
  [$E(k_1)$ followed by $E(k_2) = E(k_1 k_2)$].

Discussion

Camellia enlarged this shape by a scale factor of $\frac{1}{3}$.
She said 'To enlarge this shape I use multiplication.'

What did she mean by this? **Discuss.**
If you are asked to enlarge a shape by a scale factor of k, how does this relate to multiplication? **Discuss.**

Luke said Camellia should have said 'To reduce this shape I use division.'
Is he right? **Discuss.**

Once Camellia had drawn the enlargement, she threw out the original.
She then decided she wanted the original shape back again.
What scale factor should she use to enlarge her image to get the original shape? **Discuss.**
What should she **multiply** each length on the image by?

We use the fact that the **scale factor** is equal to the ratio of corresponding lengths to **solve problems**.

Worked Example

P′ is an enlargement of P.

a What is the scale factor?

b Find the length of x.

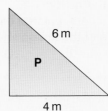

6 m

P

4 m

x

P′

3 m

Not to scale

Answer

a $k = \dfrac{\text{length on image}}{\text{corresponding length on original}}$

$= \dfrac{3}{4}$

b $\dfrac{\text{length on image}}{\text{corresponding length on original}} = k$

$\dfrac{x}{6} = \dfrac{3}{4}$

$x = 4\frac{1}{2}$ **or 4·5 m**   multiplying both sides by 6

$4 \to 3 = 6 \to x$ also $4 \to 6 = 3 \to x$

$4 \xrightarrow{\times \frac{3}{4}} 3$ $4 \xrightarrow{\times \frac{6}{4}} 6$

$4 \underset{\div 4}{\searrow} \overset{}{\underset{1}{\nearrow}} {\times 3}$ $4 \underset{\div 4}{\searrow} \overset{}{\underset{1}{\nearrow}} {\times 6}$

so $6 \to x = 6 \times \frac{3}{4} = 4\frac{1}{2}$ so $3 \to x = 3 \times \frac{6}{4} = 4\frac{1}{2}$

Exercise 2

1 a Each of the purple shapes has been enlarged to a red shape. What is the scale factor for each of these enlargements?

b Each of the red shapes has been enlarged to a purple shape. What is the scale factor for each of these enlargements?

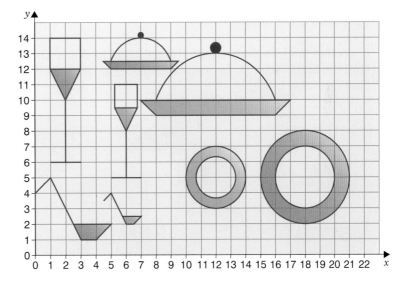

2 Use a copy of this. Enlarge each shape, centre P, by the given scale factor. Give the coordinates of the vertices of the image shapes.

a

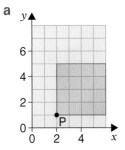

Scale factor 1·5

b

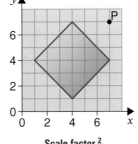

Scale factor $\frac{2}{3}$

∗c

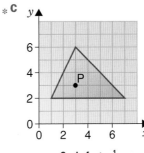

Scale factor $\frac{1}{2}$

Shape, Space and Measures

3 For each enlargement you did in question **2**, are the image and original shapes similar? Justify your answer.

4 Draw each of the following rectangles on axes.
Using (2, 3) as the centre of enlargement, enlarge each by the given scale factor.
Give the coordinates of the vertices of the image shapes.

a	A (2, 3),	B (5, 3),	C (5, ⁻3),	D (2, ⁻3)	scale factor $\frac{1}{3}$
b	E (⁻4, 5),	F (4, 5),	G (4, 1),	H (⁻4, 1)	scale factor $1\frac{1}{2}$
c	I (2, 0),	J (2, ⁻3),	K (⁻4, ⁻3),	L (⁻4, 0)	scale factor $\frac{2}{3}$
d	Q (0, 1),	R (0, 7),	S (6, 7),	T (6, 1)	scale factor $\frac{1}{2}$
e	M (7, 3),	N (7, ⁻2),	O (⁻3, ⁻2),	P (⁻3, 3)	scale factor 1·6
f	U (6, 3),	V (6, ⁻1),	W (⁻2, ⁻1),	X (⁻2, 3)	scale factor $\frac{1}{4}$

5 For each of the enlargements in question **4**, write down the scale factor for the inverse enlargement.

6 A map is an example of an enlargement by a fractional scale factor.
Give two other examples of everyday enlargements by a fractional scale factor.

7 A(2, 2), B(12, 6), C(12, 2) is enlarged to A′(1, 1), B′(6, 3), C′(6, 1).
 a What is the scale factor for this enlargement?
 *__b__ Write down the coordinates of the centre of enlargement.

8 A map, which is 24 cm wide, is reduced to $\frac{3}{4}$ of the original size.
How wide is the reduced map?

9 In each of the following, shape P is enlarged to shape Q.
Find the scale factor, k, for each enlargement.
Find the value of x.

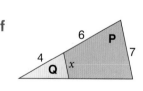

P is the large triangle.
Q is the small triangle.

10 The design on a hardcover book is to be used on the paperback version. To do this, the design is reduced to $\frac{3}{5}$ of its original size.
How high will the design be on the paperback cover if it is 18 cm high on the hardcover?

11 This diagram is reduced on a photocopier to $\frac{2}{3}$ of its original size.
If the height of the original diagram is 156 mm, how high will the reduced diagram be?

12 Here are four pictures, A, B, C and D. They are not to scale.   **[SATs Paper 2 Level 7]**

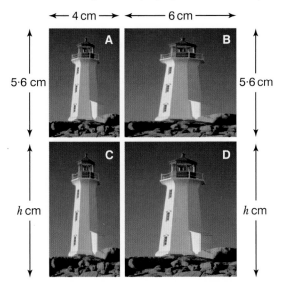

a Picture A can be stretched horizontally to make picture B.
Show that the horizontal factor of enlargement is 1·5.

b Picture A can be stretched vertically to make picture C.
The vertical factor of enlargement is **1·25**.
What is the height, h, of picture C?

c Show that pictures A and D are **not** mathematically similar.

d Picture E (not shown) **is** mathematically similar to picture A.
The width of picture E is **3 cm**.
What is the height of picture E?

13 Kent enlarged a diagram by scale factor 2 for a science project.
He decided it still wasn't large enough, so he enlarged his
enlargement by scale factor 1.5.

a What single scale factor could he have used to get the final
diagram from the original?

∗b Kent had drawn the original diagram to a scale of 1 : 10.
What would the new scale for his final enlarged diagram be?

∗**14** Priscilla enlarged a map using 85% on her photocopier.

a On the original map, Langs Road was 8 cm long.
How long will Langs Road be on her image?

b Priscilla later enlarged the **image** map by 115%.
Will the new image map be the same size as the original? Explain.

c How long will Langs Road be on the new image map?

Review 1   Shape C has been enlarged to shape C′.

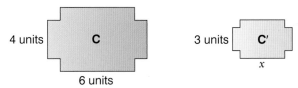

a What scale factor was used?

b What is the length of side x in shape C′?

c Are shapes C and C′ similar? Explain.

Review 2 On a set of axes, draw a triangle with vertices A($^-$4, 4), B($^-$1, 4), C($^-$2, 2).
a Using ($^-$4, 2) as the centre of enlargement, enlarge the triangle by a factor of 1·5. Give the coordinates of the vertices of the image.
b Using the same centre of enlargement, enlarge **the triangle you drew in a** by a factor of $\frac{1}{3}$. Give the coordinates of the vertices of the new image.
c What is the scale factor to enlarge the original triangle ABC to the triangle you drew in **c**?

Review 3 This diagram was made by enlarging a triangle by a scale factor of 2·5.
a Which is the original triangle?
b What is the scale factor for the inverse enlargement?
c If ∠AEB is 28°, how big is ∠ADC?
d If CD is 4·5 cm, how long is BE?

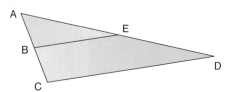

＊**Review 4** A scale on a map is 1 : 1000.
Sudi enlarges the map by scale factor $\frac{1}{2}$.

a What is the scale for the enlarged map?
On the original map, Sudi's street is 17·8 cm long.
b How long would it be on the enlarged map?
c What is the length of this street in real life?

Enlargement and area and volume

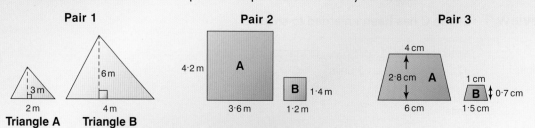

Investigation

Scale factor for area and volume

You could use ICT to help.

A You will need a copy of this table.

shapes	ratio of corresponding lengths	scale factor for length	ratio of corresponding areas	scale factor for area
Pair 1				
Pair 2				
Pair 3				

Shape B is an enlargement of shape A.
Calculate the areas for each pair of shapes below. Fill in your table.

Draw some more enlargements and fill in the table.
What is the relationship between, k, the scale factor for length and the scale factor for area?

B **You will need** a copy of this table.

shapes	ratio of corresponding lengths	scale factor for length	ratio of corresponding volumes	scale factor for volume
Pair 1				
Pair 2				
Pair 3				

3-D shape B is an enlargement of 3-D shape A.
Calculate the volumes for each pair of shapes below.
Fill in your table.

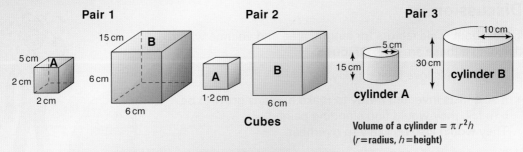

Pair 1

Pair 2

Pair 3

Cubes

Volume of a cylinder $= \pi r^2 h$
(r=radius, h=height)

Draw some more 3-D enlargements and fill in the table.

What is the relationship between, k, the scale factor for length and the scale factor for volume?

If k is the scale factor for length then
k^2 is the scale factor for **area**
k^3 is the scale factor for **volume**.

We can use this to solve problems.

Worked Example
Q is an enlargement of P.
The area of P is 12 cm².
Find the area of Q.

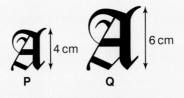

Answer
Since the scale factor for length is $\frac{6}{4}$ or $\frac{3}{2}$,
then the scale factor for area is $(\frac{3}{2})^2$ or $\frac{9}{4}$.

$\therefore$ area of Q $= \frac{9}{4} \times$ area of P
$= \frac{9}{4} \times 12$
$= \textbf{27 cm}^2$

Worked Example
a The surface area of the smaller cylinder is 126·4 cm².
   What is the surface area of the larger cylinder?
b Find the capacity of the smaller cylinder.

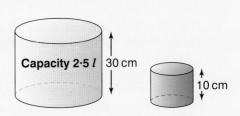

Capacity 2·5 l 30 cm

10 cm

Answer

a Since the scale factor for length is $\frac{30}{10}$ or 3, the scale factor for area is 3^2 or 9.

∴ surface area of larger cylinder = 9 × surface area of smaller cylinder

$$= 9 \times 126 \cdot 4 \text{ cm}^2$$
$$= \textbf{1137} \cdot \textbf{6 cm}^2$$

b Since the scale factor for length is $\frac{10}{30}$ or $\frac{1}{3}$, the scale factor for volume is $(\frac{1}{3})^3$ or $\frac{1}{27}$.

∴ capacity of smaller cylinder = $\frac{1}{27}$ × capacity of larger

$$= \frac{1}{27} \times 2 \cdot 5 \; \ell$$
$$= \textbf{93 m}\ell \text{ (2 s.f.)}$$

Discussion

Think about scale factors for length, area and volume when deciding if these are true. **Discuss**.

● Tiny animals lose heat quickly.

● If scientists invented a way to make humans grow to be 5 m tall, they would find standing painful.
They would also be likely to overheat.

● If a map with scale 1 : 4 is enlarged by scale factor 3, the scale on the map becomes 3 : 4.
A park which covers an area of 12 cm² on the original map will cover an area of 108 cm² on the enlarged map.

Exercise 3

1 a Enlarge this unit cube by scale factor 2.
b Find the surface area and volume of the unit cube and the enlarged cube.
c What is the scale factor for the surface area?
How is this related to the scale factor for length?
d What is the scale factor for volume?
How is this related to the scale factor for length?

cube
1 cm

2 a Find the surface area and volume of this cuboid.
b Find the surface area and volume of the cuboid if
i the length is doubled
ii both the length and width are doubled
iii the length, width and height are all doubled.
c Write a summary of the relationships between the surface area and volume of the original cuboid and the surface area and volume of each new cuboid in **bi**, **ii** and **iii**.

3 **The diagrams are *not* drawn to scale.**
These shapes are used by a manufacturer of pet food in advertising its products. One is an enlargement of the other.
a What is the scale factor of the enlargement?
b What is the scale factor for area?
c The smaller picture covers an area of 20 cm².
What area does the larger picture cover?

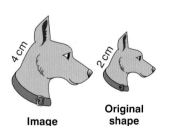

Image   Original shape

4 Cottage Craft Cooperative make hand-knotted rugs in a variety of sizes. All of the different
sizes are mathematically similar. All sell for the same price per square metre.
A rug of length 2 m and width 1·5 m sells for £72.
One of the rugs is 1·5 m long.
  a How wide will this rug be?
  b What will it sell for?

5

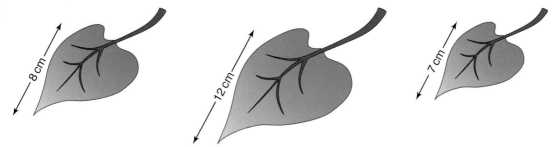

These leaves are similar. By placing a transparent grid over the smallest leaf, Adrian
estimated its surface area to be 20 cm².
Estimate the area of
  a the leaf on the left
  b the leaf in the middle.

6 One of these vases is an enlargement of the other.
The small vase holds 0·8 ℓ.
How much water does the large vase hold?

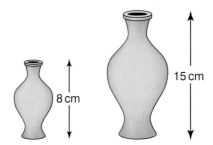

7 Two troughs on Michael's farm are mathematically similar. The larger is 1·6 m long and
can hold 150 litres of water. The smaller is 1 m long. What is the maximum amount of
water, to the nearest litre, it can hold?

8 **a** Show that triangles ABC and ADE are similar.
   b What will the ratio of their areas be?

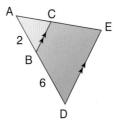

8 Two spheres, one of radius 2 cm and the other of radius 4 cm are dipped in gold. How
many times as much gold is needed for the larger sphere?

9 A slab is the shape shown.
Another slab is eight times heavier than this one.
If one of the slabs is an enlargement of the other,
find the dimensions of the heavier slab.

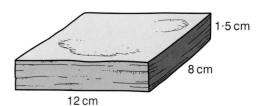

10 Eitje has a map of her town with a scale of 1 : 20 000.
Romy has another map of the same town with a scale of 1 : 10 000.
One map is an enlargement of the other.

   a Jennisie street is 14·6 cm long on Eitje's map.
How long is Jennisie street on Romy's map?

   b Eitje's map is on an A4 sheet of paper.
Show that Romy's map will be on a sheet four times as large.

*11

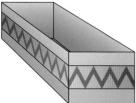

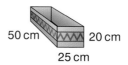

'Garden Magic' make planter boxes. They make three sizes. Each size has dimensions twice as large as the next size down. The planter boxes each have a frieze pattern painted right around the outside. The smallest size box is 50 cm by 25 cm by 20 cm. Each planter box is lined (base and sides) with polythene. The wood for the boxes costs £25 per square metre. The frieze costs £10 per metre to have painted. The polythene costs £10 per square metre. Labour costs of £10 per box are the same for all sizes.
What should 'Garden Magic' charge for each size to make a profit of 20% on each box?

*12 Two square-based pyramids of the same base length are made of the same material. One has a mass twice as much as the other. How many times greater is the height of the taller pyramid?

Review 1   Box A has been enlarged to Box B.

   a What is the scale factor of the enlargement?
   b What is the scale factor for area?
   c If the area of the top face of Box A is 5 cm^2, what is the area of the top face of Box B?
   d What is the scale factor for volume?
   e If the volume of Box B is 114 cm^3, what is the volume of Box A?
   f If the surface area of Box A is 22 cm^2, what is the surface area of Box B?

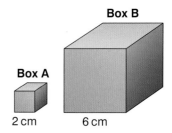

Review 2   Joanne made a three-tier wedding cake for her brother's wedding. The two bottom tiers were enlargements of the top tier.

If the middle-tier cake weighed 2 kg find the mass of the whole wedding cake. Give the answer to the nearest kg.

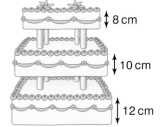

Review 3   Sara's soup comes in two sizes. The cans are similar shapes.

   a The large size holds 1000 m*l* of soup. How much does the small can hold?
   b The small can has a total surface area of 210 cm^2. What is the surface area of the large can?

Coordinates and lines

We can use similar triangles to find the **coordinates of a point that divides a line in a given ratio**.

Discussion

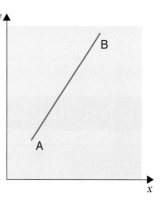

- I want to divide the line joining A and B in the ratio 1 : 2.
  The line will be divided into two lengths.
  What fraction of AB will each part be? **Discuss**.

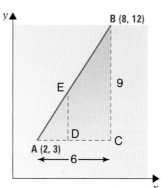

- Let E be the point that divides the line in the ratio 1 : 2.
  The coordinates of A and B are given.
  How could you use the similar triangles ABC and AED to find the coordinates of E? **Discuss**.

Worked Example

Find the coordinates of the point that divides the line joining
M($^-$2, 8) and N(6, $^-$8) in the ratio 1 : 3.

Answer

We must divide the horizontal distance and the vertical distance
between the points in the ratio 1 : 3.
We can do this using similar triangles.
Dividing the vertical side of the triangle in the ratio 1 : 3 gives 4 : 12.

Dividing the horizontal side of the triangle in the ratio 1 : 3 gives
2 : 6.
Because we are dividing MN in the ratio 1 : 3, the smaller part is
closer to M.

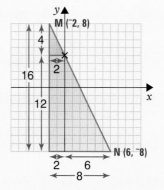

first   first

The x-coordinate is 0. $^-2 + 2 = 0$
The y-coordinate is 4. $8 - 4 = 4$
Coordinates of the point are **(0, 4)**.

Other ways of
working it out are:

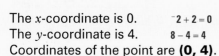

or

$$\begin{pmatrix} ^-2 \\ 8 \end{pmatrix} + \tfrac{1}{4}\begin{pmatrix} 8 \\ ^-16 \end{pmatrix} = \begin{pmatrix} 0 \\ 4 \end{pmatrix}$$

Shape, Space and Measures

Exercise 4

1 Find the coordinates of the point that divides the line joining P and Q in the given ratio.
   a P($^-$2, $^-$5)   Q(1, 1) in the ratio 1 : 2   b P($^-$2, 4)   Q(3, $^-$1) in the ratio 2 : 3
   c P($^-$6, $^-$5)   Q(12, 4) in the ratio 4 : 5   d P($^-$2, 8),   Q(5, $^-$6) in the ratio 3 : 4

*2 Derive the formula for finding the coordinates of the point which divides AB in the ratio l:m
    where A has coordinates (x_1, y_1) and B has coordinates (x_2, y_2).

Review   Find the coordinates of the point that divides the line joining R($^-$3, 2) and T(4, $^-$5) in
the ratio 2 : 5.

Length of line joining two points

Discussion

How could you use the triangle shown to find the
length of PQ? **Discuss**.

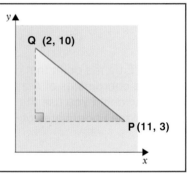

The **length of a line joining two points** P and Q can be calculated by finding the length of
the hypotenuse of a right-angled triangle.

Example   To find the length of the line joining A($^-$3, $^-$4) and
          B(2, 3)
          – plot the points on a set of axes
          – draw a right-angled triangle with the line joining
            the points as the hypotenuse
          – use Pythagoras's theorem to calculate the length
            of the hypotenuse.
          $$AB^2 = 7^2 + 5^2$$
          $$= 49 + 25$$
          $$= 74$$
          $$AB = \sqrt{74}$$
          $$= 8.6 \text{ (1 d.p.)}$$

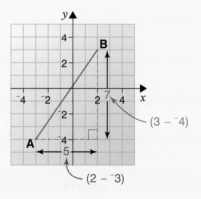

Worked Example
Flight BA852 is 10 km South and 8 km East of an airport.
Flight BA 701 is 4 km North and 5 km West of the same
airport. What is the shortest distance between the two
planes?

Answer
$$\text{distance}^2 = 14^2 + 13^2$$
$$= 196 + 169$$
$$= 365$$
$$\text{distance} = \sqrt{365}$$
$$= 19.1 \text{ km (1 d.p.)}$$

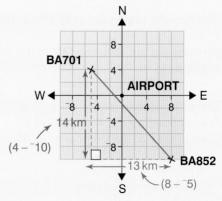

Discussion

What would the general formula be for the length, d, of a line joining two points (x_1, y_1) and (x_2, y_2)? **Discuss**.

Length of line joining two points (x_1, y_1) and (x_2, y_2) is $d = \sqrt{(x_2 - x_1)^2 + (y_2 - y_1)^2}$.

Exercise 5

1 Find the length of the line joining C and D.
  a C(2, 0) and D(5, 4)   **b** C(3, 0) and D(6, 4)   **c** C($^-$1, $^-$1) and D(0, 0)
  d C($^-$3, $^-$4) and D($^-$3, $^-$16)   **e** C($^-$6, 2) and D($^-$3, $^-$2)   **f** C(13, 8) and D(13, 10)
  g C($^-$4, $^-$4) and D($^-$6, $^-$5)   **h** C(5, $^-$3) and D($^-$2, $^-$4)

2 How far is the point ($^-$7, 13) from the origin?

3 Show that triangle MNP is isosceles if M, N and P have coordinates M($^-$1, 4), N($^-$4, $^-$2) and P(2, $^-$2).
  Which two sides are equal?

4 A yacht capsized 4 km West and 6 km North of a boat rescue station. The boat is searching for the yacht's survivors. It radios back that it is 1 km East and 4 km North of the rescue station. How far is the boat from where the yacht capsized?

5 The coordinates of the vertices of triangle STR are S($^-$4, 2), T(2, 4) and R($^-$2, 0).
  a Draw the triangle and calculate the length of the longest side.
  b Calculate the length of the line joining R to the mid-point of ST.

6 Calculate the length of the shorter diagonal of a parallelogram CDEF with coordinates C(2, 2), D(6, 3), E(7, 8) and F(3, 6).

7 Calculate the perimeter of a rectangle which has vertices at (3, 1), (9, $^-$3), (7, $^-$6) and (1, $^-$2).

8 The coordinates of point P are (5, 4).
  The x-coordinate of point Q is $^-$7.
  Line PQ is 15 units long.
  Find the coordinates of the mid-point of PQ.

Review 1   A quadrilateral has vertices A($^-$1, 3), B(3, 3), C(3, $^-$2) and D($^-$4, $^-$2).
a Calculate the length of side AD. Give your answer to 2 d.p.
b Calculate the length of the diagonal AC. Give your answer to 3 s.f.

Review 2   A yacht leaves its mooring and travels 12 km due North, then 10 km due West and finally another 4 km due South. How far does it have to travel in a straight line back to its mooring?

Summary of key points

 We can transform shapes using a **combination of reflection, rotation and translation**.

The image is always **congruent** to the original shape.

Some combinations are commutative.

Example   A has been translated 3 units left and 1 unit down, then reflected in the *x*-axis to get B. These are not commutative because doing them in a different order gives a different result.

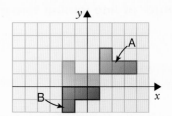

 Under enlargement, to find the distances of the image points from the centre of enlargement, multiply the distances of each point on the object from the centre of enlargement by the scale factor.

Example   ABC is enlarged to A'B'C'. The scale factor is $\frac{1}{2}$.

$$OA' = \frac{1}{2} \times OA \qquad OB' = \frac{1}{2} \times OB \qquad OC' = \frac{1}{2} \times OC$$

If the scale factor is greater than 1, the image is bigger than the original.

If the scale factor is less than 1, the image is smaller than the original.

When a shape is enlarged

- the object and its image are similar
- the ratio of any two corresponding line segments equals the scale factor k
- the scale factor for the inverse enlargement is $\frac{1}{k}$.

- Enlargement scale factor k_1, followed by enlargement scale factor
  k_1 = enlargement scale factor $k_1\,k_2$.

 We can use the scale factor of an enlargement to find unknown lengths.

Example $\dfrac{x}{5} = \dfrac{3}{2}$ **or**

$x = \mathbf{7 \cdot 5}\ \mathbf{m}$

so $x = 5 \times \dfrac{12}{8}$
$= \mathbf{7 \cdot 5}\ \mathbf{m}$

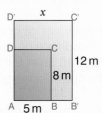

D If k is the scale factor for length, then

k^2 is the **scale factor for area**

k^3 is the **scale factor for volume**.

Example   XYZ is enlarged to X'Y'Z'.

$k = \frac{20}{10} = 2$

The scale factor for area is $k^2 = 4$.

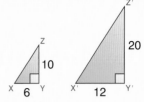
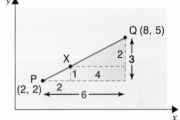

Area $\triangle$XYZ = 30

Area $\triangle$X'Y'Z' = 120 $\times$ **4**

 E We can use similar triangles to find the **coordinates of a point that divides a line in a given ratio**.

Example   X divides the line PQ in the ratio 1 : 2.

Divide the **6** in the ratio 1 : 2 to get **2** : 4.

Divide the **3** in the ratio 1 : 2 to get **1** : 2.

From the diagram it can be seen that X has coordinates (2 + **2**, 2 + **1**) or (4, 3)

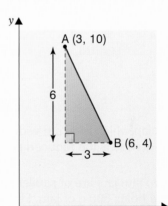

 F The **length of a line joining two points** (x_1, y_1) and (x_2, y_2) can be found using Pythagoras' theorem.

$d = \sqrt{(x_2 - x_1)^2 + (y_2 - y_1)^2}$

Example $AB^2 = 6^2 + 3^2$

$= 36 + 9$

$AB = \sqrt{45}$

$= 6\cdot7$ (1 d.p.)

Test yourself

1 These are some transformations.

A is a reflection in the x-axis.

B is a rotation 90°, centre (0, 0).

C is a rotation 180°, centre (0, 0).

I is the identity transformation.

 'Commutative' means you can do the transformations in any order.

Use this shape.

Are these pairs of transformations commutative?

a A and B   **b** B and C   **c** A and C   **d** B and I.

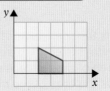

2 Draw the triangle ABC where A is the point ($^-$2, 0), B is (0, 4) and C is ($^-$2, 4).
  a Enlarge it by a scale factor 1·5 with centre of enlargement ($^-$2, $^-$2).
     Write down the coordinates of the enlarged triangle.
  b Enlarge the original triangle by a scale factor $\frac{1}{2}$ with centre of enlargement (0, 0).
     Write down the coordinates of the enlarged triangle.

3 What is the scale factor for the inverse enlargement in question **3a**?

4 The large shape is enlarged to the small shape.

  a What is the scale factor of this enlargement?
  b If the small shape was enlarged to the large
     shape what would the scale factor for this
     enlargement be?
  c Are the shapes similar? Explain.

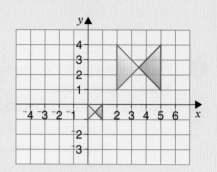

5 In each of these, M has been enlarged to N.
  Find the scale factor for each enlargement.
  Find the value of x.
  a

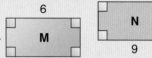

  b

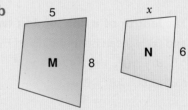

6 A plan which is 60 cm wide is reduced by a scale factor of $\frac{2}{3}$.
  How wide is the reduced plan?

7 Two jam jars are of similar shape. When full, the larger
  one holds 500 g of jam. Find the maximum amount of jam
  in the smaller jar given the heights of the jars are 8 cm
  and 12 cm. Round your answer to the nearest 100 g.

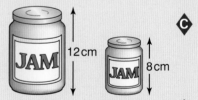

8 Two cubes have edges of 2 cm and 5 cm. What is the ratio of the total surface
  areas of these cubes? Give your answer in the form $a : b$.

9 Cuboid A has been enlarged to cuboid B.

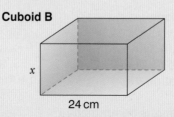

  a What is the scale factor for length in this enlargement?
  b Calculate the length marked x.
  c The area of the red face is 96 cm^2.
     What is the area of the blue face?
  d The volume of A is 900 cm^3. What is the volume of B?

*10 Ellie has a map of her district with a scale of 1 : 25 000.
Josh enlarges the map so it has a scale of 1 : 10 000.
  a A reserve on Ellie's map is 12 cm long.
     How long is it on Josh's map?
  b The reserve on Ellie's map has an area of 112·5 cm^2.
     What is its area on Josh's map?

11 Find the coordinates of the point that divides the line joining P and Q in the ratio 2 : 3.
    P($^-$2, $^-$7)   Q(3, 3)

12 Find the length of the line joining A and B.
  a A(4, 2) and B(8, 5)   b A($^-$4, 3) and B($^-$5, $^-$7)

13 a Calculate the length of the sides of a parallelogram with vertices at
     A($^-$4, $^-$2), B(2, 0), C(5, 3), D($^-$1, 1).
  b Find the lengths of its diagonals.

13 Measures, Perimeter, Area and Volume

You need to know

✓ measures   page 235
  – metric conversions
  – metric and imperial equivalents
✓ bearings   page 235
✓ perimeter, area and volume   page 235

········ **Key vocabulary** ························

> **adjacent,   angle of depression,   angle of elevation,   bearing,
> cosine (cos),   cubic centimetre,   cubic metre,   cubic millimetre,
> density,   hypotenuse,   miles per hour,   pressure,   sine (sin),
> speed,   surface area,   tangent (tan),   volume**

▶▶ Greener than green

scale 1 mm represents 5 m

Heather has a lawnmowing business.
She is asked to give a quote for mowing this circular park.
Heather doesn't know the formula for finding the area of a circle.
She draws this scale drawing of the park.

Draw a circle on a loose piece of paper with a radius of 15 mm.
Cut it into 'slices'. Rearrange these 'slices' as shown.

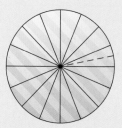
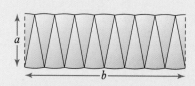

Use a piece of string to measure the circumference of the circle.

Which of *a* or *b* is equal to the radius of the circle?
Which is about the same as half the circumference?

How would Heather use what you have found out to find the area of the park?

If Heather charges £10 for every 1000 m² mowed, what should her quote be to the nearest £5?

Practical measurement

When we measure something we usually give the value to the nearest mark on the scale. No measurement is ever **exact**. A **measurement** could be **in error** by up to **half a unit** either side of this mark.

Discussion

A length given as 5 cm is presumed to be given to the nearest cm.
What is the shortest and longest it could be? **Discuss**.

What if the length was given as 5·0 cm? What accuracy would you presume it had been given to? **Discuss**.
What are the shortest and longest lengths it could be? **Discuss**.

What about the volume of a jewellery box given as 360 cm³?
What accuracy would you presume it had been given to?
What two numbers would go in the gaps if we were giving the range of values? **Discuss**.

$$\underline{\quad} \leqslant V < \underline{\quad}$$

The area of a park is given as 2·86 ha.
How would you write an inequality for the limits within which the area will lie? **Discuss**.

Worked Example
A time is given as 7·3 sec, to the nearest tenth of a second.
Suggest a possible range for this.

Answer

Any time, within the shaded region, would be given as 7·3 sec, to the nearest tenth of a second.
The shortest time could be 7·25 sec; the longest time could be up to 7·35 sec.
So the range could be given as **7·25 sec** $\leqslant t <$ **7·35 sec**.

Exercise 1

1 The length of a pencil is given as 162 mm, to the nearest mm. The range for the length of this pencil is:
    A $161 \leqslant l \leqslant 163$ B $161{\cdot}5 \leqslant l \leqslant 162{\cdot}5$ C $161{\cdot}5 \leqslant l < 162{\cdot}5$

2 The capacity of a freezer is given as 55 ℓ.
  a What accuracy would we presume this has been given to?
      A nearest $\frac{1}{2}$ ℓ B nearest litre C nearest 5 ℓ D nearest tenth of a litre
  b The capacity of the freezer could be given as:
      A $50\,\ell \leqslant c \leqslant 60\,\ell$ B $54{\cdot}5\,\ell \leqslant c < 55{\cdot}5\,\ell$ C $54\,\ell \leqslant c < 56\,\ell$

Shape, Space and Measures

3 Write the probable range these measurements lie between as an inequality.
  a The distance from Sam's place to school is 3 km.
  b A book is 123 mm long.
  c My two-year-old son weighs 17 kg.
  d It took 29·2 sec for Mem to run 200 m.
  e It is 28·3 km to the supermarket door.
  f A jug has a capacity of 1·25 ℓ
  g The floor area of a building is 7900 m^2.
  h My wardrobe is 1850 mm long.
  i Rai measured his foot to be 23·0 cm long.

4 The mass of a parcel is given as 1·4 kg, to the nearest tenth of a kilogram.
  a Find the upper limit of the mass of this parcel.
  b Find the lower limit.

5 A racing car was timed at 50·73 seconds, to the nearest hundredth of a second, for one circuit of a racing track.
  a What is the lower limit for the true time?
  b As an inequality, write down the range of possible values for the true time.

6 The mass of a box of chocolates is given as 250 g to the nearest 10 g.
  The mass of the chocolates must be at least
  A 240 g   **B** 260 g   **C** 255 g   **D** 245 g.

7 In an experiment, the temperature of a liquid was taken at 1-minute intervals.
  One of these measurements was given as 45·6 °C.
  Between what limits does the true temperature lie?

8 Sandra used an electronic balance to measure the mass of a chemical.
  Her measurement was 28·60 grams. As an inequality,
  write down the range of possible values for the mass of this chemical.

9 Deon measured the distance between two points as 46·8 cm.
  Write the true distance as 46·8 ± ... cm.

10 Beth measured the length and the width of her book.
  Her measurements, shown on the diagram, are accurate to
  the nearest mm.
  a What range must the area of the cover lie between?
  b Suggest a sensible answer for the area, given the degree of
  accuracy of the data.

11 Ellsie marked out a basketball court.
  a She makes it 84 feet long to the nearest foot.
  What is the shortest possible length of the court?
  b She makes her court 50 feet wide to the nearest
  foot.
  What is the shortest possible width?
  c Ellsie's coach makes her team run around the
  outside of the court to get fit.
  Use your answers to parts **a** and **b** to find how
  many times they should run around the court to be
  sure of running at least 4 miles.
  1 mile = 5280 feet.

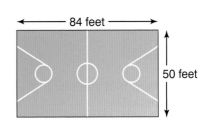

Review 1   Write the probable range these measurements lie between as an inequality.
a   The mast on Laura's boat is 2·2 m high.
b   The distance between the beds in a dormitory is 1·85 m.
c   The length of a bee's wing is 3 mm.
d   The volume of soft drink in a can is 330 mℓ.

Review 2   The doctor tells Sean he weighs 58 kg ± 500 g. Complete this sentence:
Sean weighs between _____ kg and _____ kg.

Review 3   Anya ran a 100 m sprint in 13·4 seconds to the nearest tenth of a second.
In the next race, she was timed with an electronic clock at 13·42 seconds to the nearest
hundredth of a second.
a   For Anya's first attempt write the possible time range as an inequality.
b   Write the possible time range for her second attempt as an inequality.
c   Anya said 'I ran faster in the first race.' Is Anya
        A definitely correct   **B** definitely wrong   **C** can't tell?
    Explain your choice.

Review 4   Nick is buying carpet for his bedroom and measures
the floor area he wants to cover. His measurements are shown on
the diagram.
a   What range must the floor area of the room lie between?
b   How many square metres of carpet should Nick buy, to the
    nearest hundredth of a square metre, to ensure he has enough
    carpet to cover the floor?

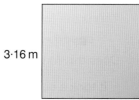

3·16 m

3·42 m

Compound measures

Discussion

Tina ran 100 m in 25 seconds.
James ran 200 m in 48 seconds.
How fast did Tina run? How fast did James run? **Discuss**.

In chemistry, Dale and Andrea both made salt solutions.
Dale used 10 g of salt to 50 mℓ of water; Andrea used 20 g of salt to 80 mℓ of water.

In each of the above, two measurements are made.
What two measurements are made to find speed?
What two measurements are made to make the salt solution?

These are called **compound** measures.

Discuss other compound measures.
Are the two quantities usually in different units?

What unit of speed might you use to measure these? **Discuss**.

    a train travelling from York to London
    an old lady walking to the shops
    a runner in a sprint race
    a deer running from a predator

Rates are a way of measuring how one quantity varies with another.

Examples   grams per cm^3, g/cm^3 newtons per cm^2, N/cm^2

g/cm^3 and N/cm^2 are called **compound measures** because the numerator and denominator have different units. .

'Per' means 'for every'. We usually use a '/' to mean 'per'.

Discussion

- A model train travels round a track at a constant rate of 0·5 m/s. How far would it travel in 5 seconds? 10 seconds? 20 seconds? What does constant rate mean? **Discuss**.
  Are the distance travelled and the time taken directly proportional? Justify your answer. **Discuss**.

Link to direct proportion, page 98.

  This is the formula for speed:

  $$\textbf{speed} = \frac{\textbf{distance}}{\textbf{time}}$$

  Does this formula only work when the speed is constant?

  > It took a car 2 hours to travel 180 km. The car made some stops and its speed varied. If we use the formula for speed,
  >
  > $$\text{speed} = \frac{180 \text{ km}}{2 \text{ hours}}$$
  > $$= 90 \text{ km/hour,}$$
  >
  > what have we worked out? **Discuss**.

- 'Yoghurt is denser than milk.'
  'Roast lamb is denser than a pork pie.' What do we mean by 'denser'? **Discuss**.

  Do you think cornflakes are denser than sugar?
  Imagine this box is filled to the top with sugar.
  Imagine an identical box filled to the top with cornflakes.

  Is the volume of both boxes the same?
  Which box is heavier, the box of sugar or the box of cornflakes?
  Is the mass of the empty boxes the same?
  What can you say about the mass of the cornflakes and the mass of the sugar in the boxes?
  What can you say about the ratios $\frac{\text{mass of cornflakes}}{\text{volume of cornflakes}}$, $\frac{\text{mass of sugar}}{\text{volume of sugar}}$? **Discuss**.

- A block of wood and a needle are pushed into a balloon with exactly the same amount of force.
  Why does the needle pop the balloon but the block of wood doesn't?
  Pressure is the force per unit area.
  $$\textbf{Pressure} = \frac{\textbf{Force}}{\textbf{Area}}$$
  How could you use this to explain why the balloon popped when the needle was pushed into it?

If one of the variables of the compound measure is directly proportional to the other, we can write a formula which connects the variables.

speed = $\frac{\text{distance travelled}}{\text{time taken}}$ $V = \frac{d}{t}$ **Usual units**
km/h or mph or m/s

Link to proportion, page 98.

Note If the speed varies, the same formula can be used to find the **average speed**.

average speed = $\frac{\text{total distance travelled}}{\text{total time taken}}$

density = $\frac{\text{mass of an object}}{\text{volume of object}}$ $d = \frac{m}{V}$ **Usual units**
g/cm^3 or kg/m^3

pressure = $\frac{\text{force on a surface}}{\text{surface area}}$ $P = \frac{F}{A}$ **Usual units**
newtons/metre^2 (N/m^2)

Worked Example

A bus travels the 14 km from Llangurig to Rhayader in 18 minutes.
Find the average speed of this bus.

Answer

We will give the average speed in km/h. We need the time in hours.

average speed = $\frac{\text{distance}}{\text{time}}$ $18 \text{ min} = \frac{18}{60}$ hours
      = $\frac{14}{0.3}$ = 0.3 hours
      = **47 km/h (to 2 s.f.)**

Worked Example

A train travels at an average speed of 108 km/h for 2 hours 13 min.
How far has the train travelled?

Answer

2 hours 13 min = $2\frac{13}{60}$ hours
       = 2.22 hours (3 s.f.)

It is better not to round until the final step of the answer.

Use $2\frac{13}{60}$ rather than 2.22 in the distance calculation.

   distance = average speed × time
       = $108 × 2\frac{13}{60}$
    = **239.4 km**   **Key** 108 × (2 + 13 ÷ 60) =

You could work out the brackets as a decimal first and store the answer in the calculator memory.

Note The fraction key a$^{b/c}$ could also be used to key $2\frac{13}{60}$.

Worked Example

The density of rock is 2·2 g/cm^3.
A large rock accidentally falls
into a pool of water with a square base of 5 m.
The rock has a mass of 580 kg.
How much does the water in the pool rise?

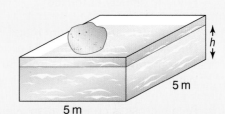

Answer

We need to find the volume of the rock.

  density = $\frac{\text{mass}}{\text{volume}}$

  so volume = $\frac{\text{mass}}{\text{density}}$

Density is given in g/cm^3 and mass in kg.

changing the subject of the formula

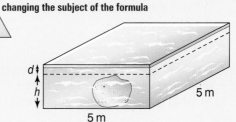

water has risen d cm

We must change the kg to g.

$$\text{volume} = \frac{580\,000}{2 \cdot 2}$$

$$= 263\,636 \cdot \dot{3}\dot{6}\ cm^3$$

580 kg = 580 × 1000 g
= 580 000 g

The rock will displace water of volume 263 636·$\dot{3}\dot{6}$ cm³.
The base of the pool is 5 m by 5 m or 500 cm by 500 cm.

$$\text{height water rises} \times 500 \times 500 = 263\,636 \cdot \dot{3}\dot{6}$$

$$\text{height water rises} = \frac{263\,636 \cdot \dot{3}\dot{6}}{500 \times 500}$$

$$= \textbf{1·1 cm (1 d.p.)}$$

The water will rise **1·1 cm**.

Exercise 2

1 Yvonne cycles a distance of 48 km in $2\frac{1}{2}$ hours. She has a 20 minute stop during this time.
Find Yvonne's average speed for the whole journey.

2 15 litres of water flows from a hose in 20 seconds.
Give this rate of flow in ℓ/sec.

3 A 30 cm³ piece of lead has a mass of 345 g.
What is the density of this lead?

4 A large safe is at rest on the floor of a bank.
The safe weighs 1200 N and the area of its base is 1·6 m².
What pressure does it put on the floor of the bank?

A weight is a force.

5 Ali ran the 800 m in 2 min 48 sec.
a What was Ali's average speed, in metres per second?
b How accurately do you think it is sensible to give your answer?

6 A piece of cork weighs 10 kg. Its volume is 0·04 m³.
Find the density of this cork.

7 The mass of a 1200 mℓ block of ice is 1104 grams.
Find the density of the ice.

8 Two models are cast from metal.
One model is made from 40 cm³ of brass. This model weighs 340 g.
The other model is made from 50 cm³ of iron. This model weighs 375 g.
Which is more dense, brass or iron?

9 An albatross is cruising at an average speed of 90 km/h.
What distance does the albatross cruise in
a 2 hours
b $\frac{3}{4}$ hour?

10 A runner sets out at midday to run to the next village, a distance of 12 km. She wants to
arrive at this village at 1330 hours.
At what average speed should she run?

11 The density of mercury is 13·6 g/cm³. How many mℓ of mercury weigh 73 grams?

12 Annabel ran a road race at an average speed of 10 km/h. She completed this race in 2 hours 15 minutes.
How long was this road race?

13 I have a weight of 750 N. The area of one of my feet is 0·013 m².
What pressure do I exert on the floor when standing?

14 The diagram shows the distance between my home, H, and two towns, A and B.
It also shows information about journey times.

[SATs Paper 1 Level 7]

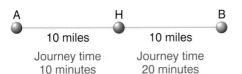

A   H   B

10 miles   10 miles

Journey time   Journey time
10 minutes   20 minutes

a What is the average speed of the journey from my home to **town A**?
b What is the average speed of the journey from my home to **town B**?
c I drive from town A to my home and then to town B.
The journey time is 30 minutes.
What is my average speed?
Show your working.

15 A train is travelling at 120 km/h.
How far does this train travel between 0720 hours and 0800 hours?

16 A statue is made from clay of density 1·4 g/cm³.
Another statue is carved from wood of density 0·8 g/cm³.

Both statues are weighed to find their masses.
The wooden one is found to have a mass of 4500 g.
The clay one has a mass of 7700 g.

Which statue has the greater volume?

17 The density of glass in an ornament is 2·5 g/cm³.
Michaela finds the volume of this ornament to be 45 cm³.
She then calculates its mass. What answer should she get?

18 Naismith's rule for mountain walkers is:
Allow 1 hour for every 3 miles you must walk.
Add $\frac{1}{2}$ hour for every 1000 feet you must climb.
Jasmine started a 17 mile walk at 0800 hours.
The path climbed 6000 feet from start to finish.
Jasmine wanted to work out about what time she would finish the walk.
If she allows $2\frac{1}{2}$ hours for stops along the way, about what time should she arrive?

19 The table given below shows the maximum speeds of some very fast animals. It also shows the distances they are able to run at these speeds.

Animal	Top speed in metres per second	Distance in metres
Cheetah	28	500 m
Racehorse	18	300 m
Antelope	15·5	6 000 m
Deer	12·5	32 000 m

Calculate how long each of these animals is able to run at maximum speed.
Round your answers sensibly.

20 A bus travels the 195 km from Cardiff to Exeter at an average speed of 75 km/h.
It then travels back to Cardiff at an average speed of 81 km/h.
If the bus spent half an hour at Exeter before returning,
how long did the total journey take?

21 The label on the outside of a bottle of turpentine is shown.
How much do the contents of this bottle weigh if the density of
the turpentine is 0·86 g/cm³?

22 A satellite passes over both the north and south poles,
and it travels **800 km above** the surface of the Earth.

[SATs Paper 2 Level 8]

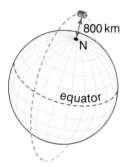

The satellite takes **100 minutes** to complete one orbit.
Assume the Earth is a sphere and that the diameter of the Earth is **12 800 km**.

Calculate the speed of the satellite, in **kilometres per hour**.
Show your working.

*23 Saturn takes 29·46 Earth years to orbit the Sun, at an average speed of 9·64 km/sec.
Assume Saturn's orbit is circular.
About how far from the Sun is Saturn?

*24 Before packaging silver necklaces they are dipped, one hundred at a time, into a liquid.
This coats them with a protective layer.
The mass of each necklace is 12·2 g.
The density of silver is 10·5 g/cm³.
The container they are dipped in is a cuboid and has a base of 3 cm by 2·5 cm.
How much will the liquid in the container rise?

Review 1

	Aberdeen					
Brighton	991	**Brighton**				
Bristol	846	253	**Bristol**			
Cardiff	877	307	74	**Cardiff**		
Dover	1025	135	320	375	**Dover**	
Exeter	967	291	135	195	451	**Exeter**
London	898	85	195	249	122	325

This distance chart gives the distances, in kilometres, between some places in Great Britain.
a How long would it take to travel from London to Exeter at an average speed of 100 km/h?
Give your answer in hours and minutes.
b Anna leaves Bristol at 9:20 a.m. and arrives at Dover at 12:50 p.m. What was Anna's
average speed?
c Steven left at 0900 hours and drove from Brighton to Bristol. He stopped for an hour then
drove to Exeter, arriving there at 1500 hours.
What was his average speed for the journey, including the hour that he stopped?
d Susan drove from Brighton at an average speed of 80 km/h for about 3 hours and
40 minutes. Where was Susan going to, Bristol or Exeter?

Review 2   The volume of a pane of glass is 1000 cm³. This pane weighs 2·4 kg.
Find the density of this glass in g/cm³.

Review 3   A circular manhole with diameter 600 mm is designed to withstand a force of
39 200 N. What pressure can it withstand?

Review 4   A piece of Cheddar cheese has a mass of 750 g.
A piece of Edam cheese has a mass of 720 g.

The density of the Cheddar cheese is 1·8 g/cm³ while that of the Edam cheese is 1·7 g/cm³.

Which piece of cheese has the greater volume?

Review 5   Joanne is mixing and pouring concrete.
Altogether she mixes and pours 0·4 m³.
a   What is the mass of 0·4 m³ of concrete if its density is 2500 kg/m³?
b   Joanne is pouring the concrete into boxing which is 3·13 m long by 75 cm wide.
How deep will the concrete be after she levels it?

We can **change between units** of compound measures.

Example   4 km/h = $\frac{4000 \text{ metres}}{3600 \text{ seconds}}$ ◄————— metres in 4 km
                                 ◄————— seconds in 1 hour
           = **1·1 m/s (1 d.p.)**

Worked Example
Convert 50 mph to metres per second.

Answer
50 mph stands for 50 miles for every hour.
We must convert 50 miles to metres and 1 hour to seconds.

50 miles ≈ 50 × 1·6 km
         = 80 km
         = 80 × 1000 m
         = 80 000 m

> See page 235 for metric
> to imperial conversions.

So 50 mph = $\frac{80\,000 \text{ m}}{3600 \text{ seconds}}$
            = **22 m/s (nearest m/s)**

Exercise 3

1   Change
    a 5 m/s to km/h   **b** 20 g/cm³ to kg/m³   **c** 50 km/h to m/s.

2   A cheetah can run 100 m in 5·4 seconds.
    A train takes 10 minutes to travel the 4 miles between stations.
    Which has the faster average speed,
    the train or the cheetah?

> Remember: 1
> mile ≈ 1600 m.

3   Jane's top running speed is 15 km/h.
    Jessie's top running speed is 10 m/s.
    Which top speed is faster?

4   Silver has a density of 10·5 g/cm³.
    Gold has a density of 19 300 kg/m³.
    Which is denser, gold or silver?

Review   Patient Smythe is given an IV drip with a flow rate of 15 mℓ/min. Patient Naimo is
given an IV drip with a flow rate of 1·2 ℓ/hour. Which patient's IV has the faster flow rate?

Area and perimeter of circles and arcs

Remember
Circumference of a circle = $2\pi r$ or πd

Area of a circle = πr^2

Part of the circumference of a circle is called an arc.
A sector of a circle is bounded by two radii and an arc.

an arc

a sector

Discussion

● What fraction of a complete circle is the dark purple section?
  What is the area of the shaded section?

What if the dark purple section is a quarter-circle?

Discuss how to find the area of each of the darker blue sectors shown below. Is the
area of each sector proportional to the angle between the radii bounding the sector?

Which arc is the arc AB? **Discuss**.

What fraction of the circumference is the arc APB? What is the length of
the arc APB?

Discuss how to find the length of the arcs APB shown below. Is the arc
length proportional to the angle between the radii that meet the ends of the arc?

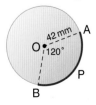

length of arc AB $= 2\pi r \times \frac{\theta}{360°}$

area of sector OAB $= \pi r^2 \times \frac{\theta}{360°}$

θ = angle between radii in degrees
r = length of radii

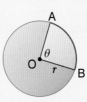

Worked Example

John made a clown's hat for his brother.
The diagram represents the pattern he used.
a Find the area of the pattern.
b Find the perimeter of the pattern.

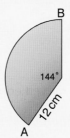

Answer

a Fraction of the whole circle $= \frac{144}{360}$
$= 0\cdot4$

Area of a pattern $= \pi r^2 \times \frac{144}{360}$
$= \pi \times 12^2 \times \frac{144}{360}$
$= \mathbf{181\ cm^2\ (3\ s.f.)}$

b Length of arc AB $= 2\pi r \times \frac{144}{360}$
$= 2 \times \pi \times 12 \times \frac{144}{360}$
$= \mathbf{30\cdot2\ cm\ (3\ s.f.)}$
Perimeter of pattern $= 30\cdot2 + 12 + 12$
$= \mathbf{54\cdot2\ cm\ (3\ s.f.)}$

Exercise 4

1 Calculate the perimeter of these shapes.
   All curves are semicircles or quarter-circles.

   a ←2·8 m→ 1·6 m

   b ←68 cm→ 50 cm

   c 5 m 5 m

   d ←4·6 cm→ 2·3 cm

   e 10 cm 10 cm 10 cm

   *f 3 cm 3 cm 4 cm 3 cm
   Each end is a semicircle

2 Find the area of each of the shapes in question **1**.

3 A clockface of diameter 16 cm is
   mounted on a rectangular board
   of dimensions 24 cm × 22 cm.
   Find the area of board not
   covered by the clockface.

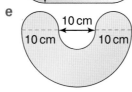

22 cm
←16 cm→
←24 cm→

Shape, Space and Measures

4 A car tyre of diameter 54 cm is marked at the bottom by a parking warden. The owner returns to the car and pushes it forward so that the mark is now at the top of the tyre. How far must the owner push the car?

5 Find the area of the sectors OPR.

a

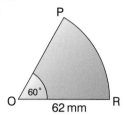

b

What proportion of a circle is OPR?

6 Find the length of the curved edge (arc) of each of the shapes in question **5**.

7 Christmas decorations are made from the pattern shown.
  a What area is this pattern?
  b Braid is sewn right around the outside of each decoration. How much braid is needed for 15 decorations?

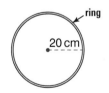

8 Karen is making a ballroom dancing skirt. Each panel of this skirt is part of a sector of a circle.
  a What fraction of a circle is this sector?
  b Find the length of the bottom of the skirt, AB.

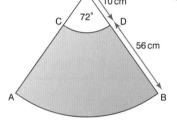

9 A path, 0·8 m wide, goes around the sides and back of a house and around a semicircular lawn in front of the house as shown. The house measures 12 m × 20 m.

Find the area of the path.

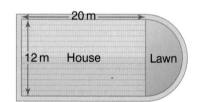

10 Pete has found a thin metal ring with radius 20 cm. He wants to bend this ring into a square.
What length will the sides of the square be?

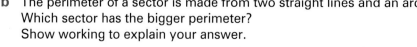

11 The diagram shows parts of two circles, sector A and sector B

[SATs Paper 2 Level 8]

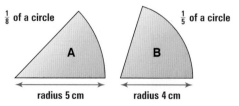

$\frac{1}{8}$ of a circle $\frac{1}{5}$ of a circle

A   B

radius 5 cm   radius 4 cm

  a Which sector has the **bigger area**?
  Show working to explain your answer.
  b The perimeter of a sector is made from two straight lines and an arc.
  Which sector has the bigger perimeter?
  Show working to explain your answer.

c A semicircle, of radius 4 cm, has the **same area** as a complete circle of radius r cm.

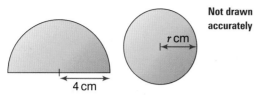

Not drawn
accurately

What is the radius of the complete circle?
Show your working.

12 Jenni made a party hat shaped like a hollow cone.
To make it, she cut a sector of a circle from card.
a Show that the arc length of the sector is 7π cm.
b Find the area of card used for the hat, leaving π in
your answer.

*13 Percy made a circle design on 1 cm squared paper.

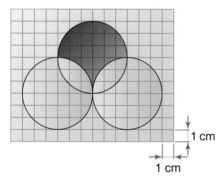

What is the area of the shaded section?

14 Here are three circles with areas as shown.
Parts of the circles have been used to make
this design.
What fraction of the design is shaded darker?

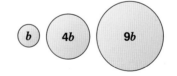

*15 What is the area of the smallest circle into which a 3 cm
by 5 cm rectangle will fit?

Review 1 A fan is the shape of a sector of a circle.
a Find the area of this fan.
b Lace is to be sewn along the curved edge of the fan.
What length of lace is needed?

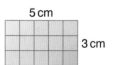

Review 2 A running rack is made in the shape below. The ends are semicircles with radius
50 m. Find the distance around the track.

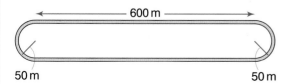

Shape, Space and Measures

Review 3

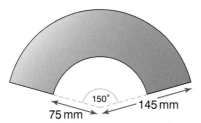

The shaded area represents the pattern piece for the neck facing of a jacket.

a How much material is used for this facing? Give your answer in cm².

b Find the perimeter of this facing. Give your answers in cm.

? Puzzle

1 The shape on the pin board has an area of two square units.
Cover this pin board with shapes of area two square units.
No two shapes may be the same.
The shape shown may or may not be one of the shapes
needed.

2 Hursad's office is very small.
It is 2·2 m by 1·75 m.
He has a desk which is 1·55 m by 85 cm.
Is it possible to move the desk, to the position
shown in pink without taking it out of the room?
He is not able to stand the desk on its end.

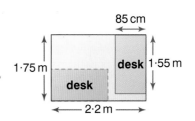

Investigation

Matthew's Pet Goat

Matthew has a pet goat.
His father tells him he can tether it so it can graze the part
of the garden shaded green. The goat must **not** be allowed
into the vegetable garden.
Matthew wants the goat to have the maximum grazing area.
He **investigates** the best place to put the stake.
Some possibilities are shown.

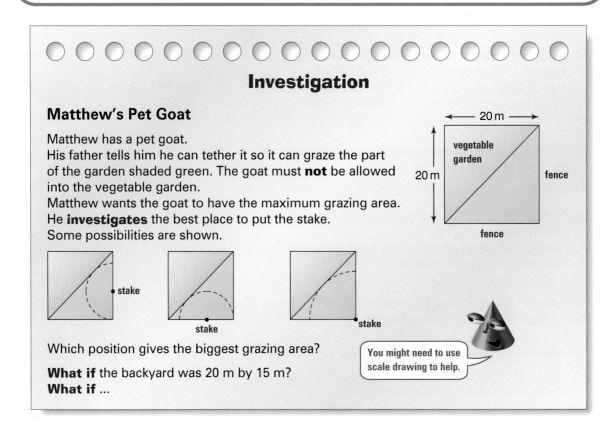

Which position gives the biggest grazing area?

What if the backyard was 20 m by 15 m?
What if ...

You might need to use
scale drawing to help.

Surface area and volume

Discussion

- What shape is the base of this cylinder? **Discuss**.
  Is a cylinder a prism?
  How could you use this to find a formula for the
  volume of a cylinder. **Discuss**.

- The net for a cylinder could be drawn like this.
  What length is the width of the rectangle? **Discuss**.
  Write a formula for the surface area of a cylinder.
  Discuss.

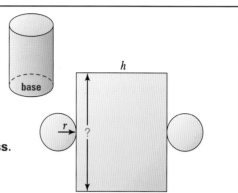

Remember
A **prism** has a uniform cross-section throughout its length. Example: a cuboid.

Volume of a prism	**Volume of a pyramid**
= area of base × height	**= $\frac{1}{3}$ area of base × height**

Volume of a cylinder = $\pi r^2 h$
Surface area of a cylinder = $2\pi rh + 2\pi r^2$

curved surface area   surface area of 2 circular faces

> See page 235 for the
> formulae for other surface
> areas and volumes.

Worked Example
In science, Tamryn was given a beaker with some copper sulphate in it.
She was asked to find the approximate volume of copper sulphate in the beaker.
She measured the diameter of the beaker as 7·1 cm.
The copper sulphate was 4·6 cm deep.

Answer
Volume of copper sulphate $= \pi r^2 h$
$$= \pi \times (\tfrac{7 \cdot 1}{2})^2 \times 4 \cdot 6$$
$$= 182 \text{ cm}^3 \text{ (nearest cm}^3\text{)}$$
$$= \textbf{182 m}\ell$$

> **Remember:**
> 1 mℓ = 1 cm^3

Worked Example
A table top is made from wood with the cross-section shown.
It is a rectangle with a semicircle on the side.
The table top is 3.8 cm deep.
What volume of wood is needed to make 15 of these table tops?

0·86 m

0·42 m

Answer
Volume of wood = area of cross-section × depth

Area of cross-section = area of rectangle + area of semicircle
$$= 0 \cdot 86 \times 0 \cdot 42 + \tfrac{1}{2} \times \pi \times 0 \cdot 43^2$$
$$= 0 \cdot 65164024 \text{ m}^2$$

**Do not round intermediate values.
Store in a calculator memory.**

Volume $= 0 \cdot 65164024 \times 0 \cdot 038$ 3·8 cm = 0·038 m
$$= 0 \cdot 024762329 \text{ m}^3$$
Volume for 15 tables $= 0 \cdot 024762329 \times 15$
$$= \textbf{0·37 m}^3 \textbf{ (2 d.p.)}$$

> When using a formula make
> sure the units are the same.

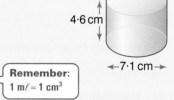

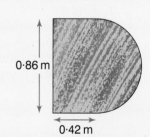

Shape, Space and Measures

Exercise 5   **Use the calculator value for π.**

1 A cylindrical tin has radius 3 cm and height 10 cm. How much metal is used to manufacture this tin?

2 Alan is going to cover a hot-water cylinder with insulating cloth. Alan measured the diameter of the cylinder as 52 cm and the height as 1·2 m. Alan then calculated the curved surface area. What answer should he get?

3 A concrete tank has diameter 2·7 m and stands 3 m high. What is the total surface area of this tank?

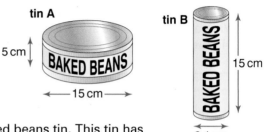

4 These tins are filled with beans.
   a Which tin holds more beans?
   b Which tin needs more sheet metal?

5 A label is to cover the curved surface of a baked beans tin. This tin has a diameter of 74 mm and a height of 115 mm. What are the dimensions of the label if it is to have an overlap of 5 mm?

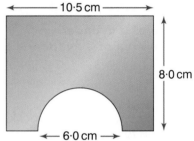

6 A 'play-tunnel' for a mouse has ends shaped as shown and is 35 cm long. Find the volume of solid plastic needed to make this.

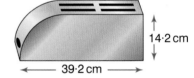

7 The cross-section of a toaster is shown.
   It is a rectangle with a quadrant (quarter-circle) on the end.
   The toaster is 13·5 cm wide.
   Find the volume of the toaster.

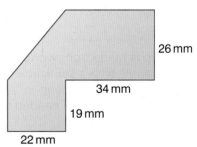

8 This diagram consists of a trapezium and a rectangle.
   a Find the total area of this shape.
   b Find the perimeter. **Hint** You will need to use Pythagoras' theorem.
   c This shape is the end of a prism of wood 52 mm long. What is the volume of this wood?

9 A partly filled lemonade bottle, of diameter 8 cm, contains lemonade to a depth of 14 cm. This lemonade is poured into ice-cube moulds each of which is a cube of side 3 cm. How many of these ice-cube moulds can be completely filled?

10 A cylindrical bucket of radius 30 cm and height 40 cm is used to fill the trough shown with water.
   Find the least number of times the bucket will need to be filled and emptied into the trough.

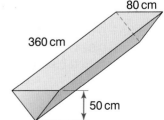

11 The diagram shows the cross-section of a 35 m long pipe.
  a Find the volume of the inside of the pipe.
  b Find the volume of PVC needed to make the pipe.
  c The water in the pipe flows at 0·02 m³/s. How long does it take water to flow from one end of the pipe to the other?

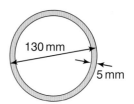

130 mm

5 mm

12 The formula for the volume, V, of a square-based pyramid is   [SATs Paper 2 Level 8]
  $$V = \tfrac{1}{3}b^2h.$$

b is the base length.
h is the perpendicular height.

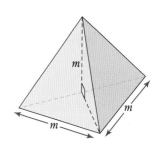

  a A square-based pyramid has base length 5 cm and perpendicular height 6 cm.
     What is its volume?
  b A different square-based pyramid has base length 4 cm.
     Its volume is 48 cm³.
     What is the perpendicular height?
  c The volume of another square-based pyramid is 25 cm³.
     Its perpendicular height is 12 cm.
     What is its base length?
     Show your working.
  d The diagram shows a triangular-based pyramid.
     The base is an isosceles, right-angled triangle.
     The perpendicular height is m.
     Write a formula, in terms of m, for the volume, V, of the pyramid.

13 400 ml of tomato sauce is in a cylindrical bottle of internal diameter 65 mm. How deep is the sauce?

14 A cylinder has a radius of 2·5 cm.   [SATs Paper 2 Level 8]
  The volume of the cylinder, in cm³, is 4·5π.

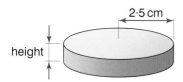

2·5 cm

height

What is the height of the cylinder?
Show your working.

∗ 15 Jam jars of diameter 6·4 cm and height 10·5 cm are used as containers in which to freeze some left-over soup. 2 cm 'head-room' must be left to allow for expansion on freezing. How many of these jars will be needed to freeze 2 ℓ of soup?

∗ 16 The outlet from the kitchen sink was blocked. Kathy decided to find out how far down the pipe the blockage was by pouring in water until no more would go down the plughole. She was able to pour in 1200 ml of water. She then put a tape measure around the pipe and found its circumference to be 13·5 cm. From these two measurements calculate how far down the blockage was.

∗ 17 a Find the height of the cylinder which has a curved surface area of 340 cm² and radius of 18 cm.
  b Find the diameter of the cylinder which has a curved surface area of 8 m² and height of 1·4 m.

339

*18 'All for Hair' have two sizes of cannisters for hair gel.
The large cannister has dimensions exactly twice that
of the smaller canister.
At Christmas each of these cannisters is packaged in
rectangular boxes.
The cannisters just fit snuggly into their respective
boxes.

a What percentage of the space in the smaller box is
air?

b What percentage of the space in the larger box is
air?

c What is the relationship between your answers to
a and b.

14 cm

7 cm

2·5 cm

5 cm

Review 1
Jude peels the label off a cylindrical jar of diameter 110 mm.
She finds the label is a rectangle 80 mm high.

a If the label had no overlap how long is the label?

b What is the area of the label in cm²?

Remember not to round
intermediate values.

Review 2
The curved surface area of a cylinder is 6·2 m². If the height of this cylinder is 1·6 m, find the
radius.

Review 3
7 litres of petrol are poured into this container.
How deep is the petrol?

40 cm

PETROL

EXTREMELY
FLAMMABLE

18 cm

*### Review 4
Ross needs to paint the inside, outside, and both ends of this cylindrical
concrete culvert before it is installed. One litre of paint will cover 10 m² per
litre. How much paint will Ross need? The outer diameter is 1·5 m and the
inner diameter is 1·2 m.

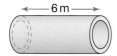

6 m

*### Review 5
A wooden bracket for shelving is shown. It was
made by removing a quarter-circle from a
rectangular block of wood, 3 cm thick. What is the
total surface area and volume of the bracket?

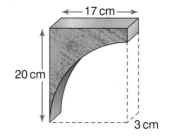

17 cm

20 cm

3 cm

Practical

Use a rain gauge to find the rainfall for a week/month at your school.
Use this to estimate the volume of water that has fallen on your school grounds
during that same time.

Investigation

Budget Buys

You will need a spreadsheet package.

a Budget Buys have always made their cans with a diameter of 8·5 cm and a height of 14·3 cm.
The mananger wants to check he is using the minimum amount of metal possible to make this can.
He wants the volume of the can to stay the same.
Find the diameter and height that give the smallest surface area.
Use a spreadsheet package if possible.

b Budget Buys want to make another can for Bulk foods with a volume of 2000 cm^3.
What should the diameter of the top and the height of the can be to use the minimum amount of metal?

Trigonometry

Practical

A You will need a calculator or graphical calculator and a copy of the table.

Using the (sin) and (cos) keys on the calculator fill in this table, e.g. key (sin) (10) (=).
Round your values to 2 d.p.
Plot the graph of sin θ versus cos θ on a larger copy of this grid.
Join the points with a smooth curve.

θ	Cos θ	Sin θ
10°		
20°		
30°		
40°		
50°		
60°		
70°		
80°		
90°		

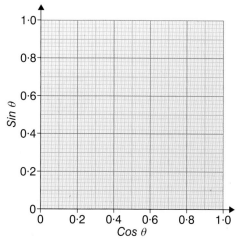

What do you notice?

B **You will need** a calculator or graphical calculator, a spreadsheet and some copies of the table.

Use the graph you just drew in **part A**.
Draw a right-angled triangle with
- hypotenuse length 1
- angle of 60°
- vertices the origin, a point on the graph and a point on the *x*-axis.

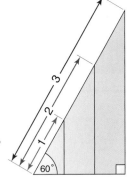

What are the lengths of the other two sides?
Compare these to the values of sin 60° and cos 60°.

Does $\quad \sin 60° = \frac{\text{side opposite } 60°}{\text{hypotenuse}}$?

Does $\quad \cos 60° = \frac{\text{side adjacent to } 60°}{\text{hypotenuse}}$?

What if you drew similar triangles with a right angle, a 60° angle and length of hypotenuse 2, 3, 4, 5, ...
Draw these triangles and measure the lengths of the vertical and horizontal sides. Fill in the table below.

Does $\quad \sin 60° = \frac{\text{side opposite } 60° \text{ (O)}}{\text{hypotenuse (H)}} \quad$ and

$\quad \cos 60° = \frac{\text{side adjacent to } 60° \text{ (A)}}{\text{hypotenuse (H)}} \quad$ for all the triangles?

	A (cm)	O (cm)	H (cm)	$\frac{O}{H}$	$\frac{A}{H}$
θ = 60°			1		
sin θ =			2		
cos θ =			3		

Use this to find the lengths of the other two sides in these triangles.

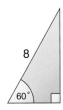

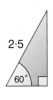

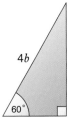

Link to similar triangles.

What if the angle changes to 70°? 50°? 45°?
Repeat **B** again and fill in a table for each of these angles instead of 60°.
Use a different table for each angle.

Write some conclusions about the values of cos θ, sin θ, $\frac{O}{H}$ and $\frac{A}{H}$.

The hypotenuse, opposite side and adjacent side have been labelled on this right-angled triangle for angle θ.

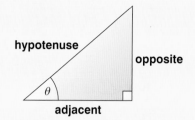

These ratios are true for any right-angled triangle, where O is the side opposite the angle and A is the side adjacent to the angle.

sine $\theta = \dfrac{\text{opposite}}{\text{hypotenuse}} \left(\dfrac{O}{H}\right)$ sin is the abbreviation for sine.

cosine $\theta = \dfrac{\text{adjacent}}{\text{hypotenuse}} \left(\dfrac{A}{H}\right)$ cos is the abbreviation for cosine.

tangent $\theta = \dfrac{\text{opposite}}{\text{adjacent}} \left(\dfrac{O}{A}\right)$ tan is the abbreviation for tangent.

Discussion

One way of remembering these trig. ratios is by using SOHCAHTOA. How can this be used to help you remember them? **Discuss**.

Make up a mnemonic to help remember SOHCAHTOA. Your mnemonic could begin '**S**cience **O**r **H**istory ...' or '**S**ome **O**ld **H**orses **C**an ...' **Discuss** your mnemonic.

Worked Example
Find the values of sin θ, cos θ, tan θ giving the answers to 2 d.p.

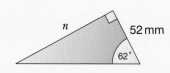

Answer

$\sin \theta = \dfrac{60}{62 \cdot 5}$ $\cos \theta = \dfrac{17 \cdot 5}{62 \cdot 5}$ $\tan \theta = \dfrac{60}{17 \cdot 5}$

   $= \mathbf{0 \cdot 96}$ $= \mathbf{0 \cdot 28}$ $= \mathbf{3 \cdot 43}$ **(2 d.p.)**

Note The calculator must be in degree mode.

In a right-angled triangle, if we are given the length of one side and the size of an angle (other than the right angle) we can use one of the trig. ratios to find the length of one of the other sides.

Worked Example
a Find m.   b Find n.

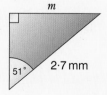

Answer

a $\sin 51° = \frac{m}{2\cdot7}$ so $\frac{m}{2\cdot7} = \sin 51°$

$m = 2\cdot7 \sin 51°$

$m = \textbf{2·1 cm (1 d.p.)}$

Key (2·7) (sin) (51) (=) .

so   *m* is a multiple of sin 51° since it is an opposite side with the hypotenuse given.

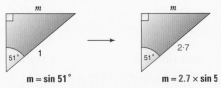

$m = \sin 51°$ $m = 2.7 \times \sin 5$

b $\tan 62° = \frac{n}{52}$ so $\frac{n}{52} = \tan 62°$

$n = 52 \tan 62°$

$n = \textbf{98 mm (to the}$

Key (52) (tan) (62) (=) . **nearest mm)**

so   *n* is a multiple of tan 62° since it is an opposite side with adjacent given.

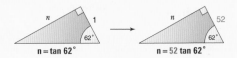

$n = \tan 62°$ $n = 52 \tan 62°$

Worked Example

Find the length of side *t* .

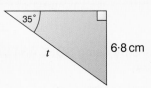

6·8 cm

Answer

t is the hypotenuse.

$\sin 35° = \frac{6\cdot8}{t}$

$t \sin 35° = 6\cdot8$ **multiplying both sides by *t***

$t = \frac{6\cdot8}{\sin 35°}$ **dividing both sides by sin 35°**

$t = \textbf{11·9 cm (1 d.p.)}$

or $\sin 35° = \frac{6\cdot8}{t}$

$\frac{1}{\sin 35°} = \frac{t}{6\cdot8}$ **take the reciprocal**

$\frac{6\cdot8}{\sin 35°} = t$

$t = \textbf{11·9 cm (1 d.p.)}$

Key (6·8) (÷) (sin) (35) (=) .

Exercise 6

1   Complete each trig. ratio giving the answers as decimals. Round to 2 d.p. if rounding is necessary.

a
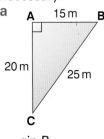

b
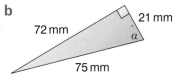

c
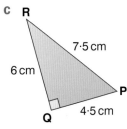

sin B = ...
cos C = ...
tan B = ...
sin C = ...

cos α = ...
tan α = ...
sin α = ...

tan R = ...
sin P = ...
cos R = ...
tan P = ...

2 Which trig. ratio (sin, cos or tan) would you use to find *d*?

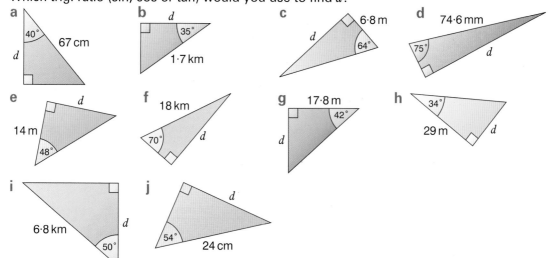

a 40° 67 cm *d*

b *d* 35° 1·7 km

c 6·8 m *d* 64°

d 74·6 mm 75° *d*

e *d* 14 m 48°

f 18 km 70° *d*

g 17·8 m 42° *d*

h 34° 29 m *d*

i 6·8 km 50° *d*

j *d* 54° 24 cm

3 For each of the triangles in question **2**, find *d*. Round your answers sensibly.

4 a A 5·2 metre ladder leans against a wall as shown.
How far is the bottom of this ladder from the wall?

b How far out would the ladder be if it is 10·4 m long?

c What if the ladder is 26 m long?

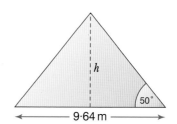

5·2 m

70°

5 This diagram shows a 4·6 m slide in a playground.

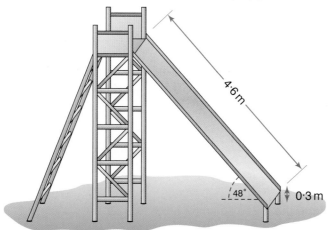

4·6 m

48° 0·3 m

How high is the top of this slide above the ground?

6 The diagram shows the side view of the loft of a garage.
The loft is symmetrical.
What is the height, *h*?

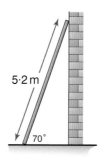

h

50°

9·64 m

Shape, Space and Measures

7 When an aeroplane is at a height of 1000 metres, it is picked up on radar. How far is the aeroplane from the radar?

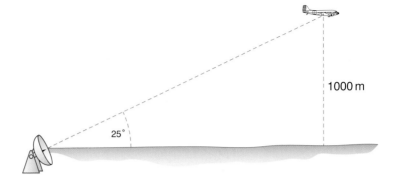

1000 m

25°

8 Find the length of a, the hypotenuse.

a

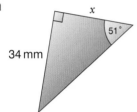

8 m

64°

a

b

6·7 cm

51°

a

c

54 mm

a

47°

d

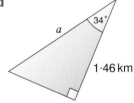

34°

a

1·46 km

9 Find x.

a

x

51°

34 mm

b

2·74 m

x

39°

c

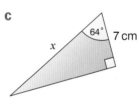

64° 7 cm

x

d

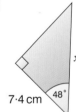

x

7·4 cm 48°

* 10 A circle has a radius of 15 cm. Calculate the length of the chord PQ.

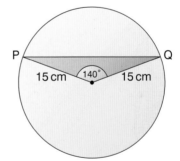

P

Q

15 cm 140° 15 cm

Review 1 Bob is pulling a trolley along the ground, as shown in this diagram. Find the height of Bob's hand above the ground.

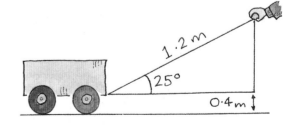

1.2 m

25°

0.4 m

Review 2 Find *d* in each of these triangles.

a
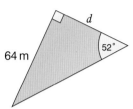

b
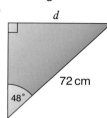

c
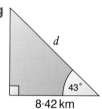

d
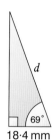

e
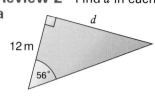

f

g

Finding angles

Discussion

If sin θ = 0·89, then θ = 62·9° to one decimal place.

What keying sequence on the calculator gives this answer for θ? **Discuss**.

What keying sequence on a graphical calculator gives this answer for θ? **Discuss**.

Worked Example
Find the size of angle A.

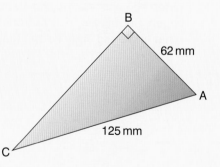

Answer
The given lengths are AB and AC. For angle A, AC is the hypotenuse and AB is the adjacent side. We use cos A to find angle A.

$\cos A = \frac{62}{125}$
$A = 60·3°$ (1 d.p.)

Key Shift cos⁻¹ (62 ÷ 125) = .

Make sure the calculator is in degree mode.

Exercise 7

1 Find, to one decimal place, the size of angle P if:

a sin P = 0·83   b cos P = 0·462   c tan P = 0·945   d tan P = 14·6
e sin P = 0·345   f cos P = 0·8236   g tan P = 56   h cos P = 0·125
i tan P = 0·82   j sin P = $\frac{1}{4}$ k tan P = $\frac{2}{5}$ l cos P = $\frac{2}{3}$
m tan P = $\frac{14}{9}$ n sin P = $\frac{8}{11}$

Shape, Space and Measures

2 Find θ.

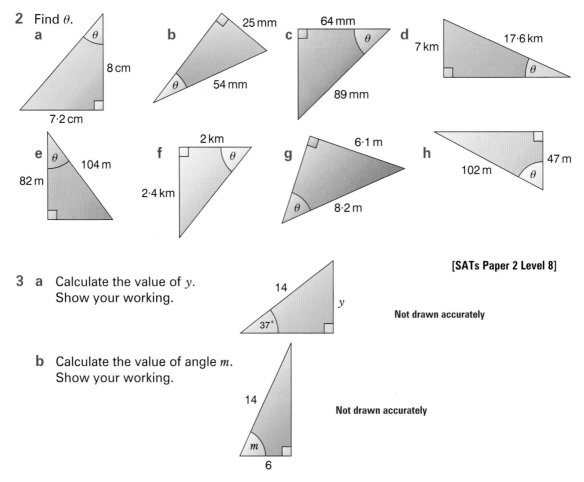

a 8 cm, 7·2 cm, θ

b 25 mm, 54 mm, θ

c 64 mm, 89 mm, θ

d 7 km, 17·6 km, θ

e 104 m, 82 m, θ

f 2 km, 2·4 km, θ

g 6·1 m, 8·2 m, θ

h 102 m, 47 m, θ

[SATs Paper 2 Level 8]

3 a Calculate the value of y.
Show your working.

14, y, 37°

Not drawn accurately

b Calculate the value of angle m.
Show your working.

14, m, 6

Not drawn accurately

4 Judy was abseiling down a building.
When she was a vertical distance of 10 m from where she began, her hips were 1 m from the side of the building.
What angle did Judy's rope make with the building?

10 m

1 m

[SATs Paper 2 Level 8]

5

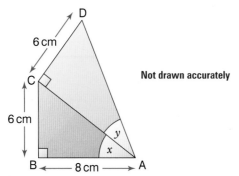

D

6 cm

C

6 cm

y

x

B ← 8 cm → A

Not drawn accurately

By how many degrees is angle x bigger than angle y?
Show your working.

6 Drainage pipes were being laid along the diagonal of a rectangular paddock, as shown in this diagram.
At what angle, to the shorter sides of this paddock, were the pipes laid?

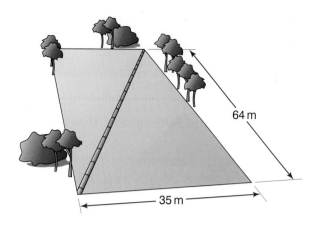

* 7 If we join the vertices of a star we get a pentagon.
If the sides of the pentagon are 16 cm long, what is the length of each of the five lines that form the star?

Review 1   Find, to the nearest degree, the size of α if
a $\cos \alpha = 0.687$ **b** $\tan \alpha = 2.7$ **c** $\sin \alpha = 0.8$ **d** $\sin \alpha = \frac{2}{7}$
e $\tan \alpha = \frac{14}{5}$ **f** $\cos \alpha = \frac{3}{8}$.

Review 2   A plane, which took off from Heathrow Airport, had gained an altitude of 2·1 km after it had travelled 8 km.
At what angle was this plane climbing?

Review 3   A wire runs from the top of a 10 m pole to the ground. The end of the wire is 15 m from the bottom of the pole.
What angle does the wire make with the ground?

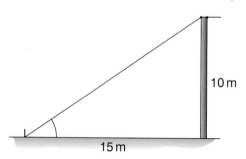

Review 4   This shows the cross-section of a swimming pool.
What angle, θ, does the sloping bottom make with the horizontal?

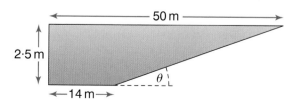

Using Pythagoras' theorem and trigonometry to solve problems

Trigonometry and Pythagoras' theorem can often be used to solve problems.

Worked Example
A ship travels due East from A to B, a distance of 74 km.
It then travels due South to C.   C is 47 km from B.
a What is the bearing of C from A?
b Use Pythagoras' theorem to find the distance from A to C.

N

A •————74 km————• B

47 km

• C

Answer
a The required bearing is given by the angle θ.
   $\theta = 90° + $ angle BAC.
   $\tan \text{BAC} = \frac{47}{74}$
   angle BAC = 32° (nearest degree)
   Key [Shift] [tan⁻¹] [( ] [47] [÷] [74] [ )] [=] .
   Then $\theta = 122°$.
   The bearing of C from A is 122°.
b $\text{AC}^2 = \text{BC}^2 + \text{AB}^2$ (Pythagoras' theorem)
       $= 47^2 + 74^2$
   AC = 87·7 km (nearest tenth of a km)
   Key [√] [( ] [47] [x²] [+] [74] [x²] [ )] [=] .

N

A [θ]————74 km————• B ⌐

47 km

• C

Note
When we look **up** at something, the angle between the horizontal and the direction in which we are looking is called the **angle of elevation**. In this diagram, θ is the angle of elevation.

When we look **down** at something, the angle between the horizontal and the direction in which we are looking is called the **angle of depression**. In this diagram, θ is the angle of depression.

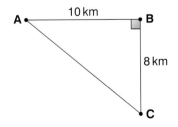

Exercise 8

1 B is 10 km East of A and 8 km North of C.
   a Use trigonometry to find the size of angle BCA.
   b What is the bearing of A from C?
   c Use Pythagoras' theorem to find the distance from A to C.

A •————10 km————• B

8 km

• C

2 A plane flies from P to Q on a bearing of 146°. Q is 229 km from P.
  a What is the size of angle RPQ?
  b Use trigonometry to find how far further East Q is than P.
  c Use Pythagoras' theorem to find how far further South Q is than P.

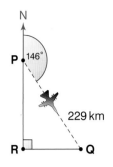

3 A helicopter leaves its base (B) and flies on a bearing of 220° to pick up an injured climber at C. It then flies the climber 16 km to the nearest hospital (H). The hospital is due South of B and due East of C, as shown in the diagram.
  a What is the size of angle CBH?
  b Use trigonometry to find how far the hospital is from B.
  c Use Pythagoras' theorem to find how far the injured climber was from the helicopter base.

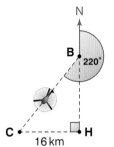

4 a What angle does the line $y = 3x - 2$ make with the positive direction of the x-axis?
  b What angle does the line $y = {}^-5x + 3$ make with the negative direction of the x-axis?
  c What angle does the line $y = \frac{1}{2}x + 1$ make with the positive direction of the y-axis?

5 At a distance of 80 m from a church tower, Donald measured the angle of elevation of the top of the tower as 24°.
How high is this church tower?

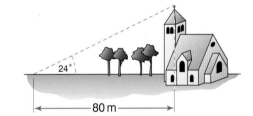

6 From the top of a 25 m high cliff, the angle of depression of a canoe is 28°. How far is this canoe from the foot of the cliff?

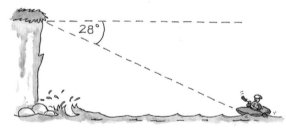

7 The guidelines for preventing occupational overuse syndrome say that you should sit with your eyes 55 cm from the centre of the screen and an angle of depression of 15°.
  a What height should your eyes be above the centre of the screen?
  b What horizontal distance should you sit from the screen?

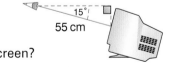

8 At the edge of a beach, which is 8 m wide, there is a 2·1 m high wall. From the top of this wall, Simon measures the angle of depression of a swimmer as 6°.
How far out to sea is the swimmer?

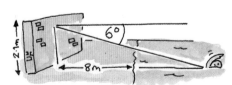

9 A flagpole is on the top of a building. From the point D, 4 m from the base of the building, Joanne measures the angles of elevation of the top, A, and the bottom, B, of the flagpole. Her measurements are shown on the diagram.
  a Find the length AC.
  b Find the length BC and the height of the flagpole.

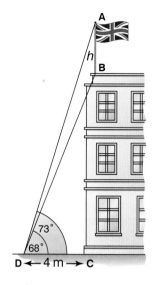

10 a A road ascends at a slope of 20%.
      What angle does the road make with the horizontal?
  b A road descends at a slope of 14%.
      What angle does the road make with the horizontal?
  c The gradient of a road up a hill is 1 in 8.
      What angle does the road make with the horizontal?

Review 1   An undersea cable runs from W to R.
R is 84 km South and 48 km West of W.
  a Use Pythagoras' theorem to find the length of this cable.
  b Use trigonometry to find the size of angle WRA.
  c What is the bearing of W from R?

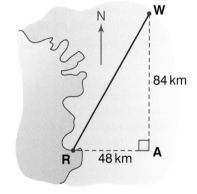

Review 2   A ship sails from B to A, a distance of 190 km. A is 100 km further West than B.

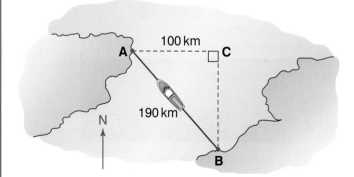

  a Use Pythagoras' theorem to find how far further North A is than B.
  b Find the size of angle ABC. Hence find the bearing on which the ship sailed.

Review 3   What angle does the line $y = \frac{-x}{2} + 3$ make with the negative direction of the x-axis?

Review 4   Valloa stood 5 m from the foot of a tree. He measured the angle of elevation of the top of the tree as 59° and the angle of depression of the foot of the tree as 20°.

a   What is the length of GH?
b   Use trigonometry to find TH.
c   What other calculations need to be made to find the height of the tree?
d   What is the height of the tree?

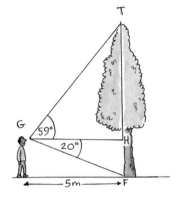
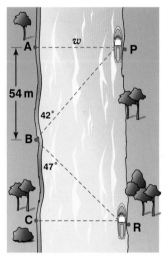

Review 5   Minami finds the distance between two boats, P and R, on the other side of the river as follows. She begins at A, opposite boat P. From A she walks to B, a distance of 54 m. She measures the angle ABP as 42° and angle CBR as 47°. From these measurements she is able to find the distance PR, between the two boats.

a   Calculate w, the width of the river, using the triangle APB.
b   Using triangle CBR, calculate the distance BC. Hence find the distance between the boats.

Practical

Use trigonometry to find the height of an object or the distance of an object from an observation point.

You could base your work on one of the examples given in the previous exercise. Use a trundle wheel or long measuring tape to measure distances. Use a theodolite or clinometer to measure angles.

Summary of key points

 A measurement can never be **exact**. We usually read to the nearest mark. Then the measurement is in error by up to half a unit either side of the mark.

Example   The length of a bike is given as 2·2 m.

The shortest the length could be is 2·15 m.

The longest the length could be is up to 2·25 m.

The range could be given as 2·15 m $\leqslant l <$ 2·25 m.

This is an inequality.

B A **rate** is used to compare how one quantity varies with another.

Examples   grams per litre, g/ℓ   kilometres per litre, km/ℓ

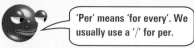

> 'Per' means 'for every'. We usually use a '/' for per.

kilometres per hour, km/h   metres per second, m/s

These measures are called **compound measures** because they involve more than one measure.

We can write a formula if the quantities are in direct proportion.

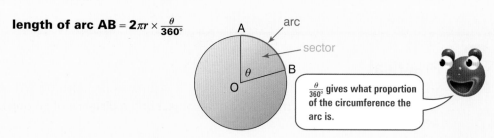

$$\textbf{speed} = \frac{\text{distance travelled}}{\text{time taken}}$$

$$\textbf{pressure} = \frac{\text{force}}{\text{area}}$$

> If speed varies this formula can also be used to find average speed.

$$\textbf{density} = \frac{\text{mass}}{\text{volume}}$$

Example   If a car travels at an average speed of 95 km/h for 3 hours 27 minutes, we can find the distance travelled.

$$\text{distance} = \text{average speed} \times \text{time}$$
$$= 95 \times 3\frac{27}{60}$$

Key ⬚95⬚ ⬚×⬚ ⬚3⬚ ⬚a^{b/c}⬚ ⬚27⬚ ⬚a^{b/c}⬚ ⬚60⬚ ⬚=⬚ **or**

⬚95⬚ ⬚×⬚ ⬚(⬚ ⬚3⬚ ⬚+⬚ ⬚27⬚ ⬚÷⬚ ⬚60⬚ ⬚)⬚ ⬚=⬚ to get **327·75 km**.

We can **change between units** of compound measures.

Example $5 \text{ km/h} = \frac{5000 \text{ metres}}{3600 \text{ seconds}}$ ⟵ number of *m* in 5 km
⟵ number of seconds in an hour
$$= 1\cdot3\dot{8} \text{ m/s}$$

C

length of arc AB $= 2\pi r \times \frac{\theta}{360°}$

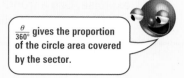

arc

sector

A

B

O

θ

> $\frac{\theta}{360°}$ gives what proportion of the circumference the arc is.

area of sector OAB $= \pi r^2 \times \frac{\theta}{360°}$

> $\frac{\theta}{360°}$ gives the proportion of the circle area covered by the sector.

Example   The perimeter of this shape is

$$\text{length of arc} + 20 \text{ cm} + 20 \text{ cm} = 2\pi r \times \frac{225°}{360°} + 20 \text{ cm} + 20 \text{ cm}$$
$$= 40\pi \times \frac{225°}{360°} + 40 \text{ cm}$$
$$= \textbf{118·54 cm (2 d.p.)}$$

The area of the shape $= \pi r^2 \times \frac{225°}{360°}$
$$= \pi \times 20^2 \times \frac{225°}{360°}$$
$$= \textbf{785·40 cm}^2 \text{ (2 d.p.)}$$

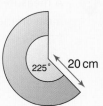

225°

20 cm

D **Volume of a prism = area of base** $\times h$

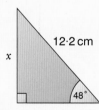 **Volume of a cylinder = area of base** $\times h = \pi r^2 h$

Surface area of a cylinder $= 2\pi rh + 2\pi r^2$

curved surface area   surface area of 2 circular faces

Example   A chimney is 4 m high and has a diameter of 60 cm.

Surface area $= 2\pi rh$

$= 2\pi \times 0.3 \times 4$ change the radius of 30 cm to metres

$= 7.5 \text{ m}^2$ (2 s.f.)

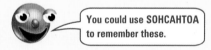

A chimney is an open cylinder with no top or bottom.

E In a right-angled triangle the sides can be labelled relative to a particular angle and ratios for the sides can be given.

$\sin \theta = \dfrac{\text{opposite}}{\text{hypotenuse}} \left(\dfrac{O}{H}\right)$ sin stands for sine

$\cos \theta = \dfrac{\text{adjacent}}{\text{hypotenuse}} \left(\dfrac{A}{H}\right)$ cos stands for cosine

$\tan \theta = \dfrac{\text{opposite}}{\text{adjacent}} \left(\dfrac{O}{A}\right)$ tan stands for tangent

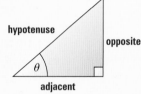

We can use these ratios to find the size of missing angles or lengths of right-angled triangles.

You could use SOHCAHTOA to remember these.

Examples

$\sin 48° = \dfrac{x}{12.2}$

$x = 12.2 \times \sin 48°$

$x = \mathbf{9.1}$ **cm (1 d.p.)**

Key 12.2 × sin 48 =

$\cos 52° = \dfrac{50}{y}$

$y = \dfrac{50}{\cos 52°}$

$y = \mathbf{81.2}$ **mm (1 d.p.)**

Key 50 ÷ cos 52 =

$\tan \theta = \dfrac{48}{37}$

$\theta = \mathbf{52.4°}$ **(1 d.p.)**

Key Shift tan⁻¹ (48

÷ 37) =

Note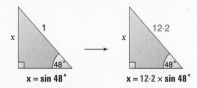

$x = \sin 48°$ $x = 12.2 \times \sin 48°$

F We can use the ratios in **E** and Pythagoras' theorem to solve problems.

Test yourself

1   Fiona and Siobhan weighed their kitten on the kitchen scales. It was 2·4 kg to the nearest tenth of a kg.
Write the possible range this measurement lies between as an inequality.

2   Charlotte measured her handspan with a ruler and found it was 18 cm to the nearest cm.
Julie measured her handspan more accurately as 178 mm to the nearest mm.
a   What are the upper and lower limits for Charlotte's handspan?
b   What are the upper and lower limits for Julie's handspan?
c   Is it possible that Julie's handspan is larger than Charlotte's? Explain your answer.

3   James applies 6 kg of fertilizer to a 200 square metre lawn.
Find the rate of application of this fertilizer in g/m^2.

4   Every hour, an ice cream manufacturer makes 120 litres of ice cream.
Give this in ℓ/min.

5   Tom ran 6 km in 24 minutes 30 seconds.
What was Tom's average speed in metres per second?

6   A block of concrete weighs 850 N.
The area of its base is 1·2 m^2.
What pressure does it exert on the ground where it rests?

Remember, weight is a force.

7   Change 50 g/cm^3 to kg/m^3.

8   The diagram shows the landing area for a shot put competition.
Find the area of this sector.

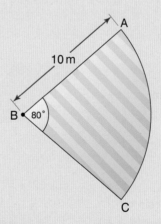

9   This diagram shows three stained-glass sectors in a window.
The angles of the sectors are 36°, 60° and 36°.
a   Calculate the total area of the three sectors.
b   Lead beading is placed along the edge of each of the sectors. Calculate the total length of lead beading required.

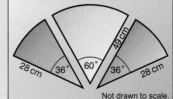

Not drawn to scale.

10 Wanita's wooden pencil case is cylindrical.
She wants to cover it with plastic, leaving one
end open.
How much duraseal will she need?

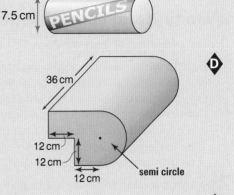

D

11 This diagram represents a brick of an usual shape.
Find the
  a surface area of the front face
  b the volume.

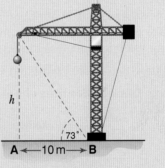

D

12 Find the lengths marked x in **a** and **b**, and find angle p in **c**.
Round sensibly.

a

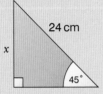
24 cm
x
45°

b

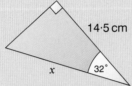
14·5 cm
x
32°

c

12·4 cm
p
8·6 cm

E

13 Ahmed stands at A, directly under the end of a crane.
From A, he walks 10 m to B. At B, he measures the
angle of elevation of the end of the crane as 73°.
How high is the end of the crane?

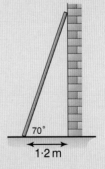
h
73°
A ←—10 m—→ B

E

14 A ladder makes an angle of 70° with the ground.
The foot of the ladder is 1·2 m from the wall.
How long is this ladder?

70°
1·2 m

E

15 Two yachts sail into a harbour at R.
One yacht sails from P, which is 112 km
West of R.
The other sails from Q, which is 54 km
South of P.
  a Find the distance from Q to R.
  b Find the bearing of R from Q.

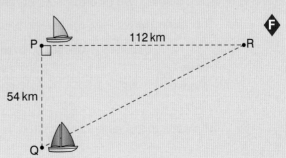
P
112 km
R
54 km
Q

F

Handling Data Support

Planning and collecting data

When **planning a survey** you need to follow these steps.

1 Decide what question you want answered and any related questions.
2 Decide what data needs to be collected.
3 Decide where to collect the data from.
   Data can be collected from primary sources such as:
      a questionnaire or survey from a sample of people
      an experiment which may use technology
      or from secondary sources such as reference books, websites ...

The **sample size** for a survey or questionnaire needs to be as large as it is sensible to make it. The sample used needs to be representative of the whole population.

Once the survey has been planned and a sample size decided on, a **collection sheet** or **questionnaire** needs to be designed. You need to decide what units to use and how accurate you want your data to be if it is measurement data.

When **writing questionnaires** you must

- allow for any possible answers
- give instructions on how you want the question answered
- avoid questions people may not be willing to answer
- word questions so your opinion is not evident
- make questions clear and concise
- keep the questionnaire as short as possible.

Once written, you should **trial** a questionnaire.

Examples **Data collection sheet**

Time for Race (min)	Tally	Frequency
$0 < t \leqslant 5$		
$5 < t \leqslant 10$		
$10 < t \leqslant 15$		
$15 < t \leqslant 20$		
$20 < t \leqslant 25$		
$t > 25$		

Questionnaire on Voting

What age, in years, do you think people should be to vote in national elections ?

younger than 16 ☐   16 ☐   17 ☐   18 ☐

19 ☐   20 ☐   21 ☐   older than 21 ☐

Practice Questions 10, 33, 34, 39, 43

Two-way tables

Data collected from surveys can sometimes be summarised in a **two-way table**. Two-way tables display two or more sets of data and are often used for comparing data.

Example   This table shows the number of males and females in some school clubs.

	Debating	Chess	Drama	Chorale	Total
Male		13	24	18	71
Female	12		28		83
Total		26			

The gaps in the table can be filled in from the data given.

Practice Question 8

Displaying data

Discrete data can only have certain values. It is usually found by counting.

Example   Number of different types of cars sold by a second-hand car yard

Continuous data can have any values within a certain range. It is usually found by measuring.

Example   Height of plants

Discrete data might be displayed on
  a pictogram
  a bar chart
  a bar line graph
  a line graph if the data is over time
  a pie chart if showing proportions

Example   Robbie collected this information about how males and females travel to work.

	Car	Bus	Train	Walk	Cycle	Other	Total
Males	42	13	36	19	6	4	120
Females	21	27	34	30	4	4	120

He drew these two pie charts after first working out the angles in each sector.
For example:

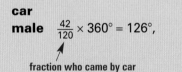

**car
male** $\frac{42}{120} \times 360° = 126°$,   **bus** $\frac{13}{120} \times 360° = 39° \dots$

  ↑ fraction who came by car   ↑ fraction who came by bus

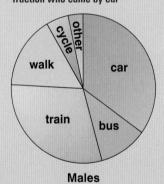

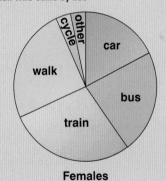

Males   Females

Grouped discrete data can be displayed in a frequency table and on a bar chart. The data must be grouped into **equal class intervals**.

Examples

e's frequency table

Number of e's	Tally	Frequency
0–2	ⅢⅢ Ⅲ	8
3–5	ⅢⅢ ⅢⅢ	10
6–8	ⅢⅢ	5
9–11	ⅢⅢ	4
12–14		0
15–17	Ⅰ	1

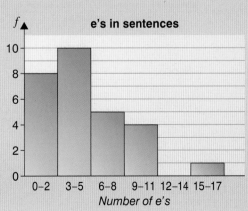

Handling Data

Continuous data is usually grouped into equal class intervals on a frequency table, and drawn on a **frequency diagram**.

Example The continuous data from the table is shown graphed as a **frequency diagram**.

There are no gaps between the bars on a frequency diagram.

Handspan of pupils	
Class interval (mm)	Frequency
$170 \leqslant l < 180$	2
$180 \leqslant l < 190$	3
$190 \leqslant l < 200$	5
$200 \leqslant l < 210$	4
$210 \leqslant l < 220$	4

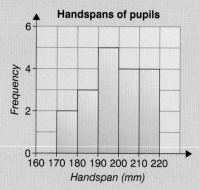

Practice Questions 4, 11, 16, 28, 30, 42

Mode, range, median and mean

The mode, range, median and mean are all ways of summarising data into a single value.

The **mode** is the most commonly occuring data value and is useful for identifying the 'most popular' or 'frequent'.

The **modal class** is the class interval with the highest frequency.

Example

Age of parents of Toby's classmates

Age (yrs)	30–34	35–39	40–44	45–49	> 49
Number	4	12	10	4	2

The modal age is 35–39.

Range = highest data value – lowest data value

The **median** is the middle value of a set of ordered data.

Example 10, 15, 18, **18**, **19**, 24, 25, 25

$$\text{median} = \frac{18 + 19}{2}$$
$$= 18 \cdot 5$$

Mean $= \dfrac{\text{sum of data values}}{\text{number of values}}$

or **mean** $= \dfrac{\text{sum of (frequency } (f) \times \text{ data value } x)}{\text{sum of frequency}}$ if the data is given in a frequency table

Example

Score in maths quiz

Score	1	2	3	4	5	6
Number of people	2	5	9	14	16	11
Total	1×2 $= 2$	2×5 $= 10$	3×9 $= 27$	4×14 $= 56$	5×16 $= 80$	6×11 $= 66$

$\text{Mean} = \dfrac{2 + 10 + 27 + 56 + 80 + 66}{2 + 5 + 9 + 14 + 16 + 11}$ ◄— sum of total scores
◄— sum of number of people

$= \dfrac{241}{57}$

$= \textbf{4·23 (2 d.p.)}$

We can find the mean using an **assumed mean**.

Example 3·2 3·5 3·1 3·6 3·5 3·4 3·2 3·5
  i Assume the mean is 3·4.
  ii Subtract 3·4 from each data value.   ⁻0·2 0·1 ⁻0·3 0·2 0·1 0 ⁻0·2 0·1
  iii Find the mean of these differences. $\frac{^-0·2}{8} = {^-}0·025$
  iv The mean is 3·4 + ⁻0·025 = 3·375.

Practice Questions 1, 3, 13, 24, 26, 37

Stem-and-leaf diagrams

We can find the median, range and mode from a **stem-and-leaf diagram**.

Example There are 35 data values.
  The median is the 18th value, 17.
  The range is 32 – 0 = 32.
  The mode is 15 and 20.

Number of Visitors

0	0 0 0 1 1 2 5 5 8
1	0 0 2 2 5 5 5 5 7 9 9
2	0 0 0 0 3 3 4 7 8
3	0 1 1 2 2 2

stem = tens
leaves = units

Practice Question 41

> Start at the lowest value and count from the left end of each row to find the median.

Scatter graphs

A **scatter graph** displays two sets of data.
A scatter graph sometimes shows a relationship or **correlation** between the data.

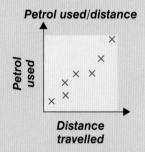

Petrol used/distance

Positive correlation
The greater the distance travelled, the more the petrol used.

Price of second–hand cars

Negative correlation
The older the car, the less it is sold for.

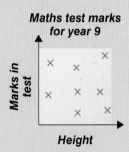

Maths test marks for year 9

No correlation
between marks in maths test and height.

Practice Questions 5, 12, 15, 31

Interpreting and comparing data

To **compare** data we can use the range and one or more of the mean, median or mode.

Example Two classes went on camp to the same place at different times.
  They recorded the number of hours of sleep they each got each night.

  Class A   **mean** 8 hours   **range** 4 hours
  Class B   **mean** $7\frac{1}{2}$ hours   **range** 1 hour

  Class B had a lower mean number of hours of sleep per night, but the number of hours per night did not vary much.
  Class A had a higher mean number of hours of sleep but the bigger range shows that the number of hours was much less consistent than for Class A.

Practice Questions 2, 14, 17, 19, 25, 27, 32

Probability

Probability is a way of measuring the chance or likelihood of a particular outcome. It can be shown on a **probability scale**.

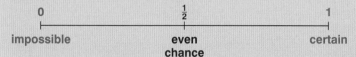

0 — impossible
½ — even chance
1 — certain

Choosing **at random** means every item has the same chance of being chosen.
The outcomes are equally likely.

If the outcomes of an event are **equally likely**:

$$\text{probability of an event} = \frac{\text{number of favourable outcomes}}{\text{number of possible outcomes}}$$

An event will either happen or not happen, so:

probability of an event not happening = 1 − probability of it happening

Mutually exclusive events are events that cannot happen at the same time.

Example   It is not possible to get a 5 and a number less than 3 on a dice with the same throw.

The sum of the probabilities of all the mutually exclusive outcomes of an event is 1.

Example   This spinner is spun.
The probabilities of spinning the letters are:
   A 0·3   **B** 0·2   **C** 0·4   **D** 0·1
   0·3 + 0·2 + 0·4 + 0·1 = **1**

To **calculate a probability** we often record all the possible outcomes.
The set of all possible outcomes is called a **sample space**.

Example   Two dice are rolled and the scores multiplied.

×	1	2	3	4	5	6
1	1	2	3	4	5	6
2	2	4	6	8	10	12
3	3	6	9	12	15	18
4	4	8	12	16	20	24
5	5	10	15	20	25	30
6	6	12	18	24	30	36

◄—— This is the sample space.

The probability of getting a number bigger than 20 is $\frac{6}{36} = \frac{1}{6}$.

number of ways of getting a number greater than 20.
total number of outcomes

We can find the **relative frequency** of an event from experimental results.

$$\text{Relative frequency} = \frac{\text{number of times an event occurs}}{\text{number of trials}}$$

We can use the relative frequency as an **estimate of probability**.

Example   Jon tossed 3 coins together 100 times. He got HHH 24 times.
The relative frequency of getting 3 heads is $\frac{24}{100} = \frac{6}{25}$.

When an experiment is repeated, the outcomes will usually be different because of **random variation**.
The more times an experiment is repeated, the better will be the estimate of probability, and the closer the **experimental probability** will be to the **theoretical probability**.

Practice Questions 6, 7, 9, 18, 20, 21, 22, 23, 29, 35, 36, 38, 40

Practice Questions

1 **a** There are four people in Sita's family.   [SATs Paper 1 Level 5]
Their shoe sizes are 4, 5, 7 and 10.
What is the **median** shoe size?

b There are **three** people in John's family.
The **range** of their shoe sizes is **4**.
Two people in the family wear shoe size 6.
John's shoe size is **not 6** and it is **not 10**.
What is John's shoe size?

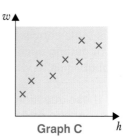

See support questions 3, 13 and 37 for more practice at this level.

2 The diagrams show the amount of rain that fell each day in two different months.

Number of mm of rain month 2

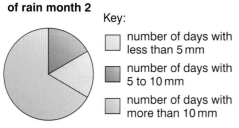

Key:
- number of days with less than 5 mm
- number of days with 5 to 10 mm
- number of days with more than 10 mm

Number of mm of rain month 1

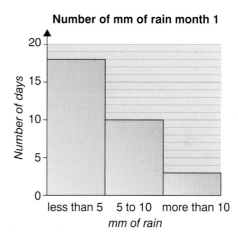

a How many days are there in month 1?
    A 28   **B** 30   **C** 31   **D** Can't tell

b How many days are there in month 2?
    A 28   **B** 29   **C** 30   **D** 31   **E** Can't tell

c Which month had more mm of rain?
Explain how you know.

3 Find the mean, median and range of each of these sets of data.
    a 14, 17, 15, 42, 27, 32, 28
    b 144 g, 179 g, 230 g, 98 g, 273 g, 87 g, 135 g, 190 g
    c 11·2 cm, 7·2 cm, 6·9 cm, 14·6 cm, 16·5 cm, 12·6 cm, 5·8 cm, 15·3 cm, 8·7 cm, 4·2 cm

4 Which of the following data is discrete and which is continuous?
    length of paper   number of books borrowed from the library
    mass of fruit   time to run a cross-country course
    number of entrants in a race

5

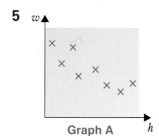

Graph A

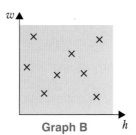

Graph B

Graph C

a Which of these graphs shows positive correlation between the variables h and w?
b Which shows negative correlation?
c Which shows no correlation?

T

6 Tom spins his spinner.

  a Use a copy of the probability scale.

0    $\frac{1}{2}$ 1

  Show, on the probability scale, the probability of spinning each of these.
  i 1   **ii** 3   **iii** an even number   **iv** an odd number

  b Tom now spins the above spinner at the same time as he spins this one.
  i Tom started writing this list of possible outcomes.
  Copy and complete this list.
      R1, R2, R2, R3, R3, ...

  ii Show on the probability scale the probability of getting red and an even number.

7 a A box has 30 marbles in it. 12 of them are blue.
  Samuel takes one marble out of the box randomly.
  Which of these values show the probability of taking out a blue marble?

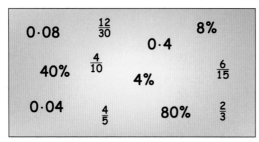

0·08 $\frac{12}{30}$ 8%   0·4

40% $\frac{4}{10}$ 4% $\frac{6}{15}$

0·04 $\frac{4}{5}$ 80% $\frac{2}{3}$

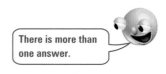

There is more than one answer.

  b Another box has 40 marbles in it. 15 of them are green.
  If you take one marble out of the box randomly, what is the probability the marble is **not** green?
  Give your answer as a fraction and a percentage.

8 This table shows the entries in some distance running events.

	Marathon	Half-marathon	10 km	Total
Male	240			
Female		96	125	
Total	425		241	846

  a How many males entered the 10 km event?
  b How many entered the half-marathon altogether?
  c How many females entered altogether?
  d How many males are entered altogether?

9 A shop sells bunches of 12 roses.
  If a rose is pulled from the bunch at random, this table shows the probability of getting each colour.
  Use the probabilities to work out how many of each colour were in the bunch.

Colour	red	pink	white	yellow
Probability	$\frac{7}{12}$	$\frac{1}{4}$	$\frac{1}{6}$	0

10 For each of these questions, how would you collect the data?
      A Questionnaire or data collection sheet
      B Experiment
      C Secondary source such as website, book, newspaper, CD-Rom, ...
  a How much does a plant grow over a few weeks?
  b What is the average rainfall of the 10 largest cities in the U.K.?
  c Do boys or girls in your class read more books?

11 This pie chart shows the proportion of passengers who flew Economy, First Class and Business Class from London to Tokyo.
Out of every 2000 passengers, how many flew
a First Class **b** Economy?

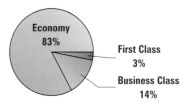

12 Would you expect there to be positive correlation, negative correlation or no correlation between the following?
a the mass of suitcases filled with clothes and the volume of the suitcases
b the distance cars travel on 10 litres of petrol and the mass of the cars
c the time students take to run 100 m and the time they take to run 200 m
d the time students take to run 100 m and the time they take to swim 100 m
e the height of the oldest child in a family and the number of children in the family

13 Julian's family own an apple orchard. Julian conducted a survey for his family.
a Julian counted the number of trees in each row. His results are shown on this frequency table.

Number of trees	15	16	17	18	19	20	21	22	23
Frequency	1	0	1	3	1	3	0	1	1

    i How many rows of trees are there on this orchard?
    ii How many trees are there?
    iii What is the mean number of trees per row? (Answer to 1 d.p.)
    iv What is the mode?
    v Find the median number of trees per row.
    vi Find the range.
b Julian collected data on the number of apples produced last season by each of the trees. This table shows his data.

Apples produced	600–799	800–999	1000–1199	1200–1399	1400–1599	1600–1799
Frequency	8	14	10	34	68	76

Julian then displayed this data on a frequency diagram. Draw this graph.

14 This table gives the mean and range for times taken by pupils in 9L and 9T to run the cross-country.

	mean (min)	range (min)
9L	15·5	3
9T	14·6	6

Write a sentence comparing the speed of the pupils in 9L and 9T.

15

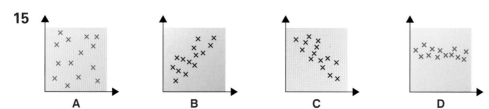

Which of the above scatter graphs **A**, **B**, **C**, or **D**, would best describe the following situations.
a The amount of practice goal shooting compared with the number of goals scored
b The length of hair compared with age of teenage girls

16 a As part of Dean and Joanne's survey on traffic, Dean decided to display this data on a bar graph. Draw this bar graph.

b Joanne decided to display the data on a pie chart. What angle should Joanne have for each of the five categories? Draw the pie chart.

Vehicle	Frequency
Lorry	5
Bus	3
Car	28
Motorbike	7
Bicycle	17

17 Freddie used this graph in his project on winter. Freddie wrote the following in his conclusion.

The number of cases of flu was three times higher in 1996 than in 1995.

Explain why Freddie was wrong.

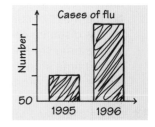

18 Jamie made up a party game.
He made these three boards and put a chocolate on each blue square.
If a ring, when dropped onto a board, landed on a chocolate square, that person got the chocolate.
On which board are you most likely to get a chocolate? Explain.

A  **B**  **C**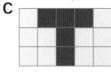

19 The school bags of pupils in class 9PQ were weighed and the results shown on this frequency diagram.
a How many pupils are in class 9PQ?
b Is the following statement true?
 'Most pupils' bags weigh between 4 kg and 6 kg.'
Explain.
c Jane's bag weighs 6 kg. Can you tell from the graph which class interval it would be in? Explain.

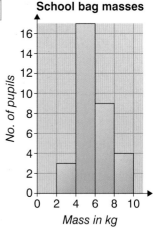

20 These spinners are spun at the same time and the numbers are added.

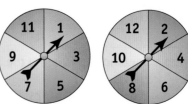

Use a copy of this table.
a Fill in the sample space to show all the possible outcomes.
b Which is the most likely total? Why?
c Use the sample space to find the possibility of getting
 i a total of 3
 ii a total of less than 9
 iii both spinners stopping on their highest number
 iv a total of more than 11.

+	1	3	5	7	9	11
2						
4						
6						
8						
10						
12						

21 These four discs are placed in a bag. The discs are all the same shape and size. Melanie chooses two of these discs at random. She replaces the first disc before choosing the second. What is the probability that the two discs she chooses, total less than 30p?

22 An unfair dice is tossed many times. The results are shown on the table.
  a What is the relative frequency of tossing a 6?
  b Estimate the probability of getting an even number the next time this dice is tossed.

Number on dice	1	2	3	4	5	6
Frequency	72	69	82	221	203	353

23 In which of these are the events M and N mutually exclusive?
  a **Event M:**   An even number is rolled with a dice.
     Event N: A three is rolled with a dice.
  b **Event M:**   A black card is drawn from a pack of cards.
     Event N: An ace is drawn from a pack of cards.
  c **Event M:**   Samuel goes to a movie.
     Event N: Joanna goes to the same movie.

24 Three friends entered a quiz.
Their median score was 24, their mean score was 24, the range of their scores was 6 and one score was 27.
What are the other two scores?

25 The club hockey team has a final coming up.
The coach has to choose which of the goalies, Tim or Dean, to play.
This shows how many goals each stopped in their last eight games.

Tim	1	1	2	2	1	4	1	3
Dean	3	3	1	5	2	0	1	2

  a Find the mean, median and range for each goalie.
  b Which goalie do you think the coach should choose?
     Explain your choice.

26 Find the mean of these masses using the assumed mean method. Show your working.
   3·9 g, 3·6 g, 4·2 g, 5·6 g, 4·6 g, 4·3 g, 4·2 g, 4·8 g

27 These line graphs show the midday temperatures at Carlisle and Calais for one week in June.
Compare these temperatures. In your comparison, refer to the ranges and the means.

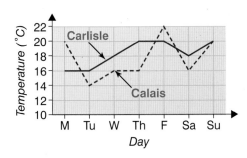

Handling Data

T

28 This table shows the masses of 25 new-born kittens.

Mass (g)	75–	100–	125–	150–	175–	200–225
Frequency	7	8	7	4	2	1

a Joel thinks that a frequency diagram is best to display this data. Explain why he is correct.

b Use a copy of this grid.
Draw a frequency diagram for the data.

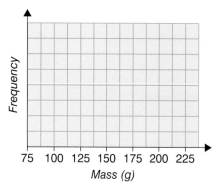

c Is the shape what you would expect? Explain.

d How many of the kittens weighed less than 150 g?

e The next kitten born had a mass of 149·5 g. This was rounded to 150 g. Which class interval should this kitten be put into? Justify your answer.

29 These two spinners are spun at the same time.

a Write down all the possible outcomes.

b What is the probability of getting

i two As   **ii** at least one A   **iii** an A and a B?

30 This table gives the heights of some Year 9 pupils.

a Draw a frequency diagram for the data.

b How many pupils were taller than 165 cm?

c How many are 150 cm or shorter?

d How many heights are recorded altogether?

e Jeanette is 142 cm.
She thinks she is the third shortest in Year 9. Is this possible? Explain.

*f The PE teacher said fewer than half these Year 9 pupils are shorter than 150 cm. Explain how you can tell that this is not correct.

Height in cm	Frequency
$135 < h \leqslant 140$	2
$140 < h \leqslant 145$	2
$145 < h \leqslant 150$	3
$150 < h \leqslant 155$	5
$155 < h \leqslant 160$	13
$160 < h \leqslant 165$	8
$165 < h \leqslant 170$	5
$h > 170$	2

31 The data in the table gives the heights of some 16-year-old boys and their heights as 3-year-olds.

	Jon	Ian	Tim	Evan	Helal	Luke	Brett	Ryan	Mark	Naim
Height at 16 (m)	1·85	1·72	1·74	1·88	1·77	1·79	1·63	1·80	1·61	1·73
Height at 3 (m)	1·03	0·93	0·94	1·04	0·96	0·99	0·85	0·95	0·84	0·93

a Plot this data on a scatter diagram. (Have height at 16 on the horizontal axis and height at 3 on the vertical axis.)

b Does the scatter graph show that boys who were taller at age 3 tended to be taller at 16 years? Is the correlation positive, negative or is there no correlation?

c At 16, Matthew is 1·68 m tall. Estimate how tall Matthew was at 3.

d When Isra's younger brother was 3, he was 1·01 m tall. Estimate his height at 16.

32 The number of fatal road accidents occurring in two counties of similar size are shown on the graph.

a What is the most obvious difference between the road deaths in the two counties?

b One county ran an aggressive road-safety campaign in 2002. Which county was it?

c Comment on the general trend of road deaths in the two counties.

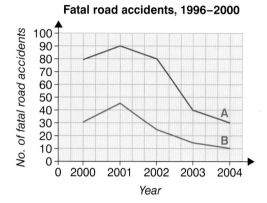

Fatal road accidents, 1996–2000

33 Karen wrote a questionnaire to survey teenagers about their leisure activities.
Two of the questions were:

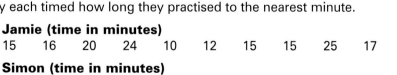

How old are you? 13 ☐ 14 ☐ 15 ☐ 16 ☐ 17 ☐ 18 ☐ 19 ☐
How long, on average, do you spend each weekday on leisure activities?
less than 1 hour ☐   between 1 and 2 hours ☐   between 2 and 3 hours ☐
more than 3 hours ☐

Write two more questions and responses that Karen could include in her survey.

34 Dina surveyed the pupils in her school to test the hypothesis 'Left-handed pupils play more sport than right-handed pupils'.
Write down two questions (with possible responses) that Dina could have included in her questionnaire.

35 A biased spinner with the numbers 1, 2, 3, 4, 5 on it is spun.
The probabilities of it stopping on a number are:

$p(1) = 0.25$ $p(2) = 0.15$ $p(3) = 0.1$ $p(4) = 0.2$

a What is the probability it will stop on 5?

b What is the probability of it landing on 1 or 3?

c What is the probability of it not landing on 1 or 2?

36 Julia repeated a probability experiment 50 times.
She tossed a drawing pin 50 times. It landed on its head 32 times.
She tossed it another 50 times.
Would she expect it to land on its head 32 times again?
Explain your answer.

37 Every day Jamie and Simon practised their cricket bowling.
They each timed how long they practised to the nearest minute.

Jamie (time in minutes)

| 15 | 16 | 20 | 24 | 10 | 12 | 15 | 15 | 25 | 17 |

Simon (time in minutes)

| 9 | 30 | 26 | 17 | 15 | 22 | 10 | 29 | 16 | 22 |

a Find the mean, median and range of each set of data.

b Draw a frequency table like this for Jamie and Simon.

c Draw a bar chart for Jamie and Simon. Put their data on the same chart.

d Compare the amount of time that Jamie and Simon spent practising.

Time (minutes)	Tally	Frequency
8–11		
12–15		
16–19		
20–23		
24–27		

Handling Data

38 Nick can't remember the last digit of his friend's mobile phone number.
He guesses the last digit.
a What is the theoretical probability he will be correct?
b Write an experiment you could do to find the experimental probability.
c What would you need to do for the experimental probability to be close to the theoretical probability?

39 Alison believes that students have better calculator skills than teachers.
Design an experiment to test whether this is true. Include how you propose to collect the data and how you would analyse it.

40 There are some yellow (Y) and purple (P) blocks in a toddlers toy bin.
The probability of getting a yellow block if you take a block at random out of the bin is $\frac{2}{5}$.
a What is the probability the block will be purple?
b Caleb takes one block out of the bin.
It is yellow.
What is the smallest number of purple blocks there could be in the bin?
c Caleb then takes another block out of the bin and it is also yellow.
What is the smallest number of purple blocks there could be in the bin?

41 Find the median, range and mode from this stem-and-leaf diagram.

Marks out of 50	
1 | 0 1 4 8	stem = tens
	leaves = units
2 | 3 8 6	
3 | 0 1 3 3 4 7 9 9 9	
4 | 0 1 4 5 5 6 6 7 7 8 8 9	
5 | 0 3 6 7 8	

42 Display this information on a suitable graph.

Money raised for charities

Year	Money raised (£)
7	120
8	80
9	190
10	300
11	220

43 Design a questionnaire to survey the opinion of people about one of these issues.

- putting a road through your nearest common or park
- raising (or lowering) the school leaving age

Explain how you would collect, organise and analyse the data.
Include the following in your explanation.

- how you might collect the data
- what tables and graphs you might use
- whether or not you might use the computer to help in your analysis of the data

14 Planning a Survey and Collecting Data

You need to know

✓ planning and collecting data   pages 358 and 359
                   – discrete or continuous data
                   – grouped data

⋯ Key vocabulary ⋯⋯⋯⋯⋯⋯⋯⋯⋯⋯⋯⋯⋯⋯⋯⋯⋯⋯⋯⋯⋯

bias, continuous, discrete, primary source, representative (sample), sample, secondary source

▶▶ Biggies

Big companies often carry out surveys to find out

- what customers want in new products
- if customers are happy with existing products
- market conditions.

What might the companies shown above want to find out in a survey?
What data might they collect?

Planning a survey

A The problem

The first thing to do when **planning a survey** is to decide on the **problem** you want to explore.

Example 1   The PE teacher wants to explore what factors affect pupil performance on an exercise circuit.

Example 2   A social scientist wants to know if development indicators such as GNP, life expectancy, % of population with access to safe water are accurate measures of development.

It is helpful to formulate a **conjecture**.
These could be the conjectures for **Example 1**.

> A conjecture is a statement that you want to test the truth of.

1 A pupil's height affects performance on the jumping part of the circuit.
2 Year 9 pupils complete the course faster than Year 7 pupils.
3 The length of the mat given for the standing start jump affects the distance pupils jump.
4 Practice improves performance on the course.

For **Example 2** conjectures might be:

1 the development measures '% of population with access to safe water' and 'infant mortality' are correlated.
2 the measure of development that will have the greatest difference between MEDCs (more economically developed countries) and LEDCs (less economically developed countries) is % of population with access to safe water.

Exercise 1

1 For each of these write three possible conjectures.
   a What effect does mass, both inside the car and on towed objects, have on the time a car take to accelerate from 0 to 50 km/h?
   b What factors affect the distribution of weeds in the school grounds in both grassed areas and gardens?
   c Jan wants to know if the claim made by this advertisement is true.

> **94%**
> of people breathe 2nd hand smoke from someone smoking a cigarette at least once a week.

2 Suggest a problem you could explore using a statistical survey.
   Write three possible conjectures for your problem.
   Keep your conjectures for the next exercise.

Review   Write three possible conjectures for each of these.
a A horticulturist wants to know the effects of playing classical music and rock music to plants in a greenhouse.
b A city council wants to determine whether decreasing the number of rubbish bags provided each year to households would encourage people to recycle more or would have other effects.

B The data

The next thing to do when planning a survey is to decide **what data** needs to be collected and possible **sources of data**.
Sometimes there are related questions you need to explore.
If so, you may need to collect data for these as well.

Remember
Data can be gathered from

1 and 2 are called **primary sources**.

1 a questionnaire or survey of a sample of people
2 an experiment which may use technology, such as a data logger, graphical calculator or computer
3 **secondary sources** such as reference books, websites, printed tables or lists, CD-ROMs, newspapers, historical records, **interrogating** a database.

Example   If you are exploring how height and length of run up affects pupil performance in the jumping section of an exercise circuit, you will need to collect data about pupil height, length of run-up and the score given for the jumping section.
A related question might be 'Is there any difference between performance on the standing high jump and standing long jump sections?'
You will need to collect individual data for each jump.

Once you have decided what data is needed, you must decide how you are going to collect the data.

Example   If you are collecting data about development indicators for more developed and less developed countries you will need to use secondary sources such as the Intenet, CD-ROMs or printed reference material.

Example   If you are collecting data about the availability of, choice of and ingredients of vegetarian food in your local area, you will need to carry out a survey with a questionnaire.

Discussion

For each of these, **discuss** possible conjectures, any related questions and the possible sources of data. Identify any extra information that might be required to pursue any further questions.
Discuss how you might collect the data.

- What factors affect pupil performance in sport?
- What factors affect the site and type of housing people live in?
- What factors affect the acceleration of a cyclist?
- What are the attitudes of smokers and non-smokers to smoking laws?
- What are the attitudes of both local customers and local shop owners to fairly traded goods?
- Which development indicators best predict death rate in a country?

C Choosing a sample and eliminating bias

The next thing to do is choose an **unbiased sample** of suitable size.
A **biased** sample is one that does not represent fairly the whole population that the survey relates to.

Remember

If we want to survey a group such as patients treated at a hospital, it is not practical to survey **every** patient. We choose a representative **sample**.

A **sample** should be as large as it is sensible to make it and free from **bias**.

Example   When investigating attitudes of smokers and non-smokers to smoking laws, if only people at a non-smoking restaurant are surveyed this would be a biased sample.
The sample chosen in this example should aim to reduce possible bias due to

        male/female differences
        socioeconomic differences
        social and full-time smokers
        age
        and so on by
        choosing

  a the same number of males and females
  b people from the full range of ages, both smokers and non-smokers
  c people from the full range of socioeconomic circumstances
  d both social and full time smokers.

Note All of these should be in the same proportion as they are in the population.

Discussion

● For each of the questions in the Discussion on page 373, **discuss** a sensible sample size, possible sources of bias and how you could eliminate these by your choice of sample.

● Some groups to be surveyed and possible samples from these are given below.
Will any of these samples represent the group well? If not, why not?
How could we choose more representative samples for each of the given groups?
Discuss.

Group	Sample
Married men	Married men aged under 40
Netball players	Netball players from Manchester
Cream packaged by a milk company	Cream packaged between 7 a.m. and 8 a.m.
Questions on a maths exam paper	Questions on algebra

● To find whether a person is HIV positive we do not test all of their blood (the population). Instead, we test a small amount of their blood (a sample). Think of other situations, in which, even if time, money etc. was available to test the whole population it would not be sensible to do this. **Discuss**.

- Hamish's group are about to carry out a survey to estimate the average time each week that people in a town spend watching sport on TV. They decided to survey a sample of 8% of the town population. They discussed ways of choosing a representative sample. All of the ways below were discussed and rejected. What reasons might Hamish's group have given for rejecting these? **Discuss**.

  Interviewing people outside the supermarket.
  Interviewing school pupils and staff.
  Interviewing by phone during the day.
  Interviewing people at the hotel.

  Discuss ways in which Hamish's group could get a representative sample.

- Emily's group are about to do a survey on the publication date of the books in the school library. They considered and rejected each of the ways given below for taking a sample of books. **Discuss** reasons for rejecting these.

  Choosing the books closest to the library door.
  Choosing the novels whose authors' names begin with R.
  Choosing the books on the bottom shelves.

  Discuss ways in which Emily's group could get a representative sample.

- Nia's group decided to ask a sample of pupils how they felt about school uniform being compulsory. They considered and rejected the following ways of choosing a sample. **Discuss** reasons for rejecting these.

  Asking the first 20 pupils who enter the school grounds.
  Asking their class.
  Asking the members of the drama club.

  Discuss ways in which Nia's group could choose an unbiased sample.

Choosing **at random** means every person, or item, has the same chance of being chosen.

For instance, if we choose, at random, 5 students from the 30 in a class then each student has the same chance of being chosen. Students who sit at the front of the class or the tallest students or the boys or ... have no greater chance of being chosen than any of the other students.

To choose at random you can, for example, pull names out of a hat, choose every fifth person on the roll, give everyone a number and then use the random number generator on your calculator to choose people, ...

Exercise 2

1 Which of the following samples are likely to be representative samples? For those that are not representative, give a reason (or reasons) why they are not.
   a Survey task: to survey opinion about building a new library.
      Sample: chosen by interviewing people in the street.
   b Survey task: the publisher of a monthly magazine wants to survey opinion about publishing the magazine weekly.
      Sample: chosen by selecting 5 readers of the magazine from each town.

c Survey task: to survey opinion about proposed changes to British Airways timetables.
Sample: chosen by giving a questionnaire to every 100th person who buys a ticket for this airline.

d Survey task: to test the effectiveness of a new drug for migraines.
Sample: chosen by giving the drug to all patients, of one doctor, who suffer from migraines.

e Survey task: to check that a packaging machine for crisps is giving the correct mass in each packet.
Sample: chosen by weighing the first 50 packets of crisps each day.

f Survey task: to survey opinion about a supermarket baking fresh bread.
Sample: chosen by interviewing every 20th shopper at the supermarket checkout on a Saturday.

g Survey task: to survey opinion about the bus service to a city suburb.
Sample: chosen by selecting, at random, 5% of the population from the city's telephone directory.

h Survey task: to check the standard of the maths investigations being done by the students at a school.
Sample: chosen by selecting, at random, 25% of the pupils taught by a particular maths teacher.

i Survey task: to survey a TV channel's viewers about the programmes shown in the past week.
Sample: chosen by asking viewers to write in for a questionnaire.

j Survey task: to check the quality of the apples an orchard is selling direct to the public.
Sample: chosen by having 50 people, chosen at random, buy a box of apples from the orchard at different times during one week.

2 For each of the conjectures given below write these down.
  i The data that needs to be collected and the degree of accuracy.
  ii Possible sources of data. Whether the source is primary or secondary.
  iii Possible sources of bias and how you would choose your sample to eliminate these.

Conjecture 1
Cats which are more than 30% above the average mass visit the vet more often for health problems than those with an average mass or less.

Conjecture 2
In general MEDCs have a lower number of people per physician than LEDCs.

> Remember MEDCs are more economically developed countries.

3 The pupils in a Year 11 class carried out a survey to find the average amount of time spent in the computer room per person each week.
The pupils decided to give the questionnaire to one of the following groups. Which of these is a good sample? Give a reason why each of the others is not a good sample.
Sample 1 the pupils who used the computer room one lunchtime
Sample 2 all the pupils in the library one lunchtime
Sample 3 the first 50 pupils to arrive at school one morning
Sample 4 every 8th pupil on the school roll

4   Describe how you might select an unbiased sample of
   a   the pupils in your class
   b   the pupils in your school
   c   the houses in your street
   d   the trees in an orchard
   e   the potatoes sold by one grower
   f   the cars assembled in a factory
   g   the shoppers at a bookshop during a one-day sale
   h   the readers of a magazine
   i   a shipment of bananas
   j   the people in your town or city or district.

Review 1   Which of the following methods of choosing adults is likely to give a representative sample?
A   choosing adults, at random, in the street
B   choosing every 20th adult at a concert
C   choosing names, at random, from a telephone directory
D   choosing names, at random, from an electoral roll
E   choosing a street, at random, then choosing all the adults who live in that street.

Review 2   Annie read this conjecture in a computing magazine. 'Most instant messaging users are under 25 years and unmarried.'
a   What data would Annie need to collect to find out if the statement is true?
b   Write down a possible primary source for the data.
c   Write down a possible secondary source for the data.
d   How could Annie choose a sample that would give unbiased results?

D Collecting the data

You must now design a **collection sheet or questionnaire** to collect the data.

Remember
If you group data, make sure the class intervals are equal and do not have gaps or overlap.

> See page 358 for more about collecting data.

Questions in a questionnaire must be worded so that they do not encourage a particular response.
This would give biased results.

> See page 358 for guidelines on writing a questionnaire.

Example   I enjoyed the delicious rich creamy chocolate flavour of Choco.
          Agree ☐   Disagree ☐

          It is obvious that the surveyor wants you to tick 'Agree'. A better question would be:

          I found the flavour of Choco
          very nice ☐   nice ☐   not very nice ☐   not very nice at all ☐

When **writing a questionnaire**, it is a good idea to include some questions you might want to pursue further at a later data.

Example   When designing a questionnaire about vegetarian food choice you might want to include some questions on attitudes and beliefs about vegetarian food or vetegarians.

When **conducting an experiment** we must vary one factor at a time keeping the other factors constant.

Example   If you are investigating which of these factors affect the growth of grass

> light, water, heat, seed type, fertiliser

you might start by varying the amount of water given to plots of grass which have the **same** amount of light and heat, are the **same** seed type and are given the **same** amount of fertiliser.

Exercise 3

1 This advert was in a newspaper.   **[SATs Paper 1 Level 6]**
It does not say how the advertisers know that 93% of people drop litter every day.
Some pupils think the percentage of people who drop litter every day is much lower than 93%.
They decide to do a survey.

a Jack says:
We can ask 10 people if they drop litter every day.
Give two **different** reasons why Jack's method might not give very good data.

b Lisa says:
We can go into town on Saturday morning.
We can stand outside a shop and record how many people walk past and how many of those drop litter.
Give two **different** reasons why Lisa's method might not give very good data.

2 a Design a data collection sheet for these. Group the data sensibly.
  i   Timing of goals scored in FA cup games one month
  ii Price of second-hand cars sold privately and by a dealer
  iii Number of weeds in eight equal-sized sections of lawn.
  b Give a reason why we group data.

∗3 Design a collection sheet(s) or questionnaire for one of the following surveys.
Include questions for later or further analysis.
  Surveys
      a survey on car performance/engine size
      a survey on attitudes and beliefs about LEDCs
      a survey on factors that affect aptitude in maths/science or some other subject
      a survey on left-handedness and its consequences
      a survey on factors that affect grass/plant growth

Review 1
a Scientists are observing three different waterholes in a game park in Africa. Design a data collection sheet to group data on how many animals visited the waterholes during the night.
b A graduate student suggests that she spend one night at each waterhole to collect the data. Give a reason why this method might not give very good data.

Review 2
Design a collection sheet or questionnaire for a survey on factors that affect how well pupils do on a maths exam. Include questions for later or further analysis.

Practical

Plan a survey.
You could choose one of the surveys already mentioned in this chapter **or** you could use one of the suggestions below **or** you could make up your own.
Check your choice with your teacher.

Follow the steps given in the previous pages.
Design a collection sheet or questionnaire. Remember you may need to group the data. Decide first if it is continuous or discrete data.
Collect the data.
Write a short report that includes

- your conjecture
- related questions
- where you collected data from
- possible sources of bias and how you chose an unbiased sample
- why you decided on the accuracy you chose for any data
- size of sample
- your collection sheets or questionnaire
- any problems you encountered and how you responded to these e.g. secondary data, for the same variables, that was presented in different forms such as on a graph, tabular or summarised.

Summary of key points

These are the stages in **planning a survey**.

 Decide on **the problem** you want to explore.

It is useful to formulate a **conjecture**, which is a statement you want to test.

Example   A health worker wants to know what factors help people get to sleep easily.
A conjecture for this may be 'A warm milky drink before bed helps people get to sleep.'

 Decide what **data needs to be collected** and possible sources of this data.

Then you must decide **how to collect the data**.

Example   For the above example you will need to carry out a survey using a questionnaire.
This is a primary data source.

You may also want to collect data for related questions.

 Choose an **unbiased sample of suitable size**.

An unbiased sample is one which represents the whole population fairly.

Every person or item in the sample must be chosen **randomly** so each has an equal chance of being chosen.

Example   For the above example, you may choose about 50 males and 50 females from a full range of ages and of socioeconomic status.

Handling Data

 D Design a **collection sheet** or **questionnaire** to collect the data.

Questions in a questionnaire must not encourage a particular response.

Often questions may be ones for later analysis.

Example

> *Do you agree that having a lovely, warm relaxing*
> *bath at night helps you go to sleep?*
> Yes ☐ No ☐

This encourages a 'Yes' answer.

A better question would be

> Does a warm bath at night help you go to sleep?
> Yes ☐ No ☐ Not sure ☐

Test yourself

1 Write three possible conjectures for this.
What effect does taking vitamin tablets have on how often a person gets sick?

2 Which of these samples are likely to be unbiased?
Explain your answers.
a Survey task: to survey opinion on banning open fires to reduce air pollution.
Sample: Surveying people that have open or enclosed fires in their homes.
b Survey task: to find out whether people think there is too much violence on television.
Sample: choosing about 50 people who buy televisions during one month.
c Survey task: to survey opinion on whether the school canteen should sell ice creams.
Sample: choosing every 10th student on the alphabetical school roll.

3 For these two conjectures, write down
i   the data that needs to be collected
ii   possible sources of data and whether it is primary or secondary data
iii how you would choose your sample to eliminate bias.
Conjecture 1 Children in higher income families get sick less often than those in low income families.
Conjecture 2 When fewer rubbish bins are placed in city parks, people will be more likely to take their rubbish home.

4 Design a collection sheet or questionnaire for one of these.
Group the data sensibly.
a Number of bottles of vitamins sold at supermarkets and sold at pharmacies one month
b Number of call-outs by dog control officers in different parts of a city one week

15 Analysing and Displaying Data

You need to know

✓ mode, median, mean, range   page 360

✓ displaying discrete data   page 359

✓ comparing data   page 361

Key vocabulary

cumulative frequency, distribution, estimate of the mean/median, interquartile range, line of best fit, population pyramid, quartile, scatter graph

Searching ....

Find each of these words in the square.

- DATA
- MEAN
- MEDIAN
- MODE
- RANGE
- CLASS
- TABLE
- FREQUENCY
- MODAL

N	P	O	M	E	D	N	M	O	D	A	L
A	D	M	M	E	D	I	R	N	R	C	S
N	A	E	E	U	R	D	A	A	A	L	S
G	T	A	M	D	E	O	N	C	N	A	T
R	P	N	E	Q	I	A	L	T	R	G	A
L	E	R	D	A	T	A	T	A	B	L	E
O	N	F	E	N	S	C	N	Y	E	D	B
M	A	A	N	S	M	G	M	I	A	N	L
W	E	M	E	A	E	D	O	M	O	C	E
H	M	T	A	F	M	O	D	S	A	L	S
M	E	F	R	E	Q	U	E	N	C	Y	F
A	N	C	L	A	S	S	A	L	C	L	A

Mean, median, mode and range

> **Remember**
>
> The **mean**, **median** and **mode** are all ways of summarising data into a single value.
>
> $$\text{Mean} = \frac{\text{sum of data values}}{\text{number of values}}$$
>
> The **median** is the middle value of a set of ordered data.
> The **mode** is the most commonly occurring data value.
> The **modal class** is the class interval with the highest frequency.
> **Range** = highest data value – lowest data value
>
> This exercise gives you practice at working with the mean, median, mode and range.

Exercise 1

1 **a** Ikram handed in four computing assignments.
       For three of them he got 8 marks each.
       In the other he got no marks.
       What was Ikram's mean mark for the four assignments?
   b Jessica only handed in two assignments.
       Her mean mark for the two was 3.
       Her range was 4.
       What did Jessica get for each assignment?
   c Sarah handed in three assignments.
       Her mean mark for these three was also 3.
       Her range was also 4 marks.
       What marks might Sarah have got in her three assignments?
       Show your working.

2   **Mean age** = 19 years 7 months
        Range = 3 years 0 months
   This gives the mean age and range of members of a football club.
   Damion, who is 20 years 7 months, joins the club.
   a Which of these is correct about the mean age of members?
       A It will increase by exactly 1 year.
       B It will increase by less than 1 year.
       C It will stay the same.
       D It is not possible to tell.
   b Which of A, B, C or D in part **a** is true about the range of ages of members once
       Damion has joined?

3 Ben measured the heights, in cm, of some of his friends. These are their heights in order.
       172,   171,   170,   170,   169,   168,   161,   158
   a Find the mean, median and mode of these heights.
   b If you were summarising the data, which of these would you use? Explain.

4 Three sisters have a mean age of 12.
   The range of their ages is 6.
   What is the lowest possible age of
   a the youngest sister   **b** the oldest sister?

*5 Rajshree picks five numbers from 0 to 20.
He says 'The range of my numbers is 4, the mode 5 and the mean is 5.'
What might his numbers be? Is there more than one possible answer?

*6 Four men and two women were on a minibus tour.
They all received letters while on tour.
The mean number of letters received by the men was 20.
The mean number of letters received by the women was 26.
Are these true or false? Explain.

a The person who received the *most* letters **must** have been a woman.

b The mean number of messages received by the six people was

Review 1
a The mean of six numbers is 7·5. If five of the numbers are 4, 6, 11, 13 and 2, what is the sixth number?
b If the median of 1, 3, 6, x, 15, 20 is 9, what is the missing number x?

Review 2 Ali, Brian, Cate and Damion read a mean number of 16 books per year. Ed reads 23 books per year. When the reading figures for Frankie are added, the mean number of books read by the six friends is 18.
a How many books did Frankie read per year?
b Is it possible that three of Ali, Brian, Cate and Damion read more books per year than Frankie? Explain.

* **Review 3** A researcher is investigating family sizes. She interviews 15 women, chosen at random. 10 of the women answer 'yes' to the question 'Do you have any children?' The interviewer then asks the 10 women how many children they have, and finds that the mean number of children per family is 2. What is the maximum number of children that any one of these 10 women could have?

Mean, median and range from tables

The statistical mode on a calculator can be used to find the **mean** of a frequency distribution.

Remember

$$\text{Mean} = \frac{\text{sum of (frequency } (f) \times \text{ data value } (x))}{\text{sum of frequency}}$$

Example

Number of movies seen last week	0	1	2	3	← This is the 'data values' row
Frequency		5	9	8	3

To find the mean of this frequency distribution key ⬚MODE⬚ ⬚2⬚ to put the calculator into statistical mode. Then key

 ⬚Shift⬚ ⬚Scl⬚ ⬚=⬚   ⬚0⬚ ⬚Shift⬚ ⬚;⬚ ⬚5⬚ ⬚DT⬚

to clear the   enters data   enters the frequency,   enters the
statistical memory   value, x 5, of the data value $f \times x$ data

 Always clear the statistical memory before every new calculation.

then key ⬚1⬚ ⬚Shift⬚ ⬚;⬚ ⬚9⬚ ⬚DT⬚ ⬚2⬚ ⬚Shift⬚ ⬚;⬚ ⬚8⬚ ⬚DT⬚ ⬚3⬚ ⬚Shift⬚ ⬚;⬚ ⬚3⬚ ⬚DT⬚ ⬚Shift⬚ ⬚x̄⬚ ⬚=⬚

to get a mean of 1·36.

Note The zero data values must be entered.

The **median** can be read from a frequency table.
For the example on the previous page, the number of data values is
found by adding the frequencies. $5 + 9 + 8 + 3 = 25$
The median is the $\frac{1}{2}(n + 1)$th value when there are an
odd number of values.
The median is the 13th value.   **13th value**
The 13th value is 1.   The data is   0, 0, 0, 0, 0, 1, 1, 1, 1, 1, 1, 1, 1, 1, 2, 2, 2, ...

Number of movies	0	1	2	3
Frequency	5	9	8	3

5 values to here   14 values to here so 13th value is in here

The **range** is the highest data value minus the lowest
data value. For the example given:
range = $3 - 0$
        = 3

Exercise 2

1 Pia made this frequency table for the number of days pupils in
her class were absent, during one week.

Days absent	1	2	3	4	5
Frequency	6	0	3	1	1

   a Find the mean, median, mode and range of the data.
   b Pia made some statements about the data, using the answers
to **a**. What statements might Pia have made? Are all of the
mean, the median, the mode and the range equally useful in analysing this data?

2 The following table gives the number of pupils in classes in a school.

Number of pupils in class	27	28	29	30	31	32	33
Frequency	3	2	3	4	5	0	1

   a Find the mean, median, mode and range of this data.
   b Make some statements about the data, using your answers from **a**. Is any one of the
mean, the median, the mode and the range more useful than others in analysing the data?

3 Owls eat small mammals.   [SATs Paper 2 Level 7]
They regurgitate the bones and fur in balls called pellets.
The table shows the contents of **62** pellets from long-eared owls.

Number of mammals found in the pellet	1	2	3	4	5	6
Frequency	9	17	24	6	5	1

   a Show that the **total** number of mammals found is **170**.
   b Calculate the **mean** number of mammals found in each pellet.
Show your working and give your answer correct to 1 decimal place.
   c There are about **10 000** long-eared owls in Britain.
On average, a long-eared owl regurgitates **1·4 pellets** per day.
Altogether, how many **mammals** do the 10 000 long-eared owls eat in **one day**?
Show your working and give your answer to the nearest thousand.

4 Draw a table like this one.
Fill it in for the number of letters in the
surnames of the pupils in your class.
Find the mean, median and range of the data.

Number of letters	3	4	5	6	7	8
Tally						
Frequency						

Review Winstone conducted a survey on hospital visiting. He included the following table as part of his analysis.

Visitors per patient	0	1	2	3	4	5	6	7	8
Frequency	6	25	14	9	6	0	2	0	1

a Winstone calculated the mean and the median number of visitors per patient. What should his answers be?

b Find the range.

c What is the mode of the frequency distribution?

d Winstone wrote a few sentences about the data, using his answers to **a**, **b** and **c**. What might he have written? Do you think the mean, the median, the mode and the range are all equally useful in analysing this data?

Mean, median, mode for grouped data

We can find an **estimate for the mean** of grouped data. We assume each item of data in a given class interval has the value of the mid-point of that class interval.

Example

Mark in test	0–9	10–19	20–29	30–39	40–49	50–59
Frequency	8	11	5	6	9	4

This table gives the marks in a test.

This is grouped discrete data. The 11 data values in the class interval 10–19 will be some or all of 10, 11, 12, 13, 14, 15, 16, 17, 18 or 19.

The mid-point of these values is 14·5. $(\frac{10 + 19}{2})$

Mark in test (x)	0–9	10–19	20–29	30–39	40–49	50–59
Mid-point (m)	4·5	14·5	24·5	34·5	44·5	54·5
Frequency (f)	8	11	5	6	9	4

$$\textbf{Mean} = \frac{\textbf{sum of (frequency } (f) \times \textbf{mid-point } (m))}{\textbf{total } f}$$

$$= \frac{8 \times 4\cdot5 + 11 \times 14\cdot5 + 5 \times 24\cdot5 + 6 \times 34\cdot5 + 9 \times 44\cdot5 + 4 \times 54\cdot5}{8 + 11 + 5 + 6 + 9 + 4}$$

$$= 26\cdot6 \text{ (to 1 d.p.)}$$

The **estimated mean mark is 26·6** (1 d.p.).

Key MODE 2 Shift Scl =
4·5 Shift ; 8 DT 14·5 Shift ; 11 DT 24·5 Shift ; 5 DT
34·5 Shift ; 6 DT 44·5 Shift ; 9 DT 54·5 Shift ; 4 DT
Shift $\bar{x}$ =

The **modal class is 10–19**.

There are 43 values, so the median is the 22nd value and is in the class interval 20–29.

Example The table below gives the time spent helping with household jobs.

The data is continuous.

The class interval   5–   means $5 \leqslant t < 10$.

The mid-point of this interval is 7·5. $(\frac{5 + 10}{2})$

Time in minutes	5–	10–	15–	20–	25–	30–	35–40
Mid-point (m)	7·5	12·5	17·5	22·5	27·5	32·5	37·5
Frequency (f)	8	12	14	17	5	3	1

$$\text{Mean} = \frac{\text{sum of } mf)}{\text{sum of } f}$$

You could use the statistical mode on a calculator to find the mean.

$$= \frac{8 \times 7\cdot5 + 12 \times 12\cdot5 + 14 \times 17\cdot5 + 17 \times 22\cdot5 + 5 \times 27\cdot5 + 3 \times 32\cdot5 + 1 \times 37\cdot5}{8 + 12 + 14 + 17 + 5 + 3 + 1}$$

$$= \textbf{18·5}$$

The mean time spent helping with household jobs is **18·5 minutes**.

Discussion

- Dale is 15.

  Discuss the following statements to do with age and Dale's
  Some of the statements are true and some are false.

  > Age is continuous, not discrete.
  > Dale could be as young as $14\frac{1}{2}$ or as old as $15\frac{1}{2}$.
  > Dale could be just 15.
  > Dale could not be 16.
  > Dale could be as young as 15 years 1 day or as old as
  > 15 years 364 days.
  > Dale could be 1 minute older than 15 or 1 minute younger than 16.

  Discuss how best to describe Dale's age.

-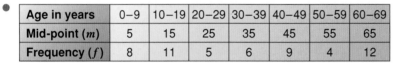

Age in years	0–9	10–19	20–29	30–39	40–49	50–59	60–69
Mid-point (m)	5	15	25	35	45	55	65
Frequency (f)	8	11	5	6	9	4	12

  The data on this table gives the ages of people in a doctor's waiting room one day.
  How were the mid-points of the class intervals found? **Discuss**.

- * **Food and non-alcoholic drink expenditure by age of head of household**

	Under 30	30 and under 50	50 and under 65	65 and under 75	75 or over	All house-holds
Average amount spent	55·00	73·60	67·10	47·00	34·70	61·90

  How could the mean spending on food and non-alcoholic drinks be calculated?
  Does it matter that the class intervals are not the same width?
  What might the mid-point of the class intervals 'under 30' and '75 or over'
  be? **Discuss**.

An **estimate of the median** of grouped data can be made.

Example   This frequency table shows the number of days of hire of vans from Daley's Van
Hire during last February.

Length of hire in days	1–4	5–8	9–12	13–16	17–20	21–24	25–28
Frequency	20	8	7	9	7	0	10

61 people hired vans. When the data is in order, the median number of days hire is
the 31st. The 31st is in the interval 9–12 days and is the 3rd in that interval. There
are 7 data values in that interval. We take the 31st to be $\frac{3}{7}$ of the way along that
interval.
The interval spans $12 - 9 = 3$ days. The 31st is then $\frac{3}{7} \times 3$ days = 1·3 days (to 1 d.p.)
from the beginning of the interval.
The estimate of the median is $9 + 1·3$ days = **10·3 days** (1 d.p.).

An **estimate of the range of grouped data** can be made.
For the example above the least number of days of hire possible is 1 and the greatest is 28.

   Range = 28 – 1 = 27

> This data is discrete
> data.

Worked Example
Estimate the range of the heights of Year 9 pupils.

Height	155–	160–	165–	170–	175–	180–185
Frequency	4	3	6	8	2	3

Answer
The first class interval, 155–, might have heights from 154·5 cm up to 160 cm. The last class interval might have heights up to 185·5 cm.
An estimate of the range is   185·5 – 154·5 = **31 cm**.

This data is continuous data.

Practical

You will need a spreadsheet package.

Ask your teacher for the **Mean from grouped data** ICT worksheet.

Exercise 3

1

Length of stay in days	1–3	4–6	7–9	10–12	13–15	16–18	19–21	22–24
Frequency	32	27	19	10	6	14	5	18

This table gives the number of days that each of 131 guests stayed in a hotel. Find
a the mid-point of the 1–3 class interval   **b** an estimate for the mean length of stay
c the modal class   **d** the class interval that contains the median
e an estimate of the range.
Make some statements about the data, based on your answers.

2 a

Test A

Number of correct answers	1–10	11–20	21–30	31–40	41–50
Frequency (*f*)	21	34	69	52	24

A multiple choice test was trialled on 200 pupils.
This table shows the number of multiple choice questions that these pupils answered correctly.
Find the mean number of correct answers.

b Another multiple choice test was trialled on these same pupils.
The following table shows the results of this test.

Test B

Number of correct answers	1–10	11–20	21–30	31–40	41–50
Frequency (*f*)	11	42	54	52	41

Find the mean number of correct answers.

∗c What other calculations could you do to help you compare the two tests in **a** and **b**?
Write a few sentences comparing these tests.

3 The pupils in Bruce's maths class organised a dart's competition. This table gives the number of darts thrown by each pupil to get a bull's eye.

Darts thrown to get bull's eye	1–5	6–10	11–15	16–20	21–25	26–30
Number of pupils	3	8	3	2	4	5

a How many pupils threw between 6 and 10 darts to get a bull's eye?
b How many pupils were in this competition?
c Estimate the median number of darts thrown to get a bull's eye.
d Estimate the range of the number of darts thrown.

4

Weight loss in kg	1–	2–	3–	4–	5–	6–	7–8
Frequency	2	0	3	7	5	4	4

This table shows the weight loss of 25 people, during their first month on a new diet.
a Find the mid-point of the first class interval.
b Estimate the mean weight loss.
c Estimate the median.
d What is the modal class?
e Estimate the range.
*f Use your answers to a, b, c, d and e to make some statements about the data.

5

Time taken in min (t)	$30 \leqslant t < 35$	$35 \leqslant t < 40$	$40 \leqslant t < 45$	$45 \leqslant t < 50$	$50 \leqslant t < 55$	$55 \leqslant t < 60$
Number of pupils	6	8	14	13	71	88

A multiple choice maths test was trialled on 200 pupils.
This table shows the time taken by these pupils to complete the test. Find
a the mid-point of the first class interval
b an estimate for the mean time taken
c an estimate of the median
d an estimate for the range.

6 52 pupils entered a school's golf tournament in Hampshire. The table shows the distance of the longest tee shot of these pupils.

Distance in metres	120–	130–	140–	150–	160–	170–	180–	190–200
Frequency	10	2	11	9	10	7	1	2

Find
a the mid-point of the first class interval
b an estimate for the mean distance
c the modal class
d an estimate of the median distance
e an estimate for the range.

7 There were 43 pupils in a golf tournament in Devon. The table shows the distance of the longest tee shot of these pupils.

Distance in metres	120–	130–	140–	150–	160–	170–	180–	190–200
Frequency	2	7	8	7	0	6	6	7

a Find the modal class and an estimate for the mean distance.
*b The median of this data is 157 m (nearest m). The median of the data in question 6 is 154 m (nearest m).
Zhaleh was comparing the tee shots of the pupils from Hampshire with those from Devon. She decided that one of the median, mean or modal class represented the data more fairly than the others. Which average measure does this? Explain your answer.

Review 1 Before releasing a new disposable razor onto the market, the manufacturers carried out market research on its performance. One of the razors was trialled by each of 400 men. Each man noted the number of satisfactory shaves from the razor he trialled. The results of the market research are shown on the table.

Number of shaves	1–5	6–10	11–15	16–20	21–25	26–30	31–35
Frequency	1	0	39	82	177	94	7

Find

a the mean number of shaves per razor
b the median class interval
c an estimate of the median
d the modal class
e an estimate of the range.

Review 2 This table shows the lengths of the cars parked in one section of a hypermarket car park.

Length in metres	$3 \leqslant l < 3.5$	$3.5 \leqslant l < 4$	$4 \leqslant l < 4.5$	$4.5 \leqslant l < 5$	$5 \leqslant l < 5.5$	$5.5 \leqslant l < 6$
Frequency	21	14	52	7	5	1

Find

a the mid-point of the first class interval
b an estimate of the mean length of the cars
c the modal class
d an estimate of the median
e an estimate of the range.

Use your answers to **a**, **b**, **c** and **d** to make some statements about the data.

Practical

Design and use a collection sheet or questionnaire to collect some grouped data, either discrete or continuous. There are some suggestions below.
Decide on a sensible number of class intervals.

Display the data on a frequency table.
Decide whether one or more of the mean, the median, the modal class would be useful in the analysis of your data. Analyse the data using a calculator or spreadsheet.

Suggested data

hours of sleep in one week for each of the pupils in a class
prices of houses advertised for sale in a newspaper
height of pupils in a year group
number of words on pages of this book
time taken to run a marathon by a group of runners

> You could collect two sets of data, e.g. for two different age ranges or year groups, and compare them.

Finding median and quartiles from cumulative frequency

Upper and lower quartiles

We may use the statistical measures, **mean**, **median**, **mode** and **range** to analyse data. Other statistical measures we may use are the **quartiles** and **interquartile range**.

For a set of data arranged in order,
– the median is the middle value,
– the **lower quartile** is the middle value of the lower half of the data
– the **upper quartile** is the middle value of the upper half of the data.

The **interquartile range** is the difference between the upper and lower quartiles.

> **Interquartile range = upper quartile – lower quartile**

Discussion

● **Discuss** how the following statements could be completed.

'The median divides a set of data in ...'
'The quartiles divide a set of data into ...'
'One half of the data values lie below the ...'
'One-quarter of the data values lie below the...'
'Three-quarters of the data values lie below the ...'
'The interquartile range is the range of the central ... of the data.'

This frequency distribution shows the time taken, by the 26 pupils in a class, to correctly solve a problem.

It is difficult to find the lower quartile or the upper quartile from this table. However, we can find these easily by drawing a **cumulative frequency graph**.

Time (sec)	Frequency
10–	1
20–	0
30–	2
40–	5
50–	6
60–	7
70–	3
80–90	2

To draw a cumulative frequency graph we first write a **cumulative frequency table**. In this table we add up the frequencies as we go along. The cumulative frequency table gives a 'running total' of the frequencies.

There are no values less than 10.
There is 1 value less than 20 in the 10– class interval.
There is 1 value less than 30 in the 10– and 20– class intervals, there are 3 values less than 40 m, 1 in the 10–class interval and 2 in the 30–class interval.

Cumulative frequency table

Time (sec)	Cumulative frequency	
less than 10	0	← 0
less than 20	1	← 1
less than 30	1	← 1+0
less than 40	3	← 1+0+2
less than 50	8	← 1+0+2+5
less than 60	14	← 1+0+2+5+6
less than 70	21	← 1+0+2+5+6+7
less than 80	24	← 1+0+2+5+6+7+3
less than 90	26	← 1+0+2+5+6+7+3+2

We draw a cumulative frequency graph by plotting the points (10, 0), (20, 1), (30, 1), (40, 3), (50, 8), (60, 14), (70, 21), (80, 24), (90, 26).
We always begin a cumulative frequency graph on the horizontal axis.

Note We plot the frequencies at the upper boundary of each class interval.

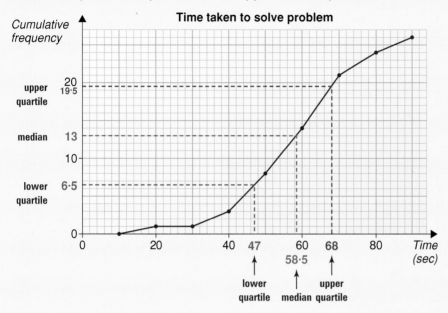

We can find the median, the lower quartile and upper quartile from the cumulative frequency graph, as shown.

There are 26 values in total.

> For 26 values, the median is in fact the $13\frac{1}{2}$th value. By convention for a cumulative frequency the median is taken as the $\frac{1}{2}n$th value when there are an even number of values

Median   The **median** value is the 13th value ($\frac{1}{2}$ of 26).
                    The red dashed line shows the median time is about **58·5 seconds**.

Lower quartile The **lower quartile** value is 6·5 ($\frac{1}{4}$ of 26).
                    The green dashed line shows the lower quartile is about **47 seconds**.

Upper quartile The **upper quartile** value is 19·5 ($\frac{3}{4}$ of 26).
                    The purple dashed line shows the upper quartile is about **68 seconds**.

We often use the median and interquartile range to compare data. The median tells us the middle value. Half the values are above and half below this. The interquartile range tells us how spread out the middle half of the values are.

Example The median and interquartile range for the ages of cats and dogs at a pet show are given. On average, the dogs tended to be older, shown by the higher median. The dogs had more variation in age, shown by the higher interquartile range.

	Cats	Dogs
Median age (years)	2.3	2.9
Interquartile range	2.1	4.7

Exercise 4

1

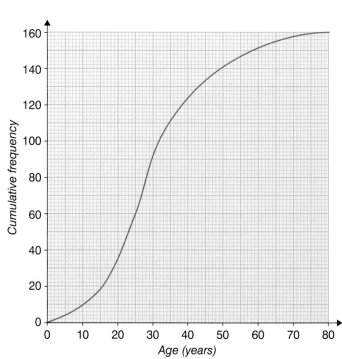

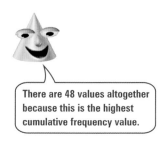

There are 48 values altogether because this is the highest cumulative frequency value.

This graph is the cumulative frequency graph for the ages of people in the United Kingdom who are under 65.

a What is the median age of these people?

b Find the lower and upper quartiles.

c What is the interquartile range?

d How many people are aged under 20?

e How many people are aged under 15?

f How many people are aged between 15 and 20?

g How many people are aged between 60 and 65?

2 Tom did a survey of the age distribution of people at a theme park.   [SATs Paper 1 Level 8]
He asked **160 people**.
The cumulative frequency graph shows his results.

a Use the graph to estimate the **median** age of people at the theme park.

b Use the graph to estimate the **interquartile range** of the age of people at the theme park. Show your method on the graph.

c Tom did a similar survey at a flower show.

Results
The **median** age was **47 years**.
The **interquartile range** was **29 years**.

Compare the age distribution of the people at the flower show with that of the people at the theme park.

3 The first 'Thomas the Tank Engine' stories were written in 1945. [SATs Paper 1 Level 8]

In the 1980s, the stories were rewritten.

The cumulative frequency graph shows the numbers of words per sentence for one of the stories.

> There are **58 sentences** in the old version.
> There are **68 sentences** in the new version.

a Estimate the **median** number of words per sentence in the old version and in the new version. Show your method on the graph.

b What can you tell from the data about the number of words per sentence in the old version and in the new version?

c Estimate the percentage of sentences in the **old** version that had more than 12 words per sentence. Show your working.

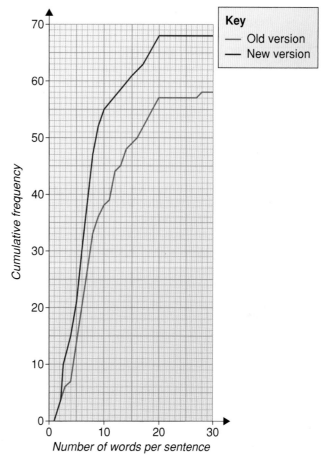

4

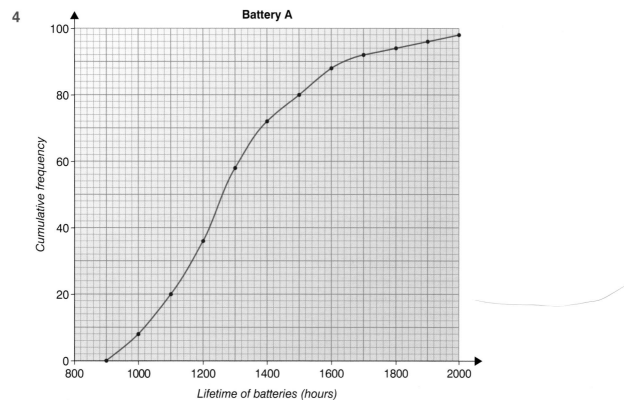

Battery A

Handling Data

a Testing was carried out on a new type of calculator battery (Battery A). One hundred of these batteries were tested. The lifetime of these batteries is shown on the cumulative frequency graph on the previous page.

  i How can you tell from this graph that 2 of the batteries were still working after 2000 hours of use?

  ii What is the median lifetime of the 100 batteries tested?

  iii What is the interquartile range?

*b One hundred calculator batteries of another type (Battery B) were also tested. The results are shown in the table below.

Lifetime (hours)	1000–	1100–	1200–	1300–	1400–	1500–	1600–	1700–1800
Frequency	3	5	9	21	33	16	7	6

  i Draw the cumulative frequency graph for this data.

  ii Use the cumulative frequency graph to find the median, the quartiles and the interquartile range.

*c Write a sentence or two comparing the two different types of calculator batteries.

Review

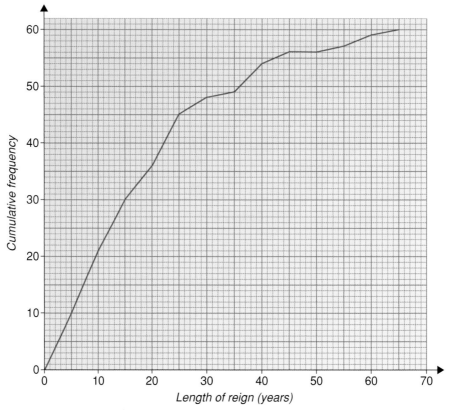

Length of reign (years)

This cumulative frequency graph shows the length of the reign of English monarchs.

a How many monarchs reigned for less than 5 years?

b How many monarchs reigned for between 5 and 10 years?

c How many monarchs has England had?

d What is the median length of reign?

e Find the lower and upper quartiles.

f What is the interquartile range?

g What percentage of monarchs reigned for more than 25 years?

h *5% of reigning monarchs reigned for more than x years.* What is the value of x?

Practical

Collect some data. Some suggestions follow.
You could carry out a survey or an experiment or you could collect your data from secondary sources such as the Internet, CD-ROMS, Books, ...
Draw graphs to display your data. Include cumulative frequency graphs.
Analyse your data. Include the median, quartiles and interquartile range as part of your analysis.

Suggested data

prices of cars advertised for sale
wrist measurements of pupils
comparison of time spent playing sport by the pupils in two different year groups
comparison of pulse rates of 10-year olds and 20-year olds

Note You could work as a group. If you do this, choose a theme such as sport. Gather, display and analyse data on more than one aspect of this theme.

Frequency polygons

A **frequency polygon** is usually drawn from a frequency diagram.
Join the mid-points of the tops of the bars.
Join the mid-points of the tops of the first and last bar to the axis at half a class interval distance from the bar (the mid-points of the zero height bars on either side).

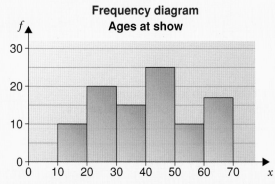

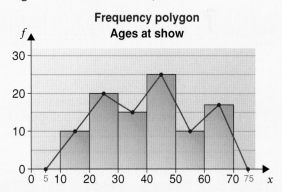

We can **estimate the mean** from a frequency polygon, using the mid-points of the intervals.
For the frequency polygon shown above

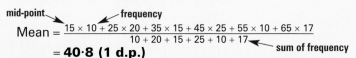

$$\text{Mean} = \frac{15 \times 10 + 25 \times 20 + 35 \times 15 + 45 \times 25 + 55 \times 10 + 65 \times 17}{10 + 20 + 15 + 25 + 10 + 17}$$

← sum of frequency

$$= \mathbf{40 \cdot 8 \ (1 \ d.p.)}$$

See page 383.

We can estimate the median from a frequency polygon.
There are 97 values.
The 49th value is the median.
This lies in the class interval 40–50. It is about $\frac{4}{25}$ of the way along. $\quad \frac{4}{25} \times 10 = 1 \cdot 6$
So the median is about $40 + 1 \cdot 6 = \mathbf{41 \cdot 6}$.

Discussion

● How can a frequency polygon be drawn for data without first drawing a frequency diagram? **Discuss.** Use the data given below in your discussion.

Boys' test marks

Class interval	f
0–9	3
10–19	2
20–29	5
30–39	9
40–49	18
50–59	22
60–69	25
70–79	8
80–89	7
90–99	1

Girls' test marks

Class interval	f
0–9	1
10–19	0
20–29	12
30–39	13
40–49	13
50–59	21
60–69	15
70–79	11
80–89	9
90–99	5

● Might it be useful to draw frequency polygons for the above data on the same set of axes? **Discuss**.

Discuss the advantages and disadvantages of data being graphed as a frequency diagram or a frequency polygon. As part of your discussion, draw the frequency diagrams and frequency polygons for the two sets of data given above.

●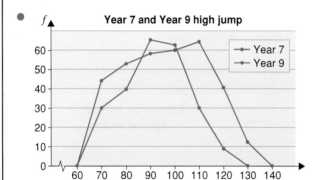

Link to PE.

These frequency polygons show the Year 7 and Year 9 high jump results.
How useful are the frequency polygons for comparing two sets of data? **Discuss**.

We can draw a frequency polygon without first drawing a frequency diagram.
We do this by – finding the mid-point of each interval
  – plotting the frequency against the mid-points
  – joining these points with a line and then joining the first and last points to the axes.

Example This shows the frequency polygon for the girls' test marks from the discussion above.

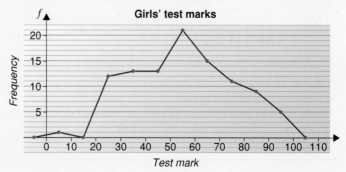

Exercise 5

1 Draw a frequency diagram and a frequency polygon for each of the following sets of data.

a

Score	10–19	20–29	30–39	40–49	50–59	60–69	70–79	80–89
f	3	4	7	5	7	4	2	1

b

Length in metres (*l*)	Frequency
$0 \leqslant l < 1$	5
$1 \leqslant l < 2$	0
$2 \leqslant l < 3$	3
$3 \leqslant l < 4$	5
$4 \leqslant l < 5$	2
$5 \leqslant l < 6$	4
$6 \leqslant l < 7$	6

2 Use a copy of these frequency diagrams, which show the money spent at 'Just T-shirts' by males and females last week.

a Draw the frequency polygon for each.

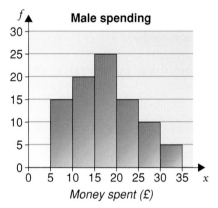
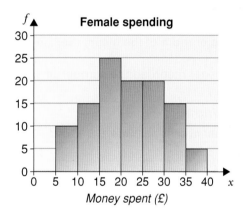

b Estimate the mean and median for male spending.

c Estimate the mean and median for female spending.

d What conclusions can you draw from your answers to **b** and **c**?

3

These frequency polygons show the points scored by the Fastfeet Netball A and B teams in their matches last season.

Comment on any similarities and differences.

4 Fifty 20-year-old males and fifty 20-year-old females were interviewed about the way they spent their incomes. The percentage of income that each group saved is shown in the tables.

Males

% saved	Frequency
0–	2
10–	0
20–	19
30–	10
40–	11
50–60	8

Females

% saved	Frequency
0–	3
10–	3
20–	14
30–	12
40–	9
50–	6
60–70	3

a On the same set of axes, draw frequency diagrams and frequency polygons for both of these frequency distributions.

b Write a few sentences about the differences or similarities between these frequency polygons.

c Estimate the mean percentage saving of the males and of the females.

d Is the modal class interval the same for each?

e In what class interval is the median for each?

f Do you think the data is easier to compare using the frequency polygons rather than one or more of the mean, median and mode and the range?

∗5 A manager recorded the number of sales calls per week made by two sales representatives over a year.

John Wright

Number of calls per week	0–4	5–9	10–14	15–19	20–24	25–29	30–34	35–39
Frequency	1	0	3	18	22	5	2	1

Fred Smith

Number of calls per week	0–4	5–9	10–14	15–19	20–24	25–29	30–34	35–39
Frequency	0	6	5	12	3	5	6	13

a Draw frequency polygons to show this data.

b Compare the number of calls made per week by each sales representative.
You could use the mean, median, mode, range and/or frequency polygons you drew to make some conclusions.

Review For one week, Mr Jones and Mr Smyth timed all the phone calls their children made. The frequency distributions are shown in these tables.

Jones

Time in min (t)	Frequency
0–	6
5–	7
10–	2
15–	2
20–	0
25–30	1

Smythe

Time in min (t)	Frequency
0–	2
5–	9
10–	3
15–	0
20–	0
25–	4
30–	3
35–40	2

a On the same set of axes draw frequency polygons for these distributions.

b Write a few sentences about the similarities and differences between these frequency polygons.

c Estimate the mean and the median time spent on each phone call for each family.

d Estimate the range of calls for each family.

e Which do you think is easier, comparing sets of data using frequency polygons or using the median or mean and the range?

⭐ Practical

You will need tangram pieces. You could make these from a pattern.

Give a set of tangram pieces to each person in your class.
Ask each person to time themselves to make this shape.
Record the time taken by boys and girls on separate frequency tables.
Draw frequency polygons for the data.
Use these and the mean, median and range to compare the times of boys and girls.

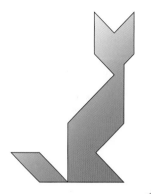

Scatter graphs

Remember
A **scatter graph** displays two sets of data.
A scatter graph sometimes shows a relationship or **correlation** between the variables.

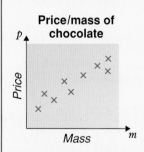

Price/mass of chocolate

This graph shows a positive correlation between price and mass of chocolate. The greater the mass, the greater the price is likely to be.

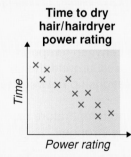

Time to dry hair/hairdryer power rating

This graph shows a negative correlation between the power rating of a hairdryer and the time to dry hair. The greater the power rating the less time it takes to dry hair.

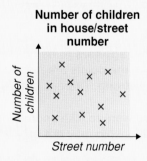

Number of children in house/street number

This graph shows no correlation. There is no relationship between the street number and the number of children living in a house.

399

Discussion

- Which of the lines $k, l, m,$ or n do you think 'fits' the data best? **Discuss**.

  Can another line be drawn that 'fits' the data even better? **Discuss**.

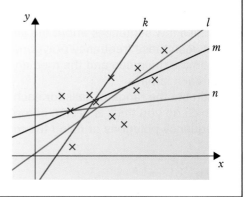

The **line of best fit** should have about the same number of values above and below the line and at about the same distance from the line.

Example

The following table gives the maximum monthly temperature and the monthly rainfall during April in a South American city.

Rainfall (mm)	42	43	54	58	50	45	47	52	59	60	53	53
Temperature (°C)	18	17	20	24	19	17	20	19	22	23	18	22

The scatter graph for this data is shown below. The line of best fit is drawn on.

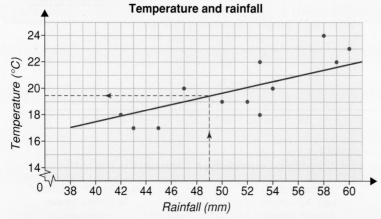

There **appears** to be positive correlation between temperature and rainfall in that city. Since the points are not clustered very closely around the line of best fit, the correlation is not very strong. There is not a very strong relationship between temperature and rainfall.

If we were to take another set of rainfall and temperature readings for the same city, the relationship may not exist at all or it may be stronger.

We can use the line of best fit to **estimate one measurement**, given another. The more closely the points are clustered around the line of best fit, the better the estimate.

Note   The estimate may not be very accurate at all.

If the rainfall in a month was 49 mm, an estimate of the maximum temperature is 19·5 °C. This estimate is not very reliable since the plotted points are not clustered very closely around the line of best fit.

Discussion

Toby investigated the lengths and wingspans of aircraft.
He collected this data and drew the scatter graph shown.

Aircraft type	Length (m)	Wingspan (m)
Boeing 747	70	60
Boeing 767	60	48
Boeing 737–400	36	29
Airbus 330	58	60
Airbus 320	38	34
Boeing 737–800	40	36

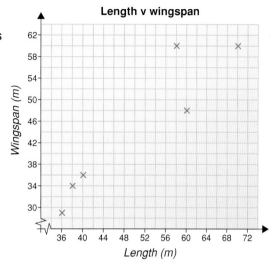

Discuss what sort of relationship exists between the length and wingspan of an aircraft?
Which aircraft data might you ignore if drawing a line of best fit? **Discuss**.
Could you estimate accurately the wingspan of an aircraft of length 50 m? **Discuss**.

Exercise 6   You could use a spreadsheet to draw the graphs.

1 The following data gives the golf scores of 14 people on two consecutive days.

Friday	74	79	71	68	81	75	72	69	78	70	81	77	82	75
Saturday	72	76	73	69	77	75	70	71	77	72	79	75	78	74

  a Use a copy of the grid.
   Draw a scatter graph for the data.
  b Does the scatter graph show that there is positive, negative or no correlation between the scores on the two days?
  c Draw a line of best fit.
  d Grace scored 68 on Friday. Estimate her Saturday score.
  e Hal scored 79 on Saturday. Estimate his Friday score.

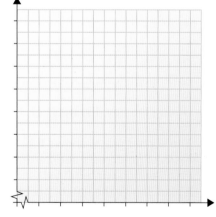

2

Long jump (m)	4·90	4·30	5·85	5·98	5·02	5·52	4·53	6·03	5·45	4·67	5·28
High jump (m)	1·55	1·60	1·86	2·03	1·83	1·74	1·54	1·30	1·88	1·78	1·73

This table gives the long jump and high jump results of the pupils who entered both events.
  a Would you expect there to be positive correlation between these results?
  b Plot the data on a scatter graph. (Have Long jump on the horizontal axis.)
  c Describe the correlation between long jump and high jump results.
  d Draw a line of best fit. Did you ignore any points? If so, explain why.
  e Meg jumped 5·32 m in the long jump. Estimate how high she jumped in the high jump. Explain why this estimate might be inaccurate.

Handling Data

3

Height (m) of pupil	1·64	1·70	1·58	1·65	1·81	1·67	1·62	1·73	1·71	1·77
Height (m) of best friend	1·72	1·69	1·82	1·70	1·74	1·64	1·58	1·68	1·60	1·70

a   Plot this data on a scatter graph. Have Height of pupil on the horizontal axis.
b   Could you use this graph to estimate the height of the best friend of a pupil who is 171 cm tall? Explain your answer.

4   The scatter graph shows information about trees called poplars.   **[SATs Paper 1 Level 6]**

a   What does the scatter graph show about the **relationship** between the diameter of the tree trunk and the height of the tree?
b   The height of a different tree is 3 m. The diameter of its trunk is 5 cm. Use the graph to explain why this tree is **not** likely to be a poplar.
c   Another tree is a poplar. The diameter of its trunk is 3·2 cm. Estimate the height of this tree.
d   **[Level 7]**
Below are some statements about drawing lines of best fit on scatter graphs.
Say whether each statement is true or false.
Lines of best fit must **always** ...
i   go through the origin
ii   have a positive gradient
iii   join the smallest and the largest values
iv   pass through every point on the graph.

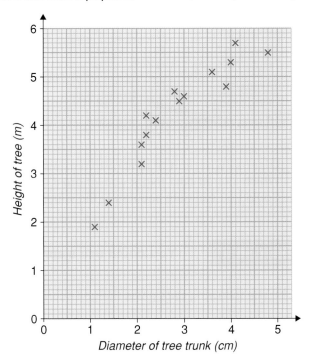

5   The goldcrest is Britain's smallest species of bird.   **[SATs Paper 2 Level 7]**
On winter days, a goldcrest must eat enough food to keep it warm at night.
During the day, the mass of the bird increases.
The scatter graph shows the mass of goldcrests at different times during winter days.
It also shows the line of best fit.
a   Estimate the mass of a goldcrest at **11:30 a.m.**
b   Estimate how many grams, on average, the mass of a goldcrest **increases** during **one hour**.
c   Which goldcrest represented on the scatter diagram is **least likely** to survive the night if it is cold?
Show your answer by giving the coordinates of the correct point on the scatter graph, then explain why you chose that point.

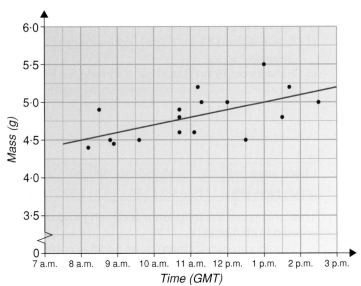

*6 **Demographic indicators of selected countries in the Commonwealth**

	Population (millions) 2000	Average annual growth rate 1995–2000 (percentage)	Percentage in urban areas 1990–99	Infant mortality rate[1]	Life expectancy at birth (years)	
					Males	Females
Australia	18·9	1·0	85	6	76	81
Bangladesh	129·2	1·7	25	79	58	58
Canada	31·1	1·0	77	6	76	82
The Gambia	1·3	3·2	33	122	45	49
India	1 013·7	1·6	28	72	62	63
Namibia	1·7	2·2	31	65	52	53
New Zealand	3·9	1·0	86	7	74	80
Sierra Leone	4·9	3·0	37	170	36	39
Singapore	3·6	1·4	100	5	75	79
United Kingdom	59·8	0·2	90	7	75	80

[1] *per 1000 live births*   *Source: Department of Economic and Social Affairs, United Nations*

a Draw a scatter graph for each of these.
   i Infant mortality rate and life expectancy
     (take the mean of male and female life expectancy).
   ii Percentage in urban areas and infant mortality.
   iii Average annual growth rate and life expectancy.
   iv Average annual growth rate and infant mortality.

Use a spreadsheet if you can.

b Which of the scatter graphs you drew appear to give a positive correlation? Does this necessarily mean there is a cause and effect relationship between the variables? Explain.

c Which of the variables show correlation with each other?

Review A group of boys had their heights measured as 3-year-olds and again as 16-year-olds. The results are given in the table.

	Jon	Imran	Tim	Evan	Adam	Luke	Brett	Ryan	Mark	Rob
Height at 16 (m)	1·85	1·72	1·74	1·88	1·77	1·79	1·63	1·82	1·61	1·73
Height at 3 (m)	1·03	0·93	0·94	1·04	0·96	0·99	0·85	0·95	0·84	0·93

a Plot the data on a scatter graph. (Put Height at 16 on the horizontal axis.)
b Describe the correlation between the heights at the different ages.
c Draw the line of best fit. Did you ignore any point when drawing this line? If so, which one?
d At 16, Matthew is 1·68 m tall. Estimate how tall Matthew was at 3.
e When Imran's younger brother was 3, he was 1·01 m tall. How tall would you expect him to be when he is 16?

Practical

Collect some data which is suitable for displaying on a scatter graph.
Some suggestions follow on the next page.
Before you collect the data, predict the likely correlation between the variables.
Display your data on a scatter graph.
Draw in a line of best fit if there seems to be any correlation.
Write a report. Include statements on how you collected the data, on any correlation between the variables and on whether your data confirmed your prediction about likely correlation.

Suggested data

armspan and handspan
circumference of head and circumference of wrist
time spent travelling to school and time spent on homework
number of brothers and number of sisters
number of letters in forename and number of letters in surname
time spent watching TV and time spent on homework
long jump distance and high jump height
time taken to do 50 calculations with and without a calculator
goals scored by home teams and goals scored by away teams
runs scored in two innings of cricket
engine size of cars and petrol consumption
birth rate and growth rate of countries
birth rate and infant mortality rate of countries

Discussion

● Sometimes in scientific conditions it is possible to show a cause and effect relationship. This does not necessarily mean one exists.

Example Data was collected on moisture content of soil when a tree was planted and rate of growth.
It appeared there was a positive correlation. Although the moisture content of the soil when a tree is planted is one factor in determining rate of growth, it does not by itself **cause** a higher growth rate.
There are many other factors affecting growth rate.

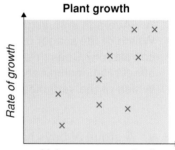

Plant growth

Discuss these graphs. Do they establish a cause and effect relationship between the variables?

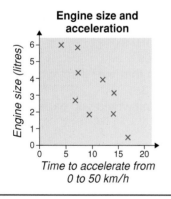

Engine size and acceleration

Salary and years of study

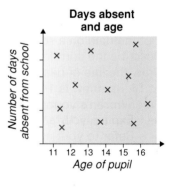

Days absent and age

Interpreting and comparing data

The next exercise gives you practice at **interpreting a range of data displays** and at **comparing data**.

We can compare data by looking at the 'shape' of distributions.

Example
Robyn drew these graphs of the Year 7 and Year 9 high jump results.

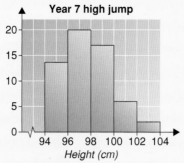
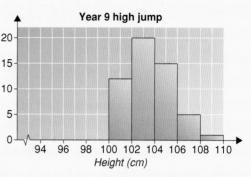

The two distributions are similar shapes with similar ranges.
This indicates that there is a similar pattern of poor, average and good jumpers in each year.
Robyn also calculated the mean of each year's jumps.
For Year 7 the mean was 97 cm and for Year 9 it was 103 cm. This indicates that, generally, Year 9 pupils jump about 6 cm higher than Year 7 pupils.

Exercise 7

1 Gareth timed the length of advertisements on two different radio stations from 9 a.m. to 12 p.m. on Saturday. The advertisements were timed to the nearest second.

Radio Station One (time in seconds)

28	15	34	23	8	19	43	26	34	54	48
19	21	18	25	16	41	30	51	46	35	29
28	18	27	33	36	26	19	9	40	43	24

Radio Station Two (time in seconds)

26	25	22	23	19	18	21	31	24	17	12
18	27	33	9	13	17	24	35	37	29	30
20	29	33	25	32	40	29	16	19	18	11

a Find the mean, median and range of each set of data.

You could use a spreadsheet.

b Draw a frequency table like this one for each radio station.

c Draw a frequency diagram for each radio station. You could put both sets of data on the same chart.

d Which radio station would you rather listen to? Explain your answer. Refer to your answers to **a** and **c**.

Time (sec)	Tally	Frequency
0–9		
10–19		
20–29		
30–39		
40–49		
50–59		

2 The percentage charts show information about the wing length of adult blackbirds, measured to the nearest millimetre. [SATs Paper 2 Level 7]

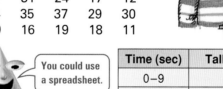
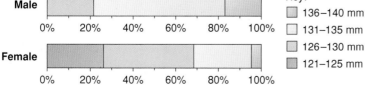

Key:
- 136–140 mm
- 131–135 mm
- 126–130 mm
- 121–125 mm

Use the data to decide whether these statements are true or false, or whether there is not enough information to tell. Explain your answers.
a The smallest male's wing length is larger than the smallest female's wing length.
b The biggest male's wing length is larger than the biggest female's wing length.

Handling Data

3 Cars more than three years old must pass a test called an MOT. **[SATs Paper 1 Level 7]**
The testers measure the right and left front wheel brakes,
and give each brake a score out of 500.
Then they use the graph.

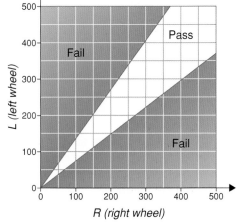

Example   A car has $R = 300$, $L = 350$.
(300, 350) is in the white region, so the
car **passes** this part of the test.

a   A man takes his car to be tested. $L = 200$
Approximately, between what values does R
need to be for his car to pass this test?

A different part of the test uses $R + L$.
To pass, $R + L \geqslant 400$.
Use a copy of the graph.

b   On the graph, draw the straight line $R + L = 400$.
Then shade the region where the car **fails**,
$R + L < 400$.

c   If $L = 200$, between what values does R need to be to pass **both** parts of the test?

4   The population pyramids for these countries are all different shapes.
a   Some other facts about each country are given.
Use these and any other facts you know about the country to give an explanation of the
shape of the pyramids.

   i   **Population Pyramid Summary for Cambodia**

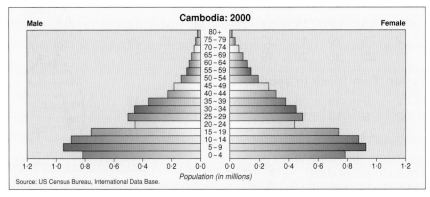

Birthrate: 35 per 1000 population   **Infant mortality rate:** 125 per 1000 born
Population growth: 3%   **Life expectancy:** 51–57 years
Death rate: 11–15 per 1000 population

   ii   **Population Pyramid Summary for Australia**

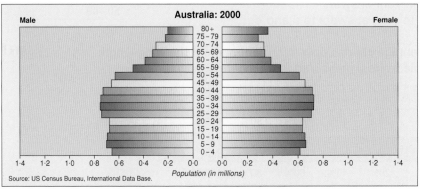

Birthrate: 15 per 1000 population   **Infant mortality rate:** 7 per 1000 born
Population growth: 1·2%   **Life expectancy:** 77 years
Death rate: 8 per 1000 population

iii Population Pyramid Summary for United Kingdom

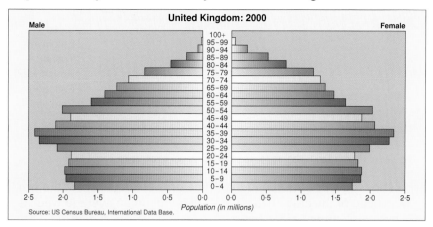

Birthrate: 12–14 per 1000 population
Population growth: 0·2%
Death rate: 10–12 per 1000 population

Infant mortality rate: 9 per 1000 born
Life expectancy: 78 years

b Compare the population pyramids for the United Kingdom and Cambodia.

5 Last year 65 000 people entered a painting in a national painting competition.
Of these, 1655 paintings were selected for a final competition.
The judges in the main competition gave each painting a score out of 50.
Paintings over a particular score, s, went through to the final.
The frequency distributions for both competitions are shown.

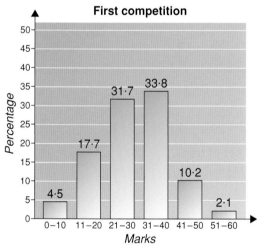

Estimate the value of s.
Show your working.

*6

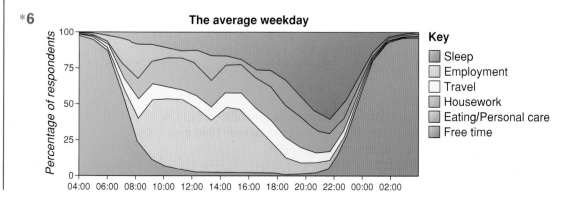

Handling Data

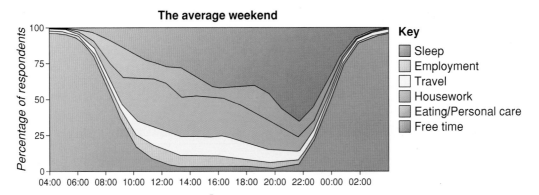

The average weekend

These graphs show how Britons spend their time on weekdays and at the weekends.

a Explain the shape of the area shaded 'employment' on an average weekday.

b Compare the employment areas for weekdays and weekend days.

c Compare the eating/personal care areas for weekdays and weekend days.

d One year a survey showed that British adults spend, on average, 12% of their time in paid work and 28% on leisure or reading.

Would a headline in the paper that said 'Britons spend too little time working and too much on leisure' be a fair comment? Explain.

Review 1 Granny Smith is experimenting with a new fertiliser for her apple orchard. She chooses an area of 40 trees and uses the fertiliser on half the apple trees (set A). The number of apples on each tree is recorded.

Set A (new fertiliser)

24	28	35	30	30	28	36	43	68	51
52	45	42	47	50	45	30	24	28	21

Set B (old fertiliser)

23	40	42	35	25	28	35	40	35	32
46	32	36	38	28	21	31	29	20	33

a Calculate the mean, median and range for each set of trees.

b Draw a frequency table like this for each set of trees.

No. of apples	Tally	Frequency
20–29		
30–39		
40–49		
50–59		
60–69		

If you put both sets on one bar chart remember to use a key.

c Draw a bar chart for each set of trees. You could put both sets of data on the same chart.

d Should Granny Smith use the new fertiliser throughout her orchard?

Review 2 A large company employed 6000 people in 1990. Graph A on the next page shows the decrease in employees between 1990 and 2000. Graph B shows the ratio of male to female employees.

Graph A

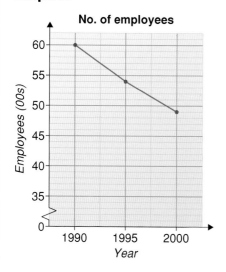

No. of employees

Graph B

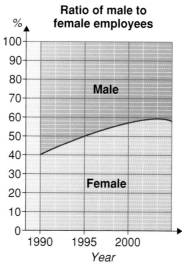

Ratio of male to female employees

a Describe what the graphs tell has been happening in the company over these 10 years.
b Use the graphs to find the number of male and female employees to the nearest 100 in 1990, 1995, 2000, by copying and completing the table

Year	Total employees	Number of females	Number of males
1990	6000		
1995			
2000			

c Is the number of female employees increasing or decreasing?

Practical

This table gives 11 indicators for each of 16 countries.

Country	Primary school enrolment (%)	Average income (USD per capita)	Energy use (kg per capita)	Vehicles (per 1000)	Phone ownership (per 1000)
Afghanistan	22	450	27	2	1
Brazil	100	2920	1068	100	78–149
China	100	370	868	1	34–86
Japan	100	26920	4070	283	488
Mexico	99	2870	1543	69	96–112
New Zealand	100	12140	4770	526	490
Nigeria	70	290	705	4	4
Rwanda	70	260	34	1	2
Singapore	100	12890	5742	105	478–593
South Africa	89	2600	2497	95	95–125
Thailand	87	1580	1169	18	59–86
Uganda	50	160	23	1	2–3
United Kingdom	100	16750	3871	367	502–567
United States	100	22560	8159	570	626–682
Uruguay	100	2850	976	116	119–271
Zimbabwe	100	620	821	29	14–21

Handling Data

Country	Population density (per km²)	Death rate (per 1000)	Maternal mortality (per 100 000)	Life expectancy	% Population with safe access to water	People per physician
Afghanistan	35	19	690	46	13	9090
Brazil	20	8	120	66	72	774
China	134	7	44	71	90	620
Japan	334	8	16	81	96	509
Mexico	50	6		72	62	613
New Zealand	13	7·9	8·3	78	87	318
Nigeria	131	15	800	51	48	3707
Rwanda	274	23	210	39	41	25–73000
Singapore	4867	5	5	80	100	720
South Africa	36	15	83	48	78	1742
Thailand	119	8	270	69	89	3461
Uganda	97	18	300	43	42	20720
United Kingdom	243	12	9	78	100	1719
United States	29	9	9	77	100	365
Uruguay	19	9	38	75	94	263
Zimbabwe	29	21	480	40	81	6904

Analyse the data to answer these questions.

Which indicators show correlation with each other?

Which are the most closely connected?

Which of these data do you think are useful as indicators of economic development?

Are there any anomalies between the indicators?

What other conclusions can you make by analysing this data?

Surveys

Remember

This diagram shows the **cycle for surveys**.

Sometimes when we evaluate the results at the end of a statistical investigation, this leads to further questions that need investigating. The cycle begins again.

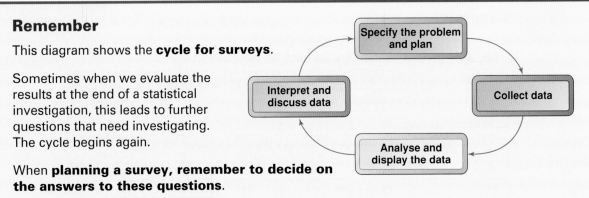

When **planning a survey, remember to decide on the answers to these questions**.

- What do you want to find out?
- Are there any related questions?
- What might you find out? Make up a conjecture to test.
- What data do you need to collect?
- How accurate does the data need to be?
- How will you collect the data and who from? You may choose a primary source or a secondary source.
- How many pieces of data do you need (sample size)?
- When and where will you collect the data?
- How will you display the data? Is using ICT appropriate or best?
- How will you interpret the data? Is finding the mean, median, mode or range appropriate?

Practical

1. Choose a problem to investigate. Suggestions follow.
2. Plan a survey using the steps above.
3. Carry out your survey.
4. Analyse and display the results using ICT if appropriate.
   Use the mean, median, mode and range if appropriate.
5. Write a report on what you found out. Include your conclusions.
   Make sure your conclusions relate to the original problem you set out to
   investigate. In your report, write about any difficulties you had and how you
   solved these difficulties.

> Use ICT to display your
> results and present your
> report.

Suggestions
- How available are fairly traded goods in local shops?
  What sort of organisations promote these goods and why?
- Is car engine size a predictor of speed of acceleration?
  Do all cars fit the same pattern and if not why not?
- What factors affect positive growth?
- Which development factors are most closely linked with other development
  factors for countries?
- What factors affect pupils' aptitude in maths?
- What factors affect people choice of holiday destination?

Summary of key points

 The **mean, median, mode and range** can be estimated for **grouped data**.

Example

Exam mark	1–20	21–40	41–60	61–80	81–100
Mid-point	10·5	30·5	50·5	70·5	90·5
Frequency	2	5	7	10	1

To find the mean, we assume all items in a class interval have the value
of the mid-point of that class interval.

$$\text{Mean} = \frac{\text{sum of (mid-point} \times \text{frequency)}}{\text{sum of frequencies}}$$

$$= \frac{2 \times 10·5 + 5 \times 30·5 + 7 \times 50·5 + 10 \times 70·5 + 1 \times 90·5}{2 + 5 + 7 + 10 + 1}$$

$$= \mathbf{52·9}$$

This can be worked out on the calculator.

The **modal class** is 61–80.

The **median** is the 13th value, so it is in the class interval 41–60. It is the 6th value
in the class interval 41–60. There are 19 values (60 − 41) in that interval.

The 13th value is $\frac{6}{7}$ of the way along the interval.

$$\frac{6}{7} \times 19 = 16·3 \text{ to 1 d.p.}$$

An estimate of the median is 41 + 16·3 = **57·3**.

The **range** can be estimated as 100 − 1 = **99**.

Example

Time (min)	10–	15–	20–	25–30
Frequency				

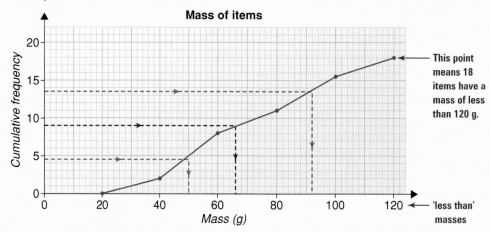

This data is continuous.

The first class interval may include a time of 9·5 min.

The last class interval may include a time of 30·5 min.

The range for this data is 30·5 – 9·5 = 21 min.

 B The **upper and lower quartiles** and **interquartile range** are best found from a **cumulative frequency graph**.

Cumulative frequency is a 'running total' of frequencies.

Example

Mass of items

This point means 18 items have a mass of less than 120 g.

'less than' masses

There are 18 values in total.

The **median** is the 9th value ($\frac{1}{2}$ of 18). It is read off the graph as about **66 g**.

The **lower quartile** is the middle value of the lower half of the data so is at 4·5 ($\frac{1}{4}$ of 18). It is read off as about **50 g**.

The **upper quartile** is the middle value of the upper half of the data so is at 13·5 ($\frac{3}{4}$ of 18). It is read off as about **92 g**.

 C **Frequency polygons** are drawn by joining the mid-points of the tops of the first and last bar in a frequency diagram to the axis at half a class interval distance from the bar.

Example

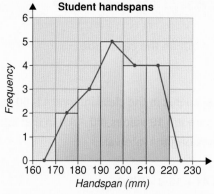

Student handspans

The mean and median can be estimated from a frequency polygon.

See page 396 for an example.

T

D A **scatter graph** displays two sets of data.

Example This scatter graph shows the marks some pupils got in maths and in science.

A scatter graph sometimes shows a **correlation** (relationship) between the variables.

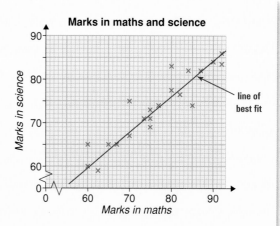

Marks in maths and science

Marks in science / *Marks in maths* — line of best fit

If data shows some correlation, we draw a line of best fit. This line should be as close as possible to the points and should go through the mean of the data values.

E We can **interpret and compare a range of data and data displays.**

F When **planning surveys** follow the cycle given on page 410.

Test yourself

1 a Julian collected data on the number of apples produced last season by each of the trees in his orchard. He organised this data onto the following frequency table.

Apples produced	600–799	800–999	1000–1199	1200–1399	1400–1599	1600–1799
Frequency	8	14	10	34	68	76

Julian then displayed this data on a frequency diagram. Draw this graph.

b Julian estimated the height of each tree.

Class interval	1 m–	2 m–	3 m–	4 m–	5 m–6 m
Mid-point					
Frequency	6	24	84	87	9

   i Use a copy of the table. Fill in the mid-points.
   ii Calculate an approximate value for the mean height.
   iii Which class interval contains the median?
   iv What is the modal class?
   v Calculate an approximate value for the median.

T

2 Ms Hassell gave her class two tests, A and B.
The frequency diagram and frequency
polygon for Test A are shown.

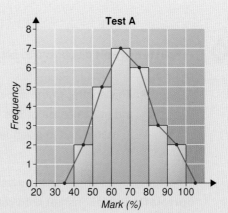

C

a Use a copy of this frequency polygon.
The marks in Test B are given in the table.

Mark range	40–49	50–59	60–69	70–79	80–89	90–99
Frequency	3	4	3	6	8	1

On the same set of axes used in **a**, draw the frequency polygon for Test B.
b Comment on any similarities or differences between the two frequency polygons.

T

3 The scatter diagram shows the heights and masses of some horses. [SATs Paper 1 Level 6] **D**
The scatter diagram also shows a line of best fit.

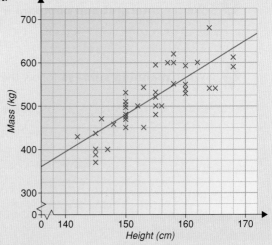

a What does the scatter diagram show
about the relationship between the
height and mass of horses?
b The height of a horse is 163 cm.
Use the line of best fit to estimate the
mass of the horse.
c A different horse has a mass of
625 kg.
Use the line of best fit to estimate the
height of the horse.

d A teacher asks his class to investigate this statement.
'The length of the back leg of a horse is always
less than the length of the front leg of a horse.'
What might a scatter graph look like if the statement
is correct?
Use a copy of the axes to show your answer.

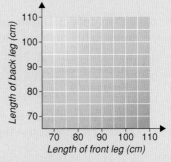

4 These line graphs show the midday
temperatures at Carlisle and Calais for one
week in June.
Compare these temperatures. In your
comparison, refer to the ranges and the
means.

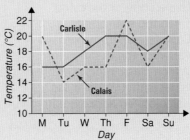

5 The pie chart shows how much time each day, on average, we spend doing different things.

[SATs Paper 2 Level 7]

a The sum of the percentages is not 100%.
Does this mean there must be a mistake in the pie chart?
Explain your answer.

b Calculate how much time in one day (24 hours) we spend on average on paid work.
Show your working and give your answer in hours and minutes.

c Most days of paid work are at least 7 hours long.
Give one reason why the average amount is less than this.

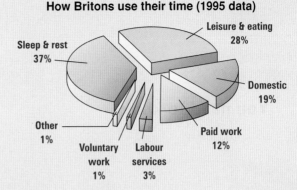

How Britons use their time (1995 data)

6 A teacher asked fifty pupils in Year 9:

[SATs Paper 2 Level 8]

How much time did you spend on homework last night?

Results

Time spent on homework (minutes)	Frequency
0 ≤ time ≤ 30	6
30 < time ≤ 60	14
60 < time ≤ 90	21
90 < time ≤ 120	9
Total	**50**

a Show that an estimate of the mean time spent on homework is **64·8 minutes**.

The teacher used the data to draw a cumulative frequency diagram.

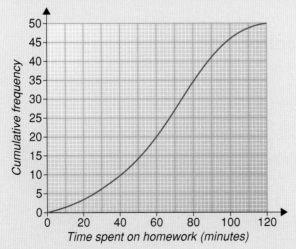

b Use the diagram to estimate the median time pupils spent on their homework.
Show on the diagram how you get your answer.

c Use the diagram to estimate how many pupils spent **more than 100 minutes** on their homework.
Show how you get your answer.

[End of SATs question]

d Estimate the range for the data.

e Use the diagram to estimate the lower and upper quartiles.

f Find the interquartile range.

You need to know

✓ probability page 362

Key vocabulary

biased, event, experimental probability, limit, mutually exclusive, $p(n)$ for probability of event n, relative frequency, sample space, theoretical probability

---

▶▶ Spiralling out of control

We can draw a spiral in two directions.

anticlockwise

clockwise

Ask about 50 people to draw a spiral. Collect the results in a table like this one.

Is it equally likely that the next person you ask to draw a spiral will draw a clockwise one or an anticlockwise one?

	Tally	Frequency
Clockwise		
Anticlockwise		

Describing and calculating probability

Worked Example

Jan likes dark chocolate.

Which box should she take a chocolate from to have the greatest probability of getting a dark chocolate?

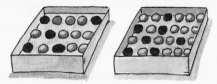

Answer

In box 1 there are 5 dark chocolates out of 15 chocolates.

In box 2 there are 7 dark chocolates out of 24 chocolates.

We need to find which box has the greater **proportion** of dark chocolates.

$$\frac{5}{15} = \frac{1}{3} = 0.\dot{3} \qquad \frac{7}{24} = 0.29 \text{ (2 d.p.) and } 0.29 < 0.\dot{3}$$

> We can use a calculator to find $7 \div 24$.

There is a greater proportion of dark chocolates in box 1 so Jan has the greater probability of getting a dark chocolate from this box.

Remember

If the outcomes of an event are **equally likely**.

> **Probability of an event** = $\dfrac{\text{number of favourable outcomes}}{\text{number of possible outcomes}}$

An event will either happen or not happen.

> **Probability of an event not happening** = **1 – probability of it happening**

If the probability of an event happening is p, the probability of it not happening is $1 - p$.

Mutually exclusive events are events that cannot happen at the same time.

Example   It is not possible to get a 3 **and** an even number on a dice with the same roll.

> **The sum of probabilities of all the mutually exclusive outcomes of an event is 1.**

Exhaustive events account for all possible outcomes. If events are exhaustive, it is certain one of them will happen.

Worked Example

a Jake estimates that the probability of winning a computer game is 0·34.
   He plays 50 of these games.
   How many should he expect to win?

b Alice played the same game.
   She won 16 games and so estimated the probability of winning each game to be 0·4.
   How many games did Alice play?

c The distributor of a different computer game claims the probability of winning it is 0·55.
   Lourdes plays this game 300 times and wins 155 times.
   She says 'The distributors are wrong.' Do you agree with her? Explain.

Answer

a $0.34 \times 50 = $ **17** He should expect to win about **17** times.

b Alice would have done this calculation.

   $0.4 \times$ games played $= 16$
   so games played $= \dfrac{16}{0.4}$
   $= \mathbf{40}$

c **No** because random variation means you wouldn't expect to win
   exactly 165 games. $0.55 \times 300 = 165$
   155 is fairly close to the 165 expected.

Handling Data

1 There are some buns in a bag.
They are either raspberry (R) or cream (C).
If you take a bun at random, the probability it will be raspberry is 0·2.
 a What is the probability that the bun will be cream? Give your answers as a decimal.
 b Jake takes a bun at random out of the bag and it is a raspberry bun.
   What is the smallest number of cream buns that could be in the bag?
 c Jess now also takes a bun at random from the bag.
   She also gets a raspberry bun.
   Now what is the smallest number of cream buns that could be in the bag?

2 A game in an arcade has a grid with some squares coloured.
When you press 'enter', one square is randomly chosen.
If a coloured square is 'hit' you get 1000 points. With each game you get three attempts.
If you get 3000 points you win ten times your money back.
A different card is displayed for each game.
 a On which of these cards are you most likely to score 3000 points? Explain.

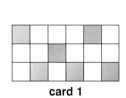

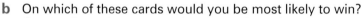

| card 1 | card 2 | card 3 |

 b On which of these cards would you be most likely to win?
   A 10 coloured squares and 54 not coloured
   B 40 coloured squares and 216 not coloured
   C 99 coloured squares and 381 not coloured

Use a calculator
if you need to.

3 Ian has a pack of 52 playing cards (no jokers).
Calculate the probability that a card chosen at random will be
 a a black card b a club c not a king d either a diamond or a heart
 e an odd numbered red card.

4 I have two bags of counters.   [SATs Paper 2 Level 7]

Bag A contains
   12 red counters and
   18 yellow counters.

Bag B contains
   10 red counters and
   16 yellow counters.

I am going to take one counter at random from either bag A or bag B.

I want to get a **red** counter.
Which bag should I choose?

Show working to explain your answer.

5 Certain conditions are hereditary, for example, left-handedness.
There is a dominant gene, R and a recessive gene, r. A child gets one gene from each parent. Children are left-handed if they inherit rr, right-handed but a carrier of left-handedness if they inherit Rr, and right-handed and not a carrier if they inherit RR.
If both of the parents are carriers, they are both Rr.

  a Copy and fill in this table to show the possible outcomes for their children.
What is the probability a child born to these parents will
    i be left-handed
    ii be a carrier of left-handedness
    iii not be left-handed nor be a carrier?

  b Repeat **a** for one parent being a carrier and the other left-handed.

Parent 1

	R	r
R		
r		

Parent 2

Parent 1

Parent 2

6 A shop has blue, green and yellow pens in a display cabinet. There are 20 pens altogether.
There is at least one of each colour.
If one is taken at random, the probability it is green is $\frac{3}{5}$.
What is the greatest number of blue pens there could be?

7 **a** The probability of Fran's school bus being late is 0·35.
On how many days in the next four weeks, should she expect the bus to be late?
  b Rajiv worked out his school bus had a probability of 0·45 of being late.
He should expect the bus to be late on 27 days out of every _____.
What goes in the gap?
  c The bus company claim another bus has a probability of 0·35 of being late.
Toby worked out that this bus was late 20 times out of 60.
He said that the bus company must have made a mistake when working out the probability. Is he right? Explain.

8 At a school there are four different house groups.
This table gives the probability of selecting a boy or girl from each house group.
  a One pupil is selected at random from the school roll to represent the school at an award ceremony.
What is the probability this pupil will be
    i from Willow house **ii** a boy?
  b There are 600 pupils at the school.
How many of these pupils are in Elm house?

House	Boy	Girl
Oak	0·15	0·15
Elm	0·15	0·1
Willow	0·2	0·05
Walnut	0·1	0·15

9 A headteacher wants to choose a pupil from Year 7, 8 or 9 to appear on television.
    [SATs Paper 1 Level 7]
The headteacher gives each pupil **one** ticket.
Then she will select the winning ticket at random.
The table shows information about the tickets used.
  a What is the probability that the winning ticket will be **blue**?
  b What is the probability that the winning ticket will show number **39**?
  c The headteacher selects the winning ticket at random.
She says:
    'The winning ticket number is 39'.
What is the probability that the winning ticket is blue?

	Colour of the ticket	Numbers used
Year 7	red	1 to 80
Year 8	blue	1 to 75
Year 9	yellow	1 to 90

10 Arshad has a bag that contains only blue, green, red and orange sweets.
Arshad is going to take one sweet at random.
This table shows the probability of each colour being taken.

	Blue	Green	Red	Orange
Probability		0·05	0·35	0·2

a Are the outcomes blue, green, red and orange exhaustive events?
b What is the probability of a blue sweet being taken?
c Explain why the number of orange sweets cannot be 10.
d What is the smallest number of each colour sweet in the bag?

11 Two classes at a school, K and Y, each have the same number of pupils.
The probability of choosing a boy at random from class K is 0·5.
The probability of choosing a boy at random from class Y is 0·3.
The classes join together to go on a class trip.
a What is the probability of choosing a boy at random from all the pupils on the class trip?
b What if class K has twice as many pupils as class Y?

12 a How many possible outcomes are there when a coin is tossed three times?
b What about if it is tossed four times?
c For each of **a** and **b**, what is the probability of getting exactly two heads?

13 At a fair, if you roll 6 dice and get 6 sixes you win £300.
What is your chance of winning?

Review 1 Molly, who is five, likes red jet planes.
At a party there are two bags of jet planes.
a In bag one there are 8 red jet planes and 10 planes of other colours.
In bag two there are 6 red jet planes and 12 of other colours.
Which bag should she pick from? Why?
b A third bag has 5 red jet planes and 6 planes of other colours.
Which bag should Molly pick from now? Why?
Show your working.

Review 2 The table shows the way two classes split into groups for an afternoon to work on three different technology topics.

	Females	Males
Computing room	7	11
Art room	5	14
Metalwork room	8	10

Sarah works in the school office and must deliver an urgent message to one of the pupils.
a What is the probability the pupil is in the metalwork room?
b Sarah can't tell from the name whether the pupil is a male or female. What is the probability that the pupil is a female in the computing room?
c If a teacher randomly sends a female pupil from the computing room and Sarah meets her in the hall, what is the probability that she's the student Sarah is looking for?
d If Sarah finds out the pupil she's looking for is male, which room should she look in first and what is the probability that he'll be in that room?

Review 3   Can you find the name of these towns in England?

a p(letter E) = $\frac{2}{5}$ p(letter D) = $\frac{1}{5}$ p(letter L) = $\frac{1}{5}$ p(letter S) = $\frac{1}{5}$

b p(letter L) = $\frac{3}{7}$ p(letter A) = $\frac{2}{7}$ p(letter W) = $\frac{1}{7}$ p(letter S) = $\frac{1}{7}$

Review 4   A bag of fruit has oranges, apples and bananas.
The probability of James taking an apple at random from the bag is $\frac{5}{8}$.

a   What is the probability that James does **not** take an apple?

b   James guesses there are 16 pieces of fruit in the bag. If he is right, what is the greatest number of oranges there could be in the bag?

James takes four pieces of fruit from the bag. None of them are apples.

c   What is the minimum number of apples in the bag?

d   What is the maximum probability that the next piece of fruit James randomly selects is an apple?

Probability of compound events

Discussion

This spinner is spun.
Are the outcomes equally likely?
Are the outcomes exhaustive?

What is the probability of the spinner stopping on 3?
What is the probability of the spinner stopping on an even number?
What is the probability of the spinner stopping on 3 **or** an even number?
Could you add the first two probabilities to get the third?

If A and B are **mutually exclusive events**:

p(A **OR** B) = p(A) + p(B)

Example   There are 50 animals in a pet show. Each owner has only one pet entered.
23 are cats and 16 are dogs.
One animal in the show is chosen at random to win a prize for its owner.
The probability it will be a cat *or* dog owner is:

$\frac{23}{50} + \frac{16}{50} = \frac{39}{50}$ The events are mutually exclusive so we add the probabilities.

Example   On a computer game, the probability of winning the jackpot in each game is $\frac{3}{100}$ and of winning a bonus is 0·35. You cannot win both.
The probability of winning the jackpot *or* a bonus is:

We must change both to fractions or both to decimals.

0·03 + 0·35 = 0·37   The events are mutually exclusive.

Discussion

If we want to know the probability of rolling a 3 or an odd number on a fair dice, we cannot add the probabilities $\frac{1}{6} + \frac{3}{6}$.

Why not? **Discuss.**

Handling Data

Exercise 2

1 A card is drawn from a pack.
Find the probability of getting an Ace or a Jack.

2 Two coins are tossed.
Find the probability of getting two heads or two tails.

3 Two dice are thrown. Find the probability of getting a total of 7 or the same numbers on both dice.

4 The probability that Belen buys 'The Mirror' is 0·2 and the probability that she buys 'The Sun' is 0·25. (Belen does not buy more than one paper.)
What is the probability that she buys either 'The Mirror' or 'The Sun'?

5 Sue drives to work in London.
The probability that she parks on Regent Street is 0·2.
The probability that she parks on Oxford Street is 0·18.
What is the probability that she parks on either Regent Street or Oxford Street?

6 A forensic pathologist finds deliberate poisoning to be the cause of a death.
The detective on the case decides that there are six suspects. The probability Karl did it is 0·15, that Max did it is $\frac{1}{4}$ and that Angeline did it is 40%.
Find the probability that
a either Karl or Max did it
b either Max or Angeline did it
c either Angeline or Karl did it
d none of these three did it.

7 John answers the phone in his office from 9 a.m. until midday and then from 1 p.m. until 4 p.m. Ruski phones this office each day. The following table shows the probability of Ruski phoning at particular times of the day.

Time	Probability
In the morning before 9 a.m.	$\frac{3}{20}$
From 9 a.m until midday	0·4
From midday until 1 p.m.	0·15
In the afternoon after 4 p.m.	$\frac{1}{20}$

a Find the probability that Ruski phones between 1 p.m. and 4 p.m.
b What is the probability that John answers Ruski's call?

8 Part of the analysis of a dentist's patients is shown on this table.
a Find the probability that a patient is either an adult or a school pupil.
b The probability that a patient is either a school pupil or female is *not* 0·3 + 0·45. Why not?
c What is the probability that a patient is not an adult?
d Find the probability of a patient having dental work other than a tooth filled.
e Of the next 150 patients, how many do you expect will be adults?

	Probability
Male	55%
Female	45%
Under school age	10%
School pupil	30%
Adult	60%
Tooth filled	38%
Reconstruction work	9%
Tooth extracti	
Teeth	

Review 1 A die is rolled.
Find the probability of getting an odd number or a six.

Review 2 The school was broken into through one of three doors; the front door, the side door or the hall door. The probability it was the front door is 30% and the hall door is 0·05.
Find the probability that the school was broken into through
a   the side door
b   either the front door or the hall door
c   either the front door or the side door.

Independent events

Event A and event B are **independent** if event A happening (or not happening) has no influence on whether event B happens. The probability of event B happening will be the same regardless of whether or not event A has happened.

Example   A coin is tossed twice. A head occurs on the first toss. This has no influence on whether or not a tail occurs on the second toss.
The events 'head on first toss', 'tail on second toss' are independent.

Example   These counters are placed in a bag.
Two counters are drawn at random, one after the other.
The first counter is not replaced in the bag.
If the first counter drawn is red then there are 4 red counters left.

So the probability the second counter is red = $\frac{4}{7}$.
If the first counter drawn is blue, the probability the second is red = $\frac{5}{7}$.
The colour of the first counter drawn affects the probability of the second counter drawn being red.

The events 'the first counter drawn is red' and 'the second counter drawn is red' are **not** independent events.

Discussion

- Two counters are drawn with the first counter being replaced before the second one is drawn.
Are the events 'the first counter is red', 'the second counter is red' independent in this case? **Discuss**.

- **Discuss** whether or not the events A and B are independent in each of the following.

Sue throws two backgammon dice, one grey and the other purple.
Event A: Sue gets a 6 on the grey dice.
Event B: Sue gets a 6 on the purple dice.

Andrew uses the alarm on his clock-radio.
Event A: There was a power failure last night.
Event B: Andrew was late for school today.

Two babies are born on Saturday at Rochford Hospital.
  Event A: The first baby born is a girl.
  Event B: The second baby born is a boy.

Nicole and Ana are the finalists in the school tennis championship. In the final, they play a three-game match.
  Event A: Ana wins the first game of the final.
  Event B: Ana wins the second game of the final.

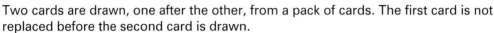

Femi and Oni are friends.
  Event A: Femi goes to the disco.
  Event B: Oni goes to the disco.

Two cards are drawn, one after the other, from a pack of cards. The first card is not replaced before the second card is drawn.
  Event A: The first card drawn is a spade.
  Event B: The second card drawn is a spade.

Two cards are drawn, one after the other, from a pack of cards. The first card is replaced before the second card is drawn.
  Event A: The first card is the King of hearts.
  Event B: The second card is a heart.

● Event A is 'it will rain on the first day of next month'.
  Event B is 'it will rain on the last day of next month'.
  Are these events independent? **Discuss**.

  Suppose $p(A) = 0.4$ and $p(B) = 0.4$. Will the probability of it raining on both the first and last days of next month also be 0·4 or more than 0·4 or less than 0·4? **Discuss**.

A green and a red dice are tossed together.
Possible outcomes are:

1, 1	1, 2	1, 3	1, 4	1, 5	1, 6
2, 1	2, 2	2, 3	2, 4	2, 5	2, 6
3, 1	3, 2	3, 3	3, 4	3, 5	3, 6
4, 1	4, 2	4, 3	4, 4	4, 5	4, 6
5, 1	5, 2	5, 3	5, 4	5, 5	5, 6
6, 1	6, 2	6, 3	6, 4	6, 5	6, 6

If event A is 'a number greater than four on the green dice' and event B is 'a five on the red dice' favourable outcomes for event A are

5, 1   5, 2   5, 3   5, 4   5, 5   5, 6   6, 1   6, 2   6, 3   6, 4   6, 5   6, 6

$p(A) = \frac{12}{36}$ or $\frac{1}{3}$

favourable outcomes for event B are

1, 5   2, 5   3, 5   4, 5   5, 5   6, 5

$p(B) = \frac{6}{36}$ or $\frac{1}{6}$

favourable outcomes for event A **and** B are

5, 5   6, 5

$p(A \textbf{ and } B) = \frac{2}{36}$ or $\frac{1}{18}$

$p(A) \times p(B) = \frac{1}{3} \times \frac{1}{6}$

$\qquad\qquad\quad = \frac{1}{18}$

> **For independent event, $p(A \textbf{ and } B) = p(A) \times p(B)$.**

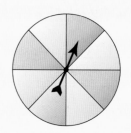

Worked Example

a The arrow on the spinner is spun once. Find the probability of the
   arrow stopping in
   i a yellow section
   ii a purple section.
b The arrow is spun twice. What is the probability of spinning yellow
   then purple?

Answer

a i $\frac{4}{8}$ or $\frac{1}{2}$ ii $\frac{3}{8}$

b We want $p(\text{Y and P}) = p(\text{Y}) \times p(\text{P})$ **events are independent**

$\qquad\qquad\qquad = \frac{1}{2} \times \frac{3}{8}$

$\qquad\qquad\qquad = \mathbf{\frac{3}{16}}$

Worked Example

The probability that it will rain on any day in May is $\frac{1}{4}$.
Find the probability that
a it will rain on both 1 May and 21 May
b it will not rain on 21 May
c it will rain on 1 May but not on 21 May.

Assume that raining on 1 May and raining on 21 May are independent events.

Answer

a P(rain *and* rain) $= \frac{1}{4} \times \frac{1}{4}$

$\qquad\qquad\qquad = \mathbf{\frac{1}{16}}$

b P(not rain) $= 1 - \text{P(rain)}$

$\qquad\qquad = 1 - \frac{1}{4}$

$\qquad\qquad = \mathbf{\frac{3}{4}}$

c P(rain *and* not rain) $= \frac{1}{4} \times \frac{3}{4}$

$\qquad\qquad\qquad\qquad = \mathbf{\frac{3}{16}}$

Discussion

For the spinner in the first **Worked Example** above, Beth said that since $p(\text{Y and P}) = \frac{3}{16}$
then the probability of spinning yellow and purple in any order is $2 \times \frac{3}{16}$ or $\frac{3}{8}$.
Is Beth correct? **Discuss**.

Worked Example

The probability of Jesse doing his homework before tea is 0·4. What is the probability he will
do his homework before tea on Monday or Tuesday but not both days?

Answer

$p(\text{before tea and not before tea}) = 0{\cdot}4 \times 0{\cdot}6$

$\qquad\qquad\qquad\qquad\qquad\quad = 0{\cdot}24$

Because he could do it before tea on Monday **or** Tuesday we multiply the answer by 2.

$p(\text{before tea and not before tea or not before tea and before tea}) = 0{\cdot}24 \times 2$ $(0{\cdot}24 + 0{\cdot}24)$

$\qquad\qquad\qquad\qquad\qquad\qquad\qquad\qquad\qquad\qquad\qquad\quad = \mathbf{0{\cdot}48}$

Handling Data

1 A coin is tossed twice.
Find the probability of
a a head on the first toss
b heads on both tosses.

2 A coin is tossed and a dice is rolled.
Find the probability of
a a multiple of 3 on the dice
b a head on the coin and a multiple of 3 on the dice.

3 A dice is thrown twice.
Find the probability of a 4 and a 6 in either order.

4 The probability that Irina is late for school is 0·3; the probability that Helena is late is 0·2.
What is the probability that both girls are late for school? Assume these girls do not know
each other. Why do you need to make this assumption?

5 In one Mercedes factory, 80% of the cars manufactured are left-hand drive and the rest are
right-hand drive. The probability that a car needs its steering adjusted before it leaves the
factory is 0·15.
One car is chosen at random from this factory. Find the probability that this car
a is a right-hand drive
b is a right-hand drive which needs its steering adjusted
c is a left-hand drive which does not need its steering adjusted.

6 Brenda plays two games of patience. The
probability that she gets all the cards out in any
one game is $\frac{1}{5}$.
a Find the probability that Brenda gets all the
cards out in both games.
b What is the probability that Brenda does not
get all the cards out on the first game?
c Find the probability that Brenda does not
get all the cards out on the first game but gets
them all out on the second.

7 On a road there are two sets of traffic lights.   **[SATs Paper 2 Level 8]**
The traffic lights work independently.
For each set of traffic lights, the probability that a driver will have to **stop** is **0·7**.
a A woman is going to drive along the road.
What is the probability that she will have to **stop** at **both** sets of traffic lights?
b What is the probability that she will have to **stop** at **only one of the two sets** of
traffic lights?
Show your working.
c In one year, a man drives **200** times along the road.
Calculate an estimate of the number of times he drives through **both sets** of traffic
lights **without stopping**.
Show your working.

8 Abbie plants two oak tree seedlings for sale in one months time.
   The probability that oak seedlings will die in the first month is 0·04.
   a Calculate the probability that both will die in the first month.
      Show your working.
   b Calculate the probability that only one of Abbie's oak tree seedlings will die in the first
      month.
      Show your working.
   c Abbie gets an order for 120 oak seedlings to be collected in one months time.
      She plants 125 oak seedlings.
      Will she have enought to fulfill the order?
      Explain your answer.

9 A robot can move N, S, E or W along the lines of
   a grid.
   It starts at the point marked ● and moves one
   step at a time.
   For each step, it is **equally likely** that the robot
   will move N, S, E, or W.
   a The robot is going to move 3 steps
      from the point marked ●.
      What is the probability that it will
      move along the path shown?
      Show your working.
   b The robot is going to move 3 steps
      from the point marked ●.
      What is the probability that it will
      reach the point marked × by any route?

[SATs Paper 1 Level 8]

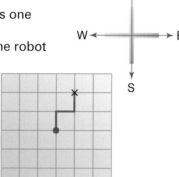

**Use the rule *p*(A *and* B) = *p*(A) × *p*(B) and the rule *p*(A *or* B) = *p*(A) + *p*(B) to do
questions 10–15.**

10 A dice is rolled twice. Find the probability of getting a
   two on just one of these throws.

Do not draw a sample space
for **10, 11, 12** or **13**.

11 Two dice are rolled. Find the probability of getting the same number on both of them.

12 A black and a red dice are rolled and the numbers obtained are added together. Find the
   probability that this sum will be 5.

13 Two coins are tossed together. Find the probability of getting
   a two tails   b no tails   c different outcomes on the coins
   d at least one tail.

14 There are two sets of traffic lights on Memorial Drive. The probability that the first set
   malfunctions is 0·02 and the probability that the second set malfunctions is 0·03. They
   operate independently.
   Find the probability that
   a both sets of lights malfunction
   b at least one of these sets of lights malfunctions
   c only one of these sets of lights functions correctly.

Handling Data

15 a At a parents' resuscitation demonstration meeting
there are 10 men and 15 women.
Two parents are chosen at random to receive prizes.
The same person cannot win both.
Copy and fill in the table.

Gender of chosen parents	Probability
Two men	
Two women	
One of each	

Dear Parents

A meeting of parents of 9P will take place this Thursday. I would be grateful if you would attend. There will be a chance for two parents to win two spot p...

Yours since...

b At a similar parents meeting for a different class there
were 8 men and 17 women.
Two parents are chosen at random to win spot prizes. The same person cannot win
both.
Copy and fill in the table in **a** for this meeting.
Which two outcomes have the same probability.

Review 1 Jamie works in a car showroom which sells British, European and Japanese cars.
65% of the cars that Jamie sells are British and 20% are Japanese.
a What is the probability that Jamie sells a European car to his next customer?
b Find the probability that the next two customers both buy Japanese cars.
Assume these customers do not know each other.
c Why do you need to make the assumption in **b**?

Review 2

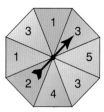

This spinner is spun twice as part of a fairground game.
Find the probability of spinning
a a 1 on the first spin
b a 1 on both spins
c a 2 on the second spin
d a 3 on the first spin and a 2 on the second
e a 3 and a 2, in any order, on the two spins.

Review 3 The probability that an RAC car is free to respond to an emergency is 0·8. Find the
probability that, of two RAC cars on duty when an emergency call comes through
a no car is available
b both cars are available
c just one car is available
d a car is available

*__**Review 4**__ The probability that a computer crashes during a week is 0·18. Room 12 has six
computers in it.
a What is the probability that none of the computers crash during the week?
b What is the probability that only computer number 4 crashes during the week?
c What is the probability that exactly one computer crashes during the week?

Estimating probabilities from relative frequency

Remember

Rani wanted to know the probability that she would shoot a goal successfully in a netball match.

In the next ten netball games, she kept a count of how many attempts at goal she had and how many of these were successful.

She successfully shot 126 out of 154 attempts.

The **relative frequency** of a successful shot = $\frac{126}{154} = \frac{9}{11}$.

The relative frequency is the number of times the event occurs in a number of trials.

> **Relative frequency** = $\frac{\text{number of times an event occurs}}{\text{number of trials}}$

We often use the relative frequency as an **estimate of probability**.

In the above example, the probability that Rani will successfully shoot a goal is estimated as $\frac{9}{11}$.

⭐ Practical

A You will need an empty matchbox or similar.

An empty matchbox when dropped could land three ways.

flat   **on side**   **on end**

Predict the probability of it landing in each of these positions.

Drop an empty matchbox 100 times and use a table to record the way it lands.

Collect four or five sets of results from other classmates or groups.

Postion	Number of trials				
	100	200	300	400	500
Flat					
On side					
On end					

Combine these with yours and make a table of the results like the one shown.

Find the relative frequency of the box landing in each position after 100 trials and then after 200, 300, 400 and 500 trials.

Draw a graph of relative frequency versus number of trials for the matchbox landing flat.

Does the relative frequency tend to a limit? If so what is it?

Use your results to make an estimate for the probability of the matchbox landing flat.

Use the experimental probabilities to predict the number of times a box would land flat if you dropped it 800 times.

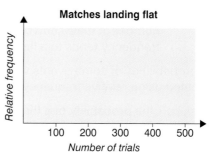

Matches landing flat

Relative frequency

100   200   300   400   500

Number of trials

B You will need computer software that can simulate rolling two dice and finding the difference.

Simulate rolling the two dice and finding the difference. Do this 100, 200, 300, 400, 500, 600, ... times.
Record the differences in a table like this one.

| | | \multicolumn{11}{c|}{Number of rolls} |
		100	200	300	400	500	600	700	800	1000	...	2000
Difference	0											
	1											
	2											
	3											
	4											
	5											

Plot a graph of relative frequency versus number of rolls for the difference 2.
What do you notice?
Calculate the theoretical probability of getting a difference of 2.
Compare this to your experimental probability.

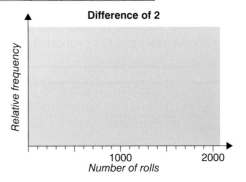

Difference of 2

C You will need a set of dominoes.

Put the dominoes in a bag or box.
Draw out a domino, record the total score of the dots and put it back.
Do this many times.
Use a tally chart to record the totals.
Draw a bar chart for your results.
Use this to calculate the relative frequency for each possible total.
Calculate the theoretical probability of getting each total.
Compare your theoretical probability with the experimental probability.

When trials are repeated, the relative frequency of an event tends to a **limit**. We can use this limit to estimate the probability of the event.

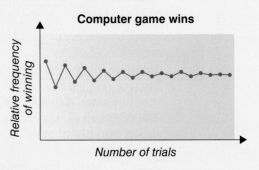

Computer game wins

Example   A computer game is played many times. The relative frequency of winning is calculated after 100, 200, 300, ... games. A graph of relative frequency versus number of trials shows the relative frequency tends to a limit.

The probabilities of some events when outcomes are not equally likely can only realistically be estimated using relative frequency.

Examples   the probability of a biased dice landing on a 6
the probability of a piece of buttered toast landing butter side up
the probability of a drawing pin landing point up
the probability of dying in a plane crash for a particular airline

Exercise 4

1 For which of these is finding an experimental probability the only realistic way to find the probability?
   a that a particular brand of bread will remain fresh for more than two days
   b that more people in London than Edinburgh exercise at least twice a week
   c that when a fair spinner is spun twice and the numbers added, the result will be greater than 6
   d that you will get to sleep in less than 7 minutes on any particular randomly chosen night

2 In his last 25 cricket games, Reece has scored these number of runs.

| 42 | 38 | 26 | 32 | 12 | 84 | 52 | 71 | 68 | 53 | 78 | 51 | 104 |
| 96 | 83 | 19 | 4 | 67 | 15 | 10 | 0 | 55 | 99 | 82 | 41 | |

Use this to estimate the probability that in his next match he will make
   a more than 20 runs   b more than 50 runs.

3 A telephone survey asked 400 people in a town out of a possible 50 000, which of the towns four fish and chip shops they used most.
      87 said Sharkies, 116 said Yangs, 35 said Top Fish and 162 said Fresh Caught.
   a Estimate the probability that a person chosen at random from the 400 chose
      i Sharkies   ii Fresh Caught
   b Assume the sample of 400 represents the population of 50 000 well.
      How many of the whole town's population would you expect to use Yang's most?
   c Sharkies did a survey themselves and said that 1162 out of 5000 liked them best so the survey must be wrong. Do you agree? Explain.

4 A drawing pin can land point up or point down.
Nelson dropped it first 10, then 20, then 30, ... times.
Which of these graphs of relative frequency of landing point up versus number of trials is most likely to show Nelson's results?

point up   **point down**

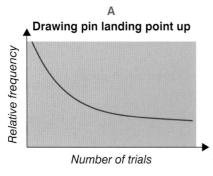

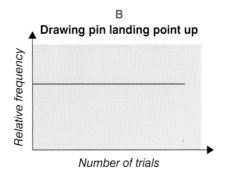

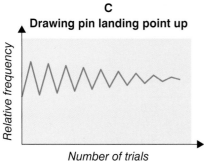

5 Design an experiment for estimating the probability that someone in your school chosen at random can run 100 m in less than 12 s.
Write the instructions down so that a classmate could follow them and do the experiment.

6 A girl plays the same computer game lots of times.   **[SATs Paper 1 Level 8]**
The computer scores each game using **1 for win, 0 for lose**.
After each game, the computer calculates her **overall mean score**.
The graph shows the results for the first **20 games**.

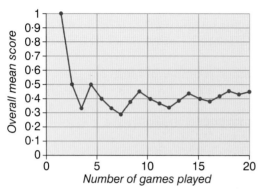

a For each of the **first 3** games, write W if she won or L if she lost.
b What percentage of the 20 games did the girl win?
The graph below shows the girl's results for the first 100 games.

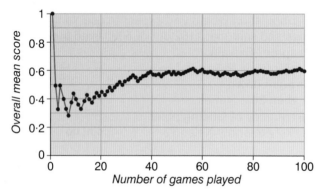

c She is going to play the game again.
Estimate the probability that she will win.
d Suppose for the 101st to 120th games, the girl were to **lose each game**.
What would the graph look like up to the 120th game?
Show your answer on a copy of the graph below.

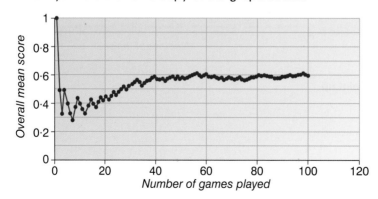

Review

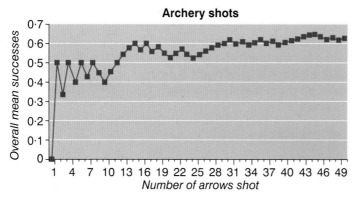

Archery shots

Overall mean successes (y-axis: 0, 0.1, 0.2, 0.3, 0.4, 0.5, 0.6, 0.7)

Number of arrows shot (x-axis: 1 4 7 10 13 16 19 22 25 28 31 34 37 40 43 46 49)

Shannon belongs to an archery club. She counts an attempt as a success if the arrow lands within the three inner rings on the target. The graph shows the relative frequency of successful shots for her first fifty shots.

a What is the largest consecutive number of successes?
b What is the estimated probability that her next attempt will *not* be a success?
c Shannon receives some coaching and her average success rate improves. Use a copy of the graph and extend it to show what the graph might look like for Shannon's next 20 attempts, if she has success with 15 of them.

Analysing games

Some games are **fair** and others are **unfair**.
If a game is fair, it means that all players have the same chance of winning.

Example In a game of Even/Odd two dice are rolled and the scores are added.
One player gets a point if the total is even and the other player gets a point if the total is odd.

To decide if this game is fair we need to look at the sample space of all the possible outcomes.
To be a fair game there needs to be the same number of even outcomes as odd outcomes.
There are 18 odd outcomes and 18 even outcomes so it **is** a fair game.

+	1	2	3	4	5	6
1	2	3	4	5	6	7
2	3	4	5	6	7	8
3	4	5	6	7	8	9
4	5	6	7	8	9	10
5	6	7	8	9	10	11
6	7	8	9	10	11	12

Exercise 5

You could do this exercise in pairs or groups and actually play the games.

1 In the game given in the example above, would the game be fair if the numbers on the dice were multiplied instead of added? Justify your answer by drawing a sample space of all possible outcomes.

You will need to draw the sample space for multiplication.

2 A game is played with three players and these two spinners.
They are both spun together.
The difference between the numbers is found.
If the result is even, player 1 gets a point, if it is odd
player 2 gets a point and if it is zero player 3 gets a point.

   a Draw a sample space of all the possible outcomes.
   b Is this a fair game? Explain.

3 Eight players play a game using these two spinners.

Eight rules on cards are put in a bag.
Each player chooses a rule at random. Then the spinners are spun together. If the rule is satisfied that player gets a point.
This shows the rule each player got.

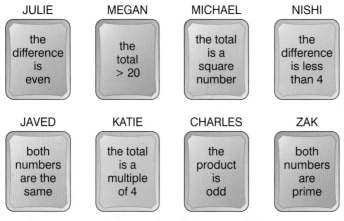

JULIE	MEGAN	MICHAEL	NISHI
the difference is even	the total > 20	the total is a square number	the difference is less than 4

JAVED	KATIE	CHARLES	ZAK
both numbers are the same	the total is a multiple of 4	the product is odd	both numbers are prime

   a Draw a sample space for all the possible outcomes of spinning the spinners.
   b Who is most likely to win?
   c Who is least likely to win?

∗4 Ralph and Sasha play a game with five coins. It costs 10p to play.
If you get exactly two heads you get 20p plus your 10p back.
If you get anything else you lose.
Justify, using mathematical reasoning, whether you are more likely to win or lose.

∗5 Ross and Robyn play Beat the Shaker. Ross is player A.
Who do you think is most likely to win? Justify your answer.

Beat the Shaker Two players, A and B, each roll a dice. A rolls first, then B. If B gets a higher number than A, then B gets 1 point. If B gets a lower number than A, or the same number, then A gets 1 point. The game continues in this manner, with A always rolling first.

∗6 Make up a game such as the one in question **3** and test the rule. Predict then test the results after 36 spins.

Review Two players play a game using two dice. The dice are thrown and the difference between the two numbers is calculated.
Player A wins if the answer is 0, 1 or 2 and Player B wins if the answer is 3, 4 or 5.
a Draw a sample space of all the possible outcomes.
b Explain why this game is not fair.
c How could you change the rules so it is fair?

Games

It is possible to develop a strategy to win each of these games.

1 **Counter game 1 – a game for 2 players**
You will need two piles of counters. There may be any number of counters in each pile.
To play
Each player, in turn, removes one or more counters from just one of the piles.
The loser is the player who is forced to take the last counter.

2 **Counter game 2 – a game for 2 players**
You will need 15 counters set out with 7 in one row, 5 in another and 3 in the other.
To play
Players take turns to remove any number of adjacent counters from any single row.
The winner is the player to remove the last counter.

3 **Circle game – a game for 2 players**
You will need the diagram shown, drawn on a piece of paper, and a blue pen for one player, a red pen for the other.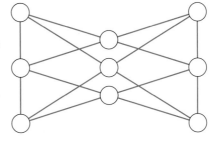
To play
The players take it in turn to colour a circle.
The winner is the first player to have coloured three connected circles.

4 **Grid game – a game for 2 players**
You will need a 4 × 4 grid and 8 counters, 4 of one colour for one player, 4 of another colour for the other player.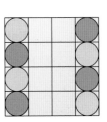
To play
The counters are placed on the grid as shown.
Players take it in turn to move their counters.
At each turn, only one counter may be moved. Counters may be moved horizontally or vertically, one square at a time. They may not be moved diagonally. Counters can not be put on top of one another.
The winner is the first player to get 3 of his or her counters in a row, either horizontally, vertically or diagonally.

Summary of key points

Mutually exclusive events cannot happen at the same time.

Example   When taking one card from a pack of cards it is not possible to take a diamond *and* a black card.

The sum of the probabilities of all mutually exclusive outcomes of an event is 1.

If A and B are mutually exclusive events:

$$p(\textbf{A } or \textbf{ B}) = p(\textbf{A}) + p(\textbf{B})$$

Example   There are 250 members in the Avon Squash Club.

The probability of a member being an adult female is 0·3 and of a student female is 0·06.

The probability a member will be an adult **or** student female is 0·3 + 0·06 = 0·36.

If event A happening has no influence on the outcome of event B, A and B are **independent events**.

$$p(\textbf{A } and \textbf{ B}) = p(\textbf{A}) \times p(\textbf{B})$$

Example   Two cards are drawn, one after the other, from a pack of cards. The first card is replaced.

Event A is drawing an Ace.

Event B is drawing a Jack, Queen or King.

A and B are independent events.

$$p(A) = \frac{4}{52} \qquad\qquad p(B) = \frac{12}{52}$$
$$= \frac{1}{13} \qquad\qquad\qquad = \frac{3}{13}$$

$$p(A \ and \ B) = \frac{1}{13} \times \frac{3}{13}$$
$$= \frac{3}{169}$$

Probability can be estimated from relative frequency.

$$\textbf{Relative frequency} = \frac{\textbf{number of times an event occurs}}{\textbf{number of trials}}$$

Example   A hockey team kept a count of how many games it won in a season.

They won 16 out of 20.

Relative frequency of winning $= \frac{16}{20}$
$$= \frac{4}{5}$$

An estimate of their probability of winning future games is $\frac{4}{5}$.

When trials are repeated many times, the relative frequency of an event tends to a **limit**. We can use this limit to estimate the probability of an event.

Example A dice is rolled many times.

The relative frequency of rolling a 5 or 6 is calculated after 100, 200, 300, ... rolls.

This graph shows the relative frequency tends to a limit.

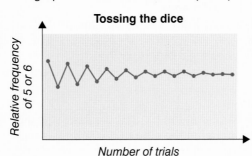

Tossing the dice

Relative frequency of 5 or 6

Number of trials

The probabilities of some events which do **not** have equally likely outcomes can only realistically be estimated using relative frequency.

Example The probability of a sunny Christmas day in York.

 Some games are **fair** and some are **unfair**.

If a game is fair it means all participants have the same chance of winning.

Test yourself

1 A school has male and female pupils from four continents. This table shows the probability of randomly choosing a pupil from the school roll.

Continent	Male	Female
America	0·05	0·1
Africa	0·05	0·05
Europe	0·25	0·15
Asia	0·25	0·1

   a If one pupil is chosen randomly, what is the probability of getting

     i a pupil from Europe

     ii a female pupil?

   b There are 400 pupils at the school. How many would you expect to be American?

2 A sports shop sells three brands of tennis racquets, A, B and C.
It has 24 racquets in stock. There are at least two of each brand.
If one racquet is taken at random, the probability it is brand A is $\frac{1}{3}$.
What is the greatest number of brand C the shop could have in stock?

3 In Jasmine's purse she has 5p, 10p, 20p and 50p coins.
She is going to take one out at random.
The probability of taking each coin is shown on this table.

Coin	5p	10p	20p	50p
Probability	0·4	0·25	0·2	0·15

a Jasmine says 'I think I have twenty 5p coins.'
Explain why she must be wrong.

b What is the smallest possible number of each coin in her purse?

4 Each week Sue borrows one book from the library.
She chooses a romance, mystery or biography.
These three actions are mutually exclusive.
a Explain what 'mutually exclusive' means.
The probability that Sue chooses a romance is 0·1 and the probability she chooses a mystery is 0·6.
b What is the probability she chooses a biography?
c What is the probability she chooses either a mystery or a biography?

5 Mrs Wild drives to school each morning.
The probability that she parks her car at the front of the school is 0·6.
The probability that she parks her car at the side of the school is 0·3.
a What is the probability that she will park either at the front or at the side of the school tomorrow morning?
b In the next 200 school mornings, approximately how many times will Mrs Wild not park either at the front or at the side of the school?

6 Sometimes Paul drives to work, sometimes he catches the bus and the rest of the time he travels by tube.
The probability that he takes the tube is 0·34.
The probability that he takes the bus is 0·26.
Find the probability that Paul
a drives to work
b either catches the bus or the tube
c either drives or takes the bus
d either drives or takes the tube.

7 Two dice are rolled. What is the probability of getting a six on one of them and an odd number on the other?

8 a A fair coin is thrown. When it lands it shows heads or tails.   **[SATs Paper 1 Level 8]**

    Game: Throw the coin three times.

          Player A wins one point each time the coin shows a head.
          Player B wins one point each time the coin shows a tail.

  Show that the probability that player A scores three points is $\frac{1}{8}$.

b What is the probability that player B scores exactly two points?
Show your working.

9 Two faults have been discovered in Mayota cars manufactured last year.
All of these cars are recalled for a check.
The probability that one of these cars has faulty seatbelts is 0·1.
The probability that the steering is faulty is 0·03.
a Find the probability that one of these cars has
   i both faults
   ii neither fault
   iii just one of the faults.
b If 355 cars were recalled for a check, how many of them are likely to have just one of the faults?

10 30 people volunteered to take part in a drug trial.
    12 were Indian and 18 were Asian.
    Two people were chosen at random to undergo a practice trial.
    Copy and fill in this table. Give the answers as fractions.

Gender of chosen people	Probability
Two Indians	
Two Asians	
One Indian and one Asian	

11 This table shows what happened to calls made
    by a telemarketer one day.
    a Find the relative frequency of each result.
    b Estimate the probability of the phone being
       answered.
    c Estimate the probability of the telemarketer not being
       able to talk with the person at home.
    d If she phones 600 people in the next week, how many
       time would she expect to get an answer machine?

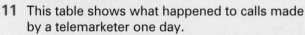

A telemarketer phones
people at home and tries
to sell them something.

Result of call	Frequency
Answered	76
Not answered	34
Answer machine	25
Out of order	3
Engaged	12

12 Angus tossed a coin 20 times, then 50 times, then 100 times,
    then 200 times and so on.
    He sketched a graph of relative frequency of tossing heads versus number of
    times tossed.
    What is his graph likely to look like?

13 Sam and Callum play a game using these spinners.
    The spinners are spun and the two numbers are
    multiplied together.
    Sam wins if the answer is 50 or less.
    Callum wins if it is greater than 50.
    a Draw a sample space of all possible outcomes.
    b Is it a fair game? Explain.
    c How could you change the rules so it is fair?

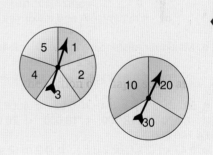

Test Yourself Answers

Chapter 1 page 30

1 **a**, **d** and **e**

2 a 3240 **b** 60 400 **c** 0·082 **d** 90 **e** 0·75 **f** 400 000 **g** 0·00000505 **h** 5·126

3 a $5·2 \times 10^1$ **b** $6·54 \times 10^2$ **c** $3·26 \times 10^1$ **d** $3·0 \times 10^0$ **e** $6·4235 \times 10^2$ **f** $7·0 \times 10^{-1}$ **g** $3·6 \times 10^{-1}$
  h $6·0 \times 10^0$ **i** $6·3 \times 10^{-2}$ **j** $2·006 \times 10^3$

4 a 1 500 000 km **b** 0·001243 cm

5 a $2·9 \times 10^2, 9·6 \times 10^2, 6·4 \times 10^3, 7·1 \times 10^3, 1·8 \times 10^4$
  b $1·6 \times 10^{-5}, 2·3 \times 10^{-5}, 6·9 \times 10^{-4}, 7·5 \times 10^{-3}, 9 \times 10^{-3}$

6 a $8·6 \times 10^8$ **b** $6·28 \times 10^{-6}$ **c** $4·32 \times 10^2$ **d** $3·21 \times 10^5$

7 a $9·03 \times 10^{14}$ **b** $3·5 \times 10^{-12}$ (2 s.f.) **c** $2·94 \times 10^0$ (2 d.p.) **d** $2·53 \times 10^{-1}$ (3 s.f.) **e** $5·96 \times 10^{-3}$ (3 s.f.)

8 a $4·0 \times 10^{-4}$ **b** $4·0 \times 10^{-5}$ **c** $\frac{1}{2500} + \frac{1}{25\,000} = 4 \times 10^{-4} + 4 \times 10^{-5}$
  $\qquad\qquad\qquad\qquad\qquad\qquad = 4·4 \times 10^{-4}$

9 a 101000 N/m^2 **b** 16·7 (1 d.p.) : 1 **c** 6×10^{10} km^3 (1 s.f.)

10 $2·93 \times 10^{-16}$ grams (3 s.f.)

11 a 312 people per km^2 (nearest whole number) **b** 110 people per km^2 (nearest whole number)

12 about 3% (nearest percent)

13 a i smallest 14 500, biggest 15 499 **ii** 14 500 ≤ number of people < 15 500
  b i smallest 15 150, biggest 15 249 **ii** 15 150 ≤ number of people < 15 250

14

Number	Nearest whole number	To 1 d.p.	To 2 d.p.
46·075	46	46·1	46·08
0·625	1	0·6	0·63
16·995	17	17·0	17·00

15 a i 205 mℓ ≤ volume < 215 mℓ **ii** 195 mℓ ≤ volume < 205 mℓ
  iii 175 mℓ ≤ volume < 185 mℓ
  b smallest 588·5 mℓ, biggest 591·5 mℓ, 588·5 mℓ ≤ volume < 591·5 mℓ

16 a Minimum distance = $1·53 \times 10^7 - 9·48 \times 10^6$
  $\qquad\qquad\qquad\qquad = 5·82 \times 10^6$ m
  or if minimise A and maximise B = $1·525 \times 10^7 - 9·485 \times 10^6$
  $\qquad\qquad\qquad\qquad\qquad\qquad\qquad = 5·765 \times 10^6$ m
  b Maximum distance = $1·53 \times 10^7 + 9·48 \times 10^6$
  $\qquad\qquad\qquad\qquad = 2·478 \times 10^7$ m
  or if maximise A and maximise B = $1·535 \times 10^7 + 9·485 \times 10^6$
  $\qquad\qquad\qquad\qquad\qquad\qquad\qquad = 2·48 \times 10^7$ m

17 a 50 **b** 360 **c** 8·4 **d** 0·02 **e** 4000 **f** 640 **g** 9·10 **h** 0·006 **i** 0·0604

18 a 18 470 **b** 18 500 **c** 18 000 **d** 20 000 **e** 20 000 **f** 18 000 **g** 18 500

19 £4 740 000

20 a 0·43 kg (2 s.f. or 2 d.p.) **b** 4·9 m (2 s.f. or 1 d.p.)

Chapter 2 page 49

1 a i $2^3 \times 3^2 \times 7$ **ii** $2^5 \times 5^2 \times 11$ **b** HCF = 8, LCM = 554 400

2 a $\frac{2}{3}$ **b** $\frac{7}{12}$ **c** $\frac{5}{8}$

3 a $\frac{11}{90}$ **b** $\frac{33}{140}$

4 a HCF = mn, LCM = m^2n^2 **b** HCF = $6abc^2$, LCM = $18a^2b^3c^3$

5 a 21·26 **b** 3·78 **c** 3·17 **d** 8·08

6 $2^{20}, 7^8, 3^{16}$

7 a a^8 **b** m^4 **c** p^{12} **d** y^{3n} **e** m^{p+q} **f** x^8 **g** x^7 **h** $81m^8n^{12}$

8 a 4 **b** 14 348 907

9 a 5 **b** 3 **c** 2 **d** 32 **e** $\frac{1}{3}$ **f** 1 **g** $\frac{1}{27}$ **h** 0·4 **i** $\frac{1}{6}$

10 a x^2 **b** x^{-3} or $\frac{1}{x^3}$ **c** $m^{6·5}$

11 a 6 **b** $3\sqrt{3}$ **c** 18 **d** $\sqrt{35}$ **e** $6\sqrt{5}$ **f** $2\sqrt{10}$ **g** 4 **h** $6\sqrt{7}$

12 a $3\sqrt{6}$ **b** $9\sqrt{5}$ **c** $21\sqrt{5}$

Chapter 3 page 70

1 a 16 m **b** 20% **c** 12 **d** 4 **e** 30 km **f** 60 cm^2 **g** $3x^3$
  h $x = 102$ **i** 23 **j** $^-5$ **k** 5·75 **l** 120 cm^3 **m** 240

2 c and d

3 a 7 **b** $^-4$ **c** 9 **d** 6 **e** 3

4 a 10×5 **b** $10 \div 0.01$ **c** 10×0.1

5 36, 38 and 40

6 A possible answer is:

	25·9		
	14·3	11·6	
	7·7	6·6	5·0
4·5	3·2	3·4	1·6

7 6 dogs, 1 cat and 1 mouse

8 $(6 - 4)^2 \times (3 + 5) = 32$

9 a 40% **b** $\frac{3}{5}$ **c** 0·375 **d** 62·5% **e** 27 **f** 54 **g** 48 **h** $8\frac{3}{4}$ or 8·75

10 £4·65

11 a C 8 **b** D 16 **c** 6 **d** 25 $\left(\frac{^530 \times 20^5}{_16 \times _14}\right)$ or 30 $\left(\frac{30 \times ^124}{_16 \times _14}\right)$

12 a Perimeter 160 cm, Area 1200 cm^2
  b Perimeter 164·8 cm, Area 1286·02 cm^2

13 a $452 \div 8$ **b** $69·4 \div 7$

14 a 0·343 **b** 0·004 **c** 0·367 **d** 0·015

15 £4

16 a $\frac{5}{2}$ **b** $\frac{7}{10}$ **c** 8 **d** $\frac{1}{6}$ **e** $\frac{5}{2}$ or $2\frac{1}{2}$ **f** $\frac{10}{17}$ **g** $\frac{y}{x}$ **h** $\frac{3}{m}$ **i** $\frac{1}{p}$ **j** $\frac{1}{3q}$

17 a 4 **b** 0·16 **c** 5·9 **d** 0·043 **e** 0·011

18 a $x = \frac{1}{4}$ or 0·25 **b** $x = 5$ **c** $x = \frac{4}{5}$ or 0·8

19 5·091 (4 s.f.)

20 a 69·4 **b** 12·6 **c** $^-95·4$ **d** $8\frac{9}{35}$ **e** $2\frac{2}{39}$ **f** 29·9 **g** 2·72 **h** $2·61 \times 10^{11}$

21 a

Circumference	3·5 m	4·0 m	4·5 m	5·0 m	5·5 m
Price (£)	89	101·71	102·99	114·43	125·87

  b 4·0 m circumference umbrella **c** 0·64 m^2 (2 d.p.) **d** 5·0 m circumference umbrella

Chapter 4 page 88

1 $\frac{12}{20}$, 0·9, 130%, $\frac{35}{25}$

2 a 28·0% (1 d.p.) **b** 45·2% (1 d.p.) **c** Chips **d** 3·5% (1 d.p.)

3 a $\frac{7}{16}$ **b** £60

4 a 46·6% (1 d.p.) **b** Percentage in 2002 was greater.

5 a $1\frac{1}{12}$ **b** $\frac{1}{24}$ **c** $\frac{1}{8}$ **d** $4\frac{7}{15}$ **e** $1\frac{33}{40}$

6 a 5 **b** $5\frac{2}{3}$

7 20 g

8 a 30 **b** $7\frac{1}{2}$ **c** $\frac{3}{20}$ **d** $4\frac{8}{15}$ **e** $2\frac{1}{4}$

9 a $3\frac{71}{84}$ **b** $\frac{19}{32}$

10 a $11\frac{3}{8}$ m^2 **b** $12\frac{7}{10}$ m

11 a 20 **b** $\frac{3}{50}$ **c** $1\frac{1}{4}$ **d** $3\frac{9}{10}$ **e** $8\frac{1}{3}$ **f** $1\frac{1}{4}$

12 a $\frac{9}{m}$ **b** $\frac{13x}{15}$ **c** $\frac{ay - bx}{xy}$ **d** $\frac{2pn + 3qm}{mn}$

13 a $\frac{71}{99}$ **b** $\frac{57}{999}$ **c** $\frac{161}{495}$ **d** $\frac{64\,759}{99\,900}$

Chapter 5 page 110

1 £835 022·69

2 6·4% (1 d.p.)

3 a 60×0.93
  b 60×1.07 could be 'What is 60 increased by 7%?' 60×0.3 could be 'What is 60 decreased by 70%?'
  60×1.7 could be 'What is 60 increased by 70%?' **c** 1·065

4 £228 ÷ 1·2 = £190

5 500 mℓ

6 Smallest mass is 2·975 kg, biggest is 4·025 kg.

7 0·8925 (1·05 × 0·85)

8 a 850 × 1·065³   **b** £1320·89 (nearest p)   **c** £470·89

9

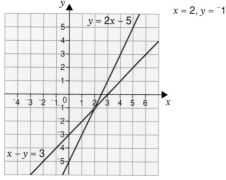

Apple Muffins

300 g flour
4 tsp baking powder
1 cup milk
1 egg
160 g sugar
1 chopped apple
60 g butter

10 a 487·5 mℓ   **b** 500 mℓ apple concentrate, 1 ℓ of orange concentrate and 5 ℓ water

11 The 2nd bar with ratio 5 : 104, because for the 1st bar the ratio is 1 : 20·75 and for the 2nd bar it is 1 : 20·8.
The 2nd bar has a smaller proportion of fat.

12 44 cm³

13 19 : 41

14 a $P \propto T$ **b** $K_{sp} \propto s^2$ **c** $R \propto \frac{1}{I}$

15 a It becomes 3 times bigger.   **b** It is halved.   **c** $A \propto h$

16 a 1·2, 6, 15, 18, 24, 25·2
   b No
   c 0·048, 6, 93·75, 162, 384, 444·528; no
   d 0·04, 1, 6·25, 9, 16, 17·64
   e $\frac{SA}{l^2} = 6$ for each set of values; yes

Chapter 6 page 153

1 Equations **c** and **e**; Formulae **b** and **d**; Functions **a** and **f**

2 a i $4(x-3) = 4(2-3) = 4(^-1) = ^-4$ **ii** $4(x-3) = 4(7-3) = 4 \times 4 = 16$
            $4x - 12 = 4 \times 2 - 12 = ^-4$ $4x - 12 = 4 \times 7 - 12 = 28 - 12 = 16$
   iii $4(x-3) = 4(^-4-3) = 4 \times ^-7 = ^-28$
            $4x - 12 = 4 \times ^-4 - 12 = ^-16 - 12 = ^-28$
   c An identity

3 a $x = 2$ **b** $x = ^-4\frac{1}{3}$ or $^-4 \cdot \dot{3}$ **c** $x = ^-26$ **d** $x = 1 \cdot 4$ **e** $y = \frac{2}{3}$ or $0 \cdot \dot{6}$ **f** $y = 1$ or $y = ^-7$

4 a $x = ^-2, y = ^-3$ **b** $x = 2, y = 0 \cdot 5$

5 a and b No because in each case the lines are parallel.
   c No because you just get the solution $x = x$.

6 $x = 1\frac{1}{2}, y = 4$

7 $a = 14, b = 3 \cdot 5$

8 a $x = 6, y = 3$ **b** $x + y = 5$ **c** $x = ^-2, y = 7$ **d** $y = 2x - 4$ and $x + y = 5$

9

$x = 2, y = ^-1$

10 20 people paid before 1 May. The two equations are $e + l = 140$ and $75e + 80l = 11\ 100$.

11 a C   **b** A

12 0, 1, 2, 3, 4

13 a

Integer solutions are ..., 2, 3, 4.

b

Integer solutions are ⁻1, 0, 1, 2, 3, 4.

c

Integer solutions are 0, 1, 2, ...

14 a i $n \leqslant 3$ **ii** $n > {}^-4$ **iii** $1 < n \leqslant 5$
b i (number line from ⁻3 to 2) **ii** ⁻2, ⁻1, 0, 1

15 a $a < 2 \cdot 6$ **b** ${}^-3 < n \leqslant 5$ **c** $x < 2 \cdot 5$ **d** $x \geqslant 2$ **e** $x < {}^-3$

16 $x \geqslant 0$, $x < 6$, $y \geqslant 0$, $x + 2y \leqslant 8$

17 a ${}^-2 \leqslant x \leqslant 2$ **b** $x \geqslant 8$ and $x \leqslant {}^-8$ **c** ${}^-3 < n < 3$ **d** ${}^-4 < n < 4$ **e** $n < {}^-1$ or $n > 1$

18

	Correct for **no** values of x	Correct for **one** value of x	Correct for **two** values of x	Correct for **all** values of x
$3x + 7 = 8$		✓		
$3(x + 1) = 3x + 3$				✓
$x + 3 = x - 3$	✓			
$5 + x = 5 - x$		✓		
$x^2 = 9$			✓	

Chapter 7 page 176

1 a $5x + 4$ **b** $6y - 2$

2 a Total number of people $= 2 \times 1 + 3 \times 5 + 4 \times x + 5 \times 7 + 6 \times 8 + 7 \times 4 + 8 \times 2$
$= 2 + 15 + 4x + 35 + 48 + 28 + 16$
$= 4x + 144$

b $27 + x$ **c** $x = 9$

3 a $n^2 - 1$ **b** 82 **c** $n^2 - 2n + 3$ **d** $n + 1$ **e** $n^2 + n + 1$

4 a $x^2 + x - 12$ **b** $y^2 - 6y + 5$ **c** $a^2 + 5an + 6n^2$ **d** $2c^2 - 5cx - 3x^2$ **e** $10n^2 - 17ny + 3y^2$
f $6 - 3d^2 - 7d$ **g** $3x^2 + 4x - 4$ **h** $5ax - 15bx + 2an - 6bn$

5 a $x^2 + 2xy + y^2$ **b** $2m^2 - 4m + 4$ **c** ${}^-3p^2 + 20p + 7$

6 a $4(x + y)$ **b** $3m(2m^2 + 1)$ **c** $14(2 - y)$ **d** $p(4p - 1 + 6p^2)$ **e** $12p(3 - 2p)$

7 a $c + 2$ **b** 2 **c** This cannot be simplified because y is not a common factor of x and $2y$.
d $\frac{2 + m}{2m}$ **e** $\frac{3 - 2y}{x}$

8 a 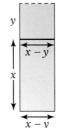 can be redrawn as It can be seen that the area of this shape
is $(x + y)(x - y)$.
So $x^2 - y^2 \equiv (x + y)(x - y)$

Area of green shape $= x^2 - y^2$

b $a^2 - b^2 = (a + b)(a - b)$ **c** $4a^2 - b^2 = (2a + b)(2a - b)$ **d** $(8 + 4)(8 - 4) = 48$

9 a $y = \frac{a}{x}$ **b** $P = \frac{k}{VT}$ **c** $l = \frac{2S}{n} - a$ **d** $h = \frac{A - \pi r^2}{2\pi r}$ **e** $b = 2s - a - c$

10 a 10 cm² **b** 5 cm **c** 8·86 m (3 s.f.)

11 17·2 (3 s.f.) or ⁻17·2 (3 s.f.)

12 $v = \sqrt{\frac{2E}{m}}$

13 a 3·8 (2 s.f.) **b** 10·5 (3 s.f.)

14 $f = 4n + 2$

15 a 3·227486122 **b** $+3 \cdot 23$ and $-3 \cdot 23$

Test Yourself Answers

Chapter 8 page 200

1 a 10, 9·5, 9, 8·5, 8, 7·5, 7, 6·5, 6, 5·5 **b** 2, 2, 4, 6, 10, 16, 26, 42, 68, 110
  c $1\frac{1}{2}, 2\frac{1}{2}, 3\frac{1}{2}, 4\frac{1}{2}, 5\frac{1}{2}, 6\frac{1}{2}, 7\frac{1}{2}, 8\frac{1}{2}, 9\frac{1}{2}, 10\frac{1}{2}$ **d** 3, 2, 3, 6, 11, 18, 27, 38, 51, 66

2 One possible answer is: **1st term** 10 **rule** subtract 0·2

3 $m + 2n, 2m + n, 3m, 4m - n, 5m - 2n, 6m - 3n$

4 a Yes, 9 **b** $x = 9$. It is the same as the number the sequence converges to.

5 a and **c** because for each of these the 2nd differences are constant.

6 a 3, 6, 11, 18, 27, 38 **b** ⁻1, 5, 15, 29, 47, 69

7 a 2 **b** 4

8 a 47, 62 **b** 73, 99 **c** 31, 44

9 a $T(n) = n^2 - 1$ **b** $T(n) = 4n^2$

10 a $T(n) = n^2 + 3$ **b** $T(n) = n^2 + n - 2$ **c** $T(n) = 3n^2 - n + 4$

11 a   **b** 4, 12, 24, 40, ... **c** $T(n) = 2n^2 + 2n$

12 a $\frac{5}{25}, \frac{6}{36}, \frac{7}{49}$; rule is **1st term** $\frac{1}{1}$ **rule** 'add 1 to the numerator of the previous term, square the numerator to get the denominator'

  b $\frac{8}{13}, \frac{13}{21}, \frac{21}{34}$; rule is **1st terms** $\frac{1}{1}, \frac{1}{2}$ **rule** 'add numerators of two previous terms and add denominators of two previous terms'

13 a $\frac{1}{3}, \frac{2}{4}, \frac{3}{5}, \frac{4}{6}, \frac{5}{7}$ **b** $\frac{2}{1}, \frac{3}{4}, \frac{4}{9}, \frac{5}{16}, \frac{6}{25}$ **c** $\frac{2}{3}, \frac{5}{4}, \frac{10}{5}, \frac{17}{6}, \frac{26}{7}$

14 The terms get closer and closer to 1.

15 a $\frac{n}{3n - 1}$ **b** $\frac{2}{7}, \frac{3}{17}, \frac{4}{31}$

16 a i Output ⁻8, ⁻4, 4, 5 **ii** Output ⁻2, 2, ⁻1; input 6
  iii Output $-\frac{1}{2}$, ⁻1·75; input 5, 8 **iv** Output ⁻1, $-3\frac{5}{8}$; input 4, 7

  b i C **ii** A **iii** D **iv** B

17

18 a $x \rightarrow \frac{x}{4} + 1$ **b** $x \rightarrow 2(x + 3)$ **c** $x \rightarrow \frac{3x}{2} - 1$ or $x \rightarrow \frac{3x - 2}{2}$ **d** $x \rightarrow \sqrt{x + 1}$

19 a i $x \rightarrow 6 - x$ **ii** $x \rightarrow x$ **iii** $y = \frac{1}{2}x$ **b** Identity function **c** $y = x$

20 a $3x + 5 = 32$ or $3(x + 5) = 42$ **b** $3(x + 5) - (3x + 5)$
          $3x = 27$ $x + 5 = 14$ $= 3x + 15 - 3x - 5$
          $x = 9$ $x = 9$ $= 10$
                                                The difference will always be 10.

21 a False **b** True **c** True

Chapter 9 page 226

1 a l_1 ⁻1, l_2 1, l_3 2 **b** l_1 $y = {}^-x$, l_2 $y = 1x + 4$ or $y = x + 4$, l_3 $y = 2x - 4$

2 a $2y = 5x$ **b** $y = 2x + 4$ and $y = -\frac{1}{2}x - 2$ **c** $y = x^2 - 2$ **d** $y = \frac{5}{2}x + 3$ **e** $2y = 5x$ and $y = \frac{5}{2}x + 3$ **f** $y = {}^-4$

3 a $y = {}^-2x + 3$ **b** $y = -\frac{1}{2}x + 3$ **c** $y = 3x - 1$

4 (⁻3, ⁻14)

5 l_1 D, l_2 B, l_3 E, l_4 A, l_5 C

6 AB

7 a C **b** F **c** H **d** B **e** E **f** A **g** D **h** G

8 a Surface area $= x \times x + 4(x \times 2x)$
$$= x^2 + 8 \times x^2$$
$$= 9x^2$$

b

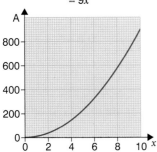

c About 9·3 cm

9 a

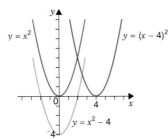

b i $x = 0$ **ii** $x = 4$

10 a

x	⁻3	⁻2	⁻1	0	1	2	3
x	⁻27	⁻8	⁻1	0	1	8	27
⁻4x	12	8	4	0	⁻4	⁻8	⁻12
y	⁻15	0	3	0	⁻3	0	15

b

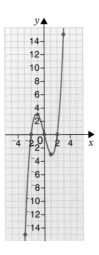

c It crosses the x-axis at (–2, 0), (0, 0) and (2, 0). It crosses the y-axis at (0, 0).

11 a B **b** C **c** D **d** A

12 a The skier skied downhill, stopped for a while then skied on down to the lodge.

b Some water was put in a kettle. After a while some water was taken out but then more water was put into the kettle straight away. Then the water level stayed the same.

c The newborn baby lost weight for the first few days then slowly started to put weight on. The weight gain slowed down a bit after a while.

13 a The sound level increased very quickly, then more slowly, then it stayed at the same level.

b A possible reason is that the CD finished.

14 a

b	1	2	3	4	5	6	7	8
a	8	4	2·7	2	1·6	1·3	1·1	1

b

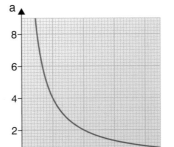

c About 1·8

15 a A – 3, B – 4, C – 1, D – 2

b A – 4, B – 3, C – 2, D – 1

Chapter 10 page 266

1 a Convention **b** Derived property **c** Convention **d** Definition **e** Derived property **f** Definition

2 A demonstration because it only deals with one triangle and does not prove the results for all right-angled triangles.

3 $a = 65°$ (corresponding angles on parallel lines)
$\text{MÔN} = 65°$ (base angle of isosceles triangle)
$x = 65°$ (corresponding angles on parallel lines)
$b = 180° - 65°$ (angles on a straight line add to 180°)
$= 115°$

4 a $a = 38°$ (alternate angles on parallel lines)
$b = 100°$ (angles in a triangle add to 180°)
$c = 80°$ (angles on a straight line add to 180°)
$d = 20°$ (isosceles triangle and angles in a triangle add to 180°)
$f = 42°$ (alternate angles on parallel lines)
$e = 62°$ (alternate angles on parallel lines)

b $a = 25°$ (angles in an equilateral triangle are 60°)
$b = 60°$ (angle in an equilateral triangle)
$y + b = 130°$ (isosceles triangle and angles in a triangle add to 180°)
$y = 130° - 60°$
$= 70°$

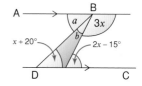

5 $a = x + 20°$ (alternate angles on parallel lines are equal)
$b = 2x - 15° - (x + 20°)$ (exterior angle of triangle = sum of interior opposite angles)
$= 2x - 15° - x - 20°$
$= x - 35°$
$a + b + 3x = 180°$ (angles on a straight line add to 180°)
$x + 20° + x - 35° + 3x = 180°$
$5x - 15° = 180°$
$5x = 195°$
$x = 39°$

6 $a = 50°$ (alternate angles on parallel lines)
$\angle\text{QTS} = 55°$ (angles on a straight line add to 180°)
$c = 105°$ (exterior angle of triangle = sum of interior opposite angles)
$b = 55°$ (alternate angles on parallel lines)
$\angle\text{PUS} = 130°$ (angles on a straight line add to 180°)
$d + 130° + 105° + 125° + 84° = 540°$ (angles in a pentagon add to 540°)
$d = 96°$

7 a 10·6 m **b** 8·6 m **c** 13·7 mm **d** 17·7 m

8 a $AC^2 = CB^2 + AB^2$ (Pythagoras' theorem)
$= 6^2 + 8^2$
$= 100$
$AC = 10$ cm

or △ABC is a 3, 4, 5 Pythagorean triple where each side is multiplied by 2.
$(3 \times 2)^2 + (4 \times 2)^2 = (5 \times 2)^2$
The length of AC is $2 \times 5 = 10$ cm.

b $AD^2 = DC^2 + AC^2$ (Pythagoras' theorem)
$= 6^2 + 10^2$
$= 136$
$AD = \sqrt{136}$
$= 11·7$ cm (1 d.p.)

9 a 13 **b** 12 **c** 75 **d** 16

10 a $26^2 = 24^2 + 10^2$ so Pythagoras' theorem is true for this triangle.
Therefore it must have a right angle opposite the longest side, 26 units.

b 1680 cm³

Chapter 11 page 296

1 a gives 1 unique triangle, **b** gives 1 unique triangle, **c** gives 1 unique triangle and **d** gives no triangles.

2 a A and B (reason SAS) **b** E and F (reason RHS)

3 By alternate angles on parallel lines, angles with same symbol are equal.
Triangles ABC and CDA are congruent (AAS)
∴ AB = CD and AD = CB.

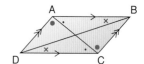

4 a **B** **b** **A** and **C**

5 20 cm

6 0·5 cm

7 a 90°   **b** 58°

9 Possible sketches are:

10 a Circle, centre A, radius 28 mm

  b A line parallel to CD and EF, and equal distances from them

  c The bisector of angle JGH

11

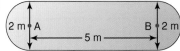

12 Diagram not drawn to scale.

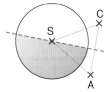

Chapter 12 page 319

1 A and B is not commutative. All the others are commutative.

2 a A'(⁻2, 1), B'(1, 7), C'(⁻2, 7)   **b** A'(⁻1, 0), B'(0, 2), C'(⁻1, 2)

3 $\frac{2}{3}$

4 a $\frac{1}{3}$ **b** 3   **c** The shapes are similar because all pairs of corresponding sides have equal ratios and the angles in each shape and its enlargement remain unchanged.

5 a $k = \frac{3}{2}$, $x = 6$ **b** $k = \frac{3}{4}$, $x = 3\cdot75$

6 40 cm

7 150 g

8 $2^2 : 5^2$ or 4 : 25

9 a 2   **b** 16 cm   **c** 384 cm^2 **d** 7200 cm^3

10 a 30 cm   **b** 703 cm^2 (3 s.f.)

11 (0, ⁻3)

12 a 5   **b** $\sqrt{101} = 10\cdot05$ (2 d.p.)

13 a $\sqrt{40} = 6\cdot32$ (2 d.p.) and $\sqrt{18} = 4\cdot24$ (2 d.p.)   **b** $\sqrt{106} = 10\cdot30$ (2 d.p.) and $\sqrt{10} = 3\cdot16$ (2 d.p.)

Chapter 13 page 356

1 2·35 kg ⩽ m < 2·45 kg

2 a Upper limit = 18·5 cm, lower limit = 17·5 cm   **b** Upper limit = 178·5 mm, lower limit = 177·5 mm

  c Yes. The upper limit of Julie's handspan is greater than the lower limit of Charlotte's handspan. That is 178·5 mm = 17·85 cm which is greater than 17·5 cm.

3 30 g/m^2

4 2 ℓ/min

5 4·1 m/s (1 d.p.)

6 708·3 N/m^2 (1 d.p.)

7 50 000 kg/m^3

8 69·8 m^2 (1 d.p.)

9 a 1749·8 cm^2 **b** 296·5 cm

10 562·5 cm^2

11 a 660 cm^2 (2 s.f.)   **b** 24 000 cm^3 (2 s.f.)

12 a 17 cm (2 s.f.)   **b** 17·1 cm (1 d.p.)   **c** 46° (2 s.f.)

13 32·7 m (1 d.p.)

14 3·5 m (1 d.p.)

15 a 124 km   **b** 064°

Chapter 14 page 380

1 Possible answers are:

Taking a mixture of vitamins prevents a person getting sick so often.

Taking vitamin C on its own has no effect on a person's health.

Taking vitamin tablets in winter prevents a person getting sick so often.

2 **c** is likely to be unbiased because the sample is chosen randomly.

a is likely to be biased as people who have fires are more likely to live in older houses. They are more likely to want to keep using their fires. The sample is not representative of everyone. It doesn't include people who only have electrical or gas heating in their houses.

b is likely to be biased. It is not likely to include any children nor young people who are not so likely to be able to afford to buy a television.

3 Possible answers are:

Conjecture 1

i Income status of family, number of days a child is sick each year, number of days a child is away from school because of sickness

ii A survey of families from a range of incomes. This is primary data.

iii Choose a sample which includes a complete range of socioeconomic backgrounds, of ethnic backgrounds, of family size and with children of all ages and with equal numbers of males and females.

Conjecture 2

i Numbers of rubbish bins in parks, amount of litter left lying round in parks

ii Data could be collected by doing a survey on the number of rubbish bins in parks and on the amount of rubbish left lying round. Primary data

iii You could collect data from a number of different parks in different areas – some in the business area, some in residential, some near schools and so on.

4 Possible answers are:

a

Retailer \ Number of bottles	<50	50–99	100–149	150–199	200–249	250+
Supermarket						
Pharmacy						

b

Number of call-outs	Area A	Area B	Area C	Area D	Area E	Area F
0–4						
5–9						
10–14						
15–19						
20–24						
25–29						
30+						

Chapter 15 page 413

1 a

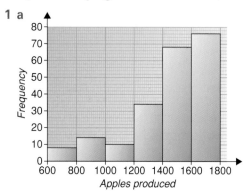

b i

Class interval	1 m–	2 m–	3 m–	4 m–	5 m–6 m
Mid-point	1·5 m	2·5 m	3·5 m	4·5 m	5·5 m
Frequency	6	24	84	87	9

ii 3·8 m (1 d.p.)

iii 3 m–

iv 4 m–

v 3·9 m (1 d.p.)

2 a

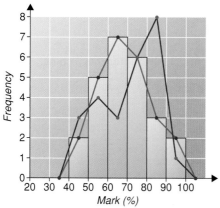

b There are more very high marks in Test B.

3 a As height increases, so does mass.
b 580–595 kg
c 166·5–167·5 cm
d All points must lie below the line $y = x$.
A possible answer is:

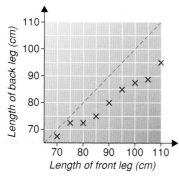

4 Carlisle: Range = 4 °C, Mean = 18·3 °C
Calais: Range = 8 °C, Mean = 17·7 °C
The range of temperatures is much greater for Calais.
The mean temperature for Calais is less than Carlisle, but its temperatures are less consistent, as shown by the range. Calais records both the highest (22°C) and the lowest (14°C) of all the temperatures.

5 a No. The percentages will have been rounded. If more were rounded up than down, this could give the 101% total that this pie chart gives.
b Any time between 2 hours 51 minutes and 2 hours 54 minutes is correct.
c Possible answers are: Days you don't work lower the average.
    or   Some days you don't work.

6 a Estimate of mean = $\dfrac{15 \times 6 + 45 \times 14 + 75 \times 21 + 105 \times 9}{50}$ = 64·8 min

b   Estimated median = 67 min

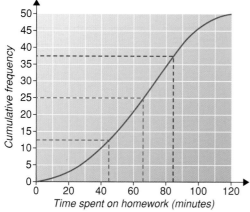

c About 4   **d** 120·5 min   **e** About 47 min for lower quartile and about 85 min for upper quartile
f 38 min

Test Yourself Answers

Chapter 16 page 437

1 a i 0·4 **ii** 0·4 **b** 60

2 14

3 a Because if she has twenty 5p coins, 0·4 represents 20 coins. So 0·1 represents 5 coins.
So 0·25 would represent 12·5 coins which is impossible.

  b 5p – 8, 10p – 5, 20p – 4, 50p – 3

4 a Mutually exclusive events cannot happen at the same time.

  b 0·3 **c** 0·9

5 a 0·9 **b** 20

6 a 0·4 **b** 0·6 **c** 0·66 **d** 0·74

7 $\frac{1}{6}$

8 a $\frac{1}{2} \times \frac{1}{2} \times \frac{1}{2} = \frac{1}{8}$

  b $p(\text{win, win, not win}) = \frac{1}{2} \times \frac{1}{2} \times \frac{1}{2} = \frac{1}{8}$. But the two wins can be in any order, that is win, win, not win, or win, not win, win, or not win, win, win.
So $p(2 \text{ wins}) = 3 \times \frac{1}{8} = \frac{3}{8}$.

9 a i 0·003 **ii** 0·873 **iii** 0·124

  b 44 to the nearest car

10

Gender of chosen people	Probability
Two Indians	$\frac{22}{145}$
Two Asians	$\frac{51}{145}$
One Indian and one Asian	$\frac{72}{145}$

11 a Answered $\frac{38}{75}$, not answered $\frac{17}{75}$, answer machine $\frac{1}{6}$, out of order $\frac{1}{50}$, engaged $\frac{2}{25}$

  b $\frac{38}{75}$ **c** $\frac{37}{75}$ **d** 100 times

12 A possible answer is:

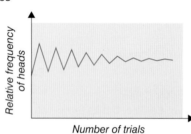

Number of trials

(vertical axis: *Relative frequency of heads*)

13 a

×	10	20	30
1	10	20	30
2	20	40	60
3	30	60	90
4	40	80	120
5	50	100	150

  b No. There are 8 results which are 50 or less and only 7 results greater than 50.
So Sam is more likely to win.

  c A possible answer is you could change the rule to:
Sam wins if the answer is less than 50.
Callum wins if the answer is more than 50.
No one wins if the answer is 50.

Index

adding 1,4, 51, 68
    fractions 5, 34, 47, 78, 83, 87
adjacent sides 343, 344
angle bisectors 230, 231, 233, 292, 295
angles 229, 230, 250
    alternate 229, 230, 248
    base 230
    corresponding 229, 250
    of elevation and depression 350
    exterior 229, 230
    interior 229, 230, 248, 265
    naming 229
    opposite 229
    made with parallel lines 229, 230, 248, 250
    at a point 229, 248
    of polygons 230, 248, 265
    on a straight line 229, 230, 248, 250
    of a triangle 229, 248
    vertically opposite 229
annulus 171, 354
answers 440–50
answers, checking 4, 60
answers, estimating 3, 4, 57, 69
arc of a circle 231, 332–3, 354
area 235, 354
    of circles 110, 235, 322, 332–3, 354
    of parallelograms 235
    of sectors 332–3, 354
    of trapeziums 112, 235
    of triangles 235
arrowheads 230
assumed mean 361
average speed 327
axes of symmetry 212, 225, 303

bacteria growth 188, 192
balancing act 124
bar charts 359
bearings 235, 350
bias 374, 377, 379, 380, 430
BIDMAS 3, 51, 68
billion 18, 21, 22
bisectors of angles 230, 231, 233, 292, 295
body mass index 61
bones, human 156
boundaries of class intervals 391
bounds for rounded numbers 22–3, 29
boxes 60
brackets 65, 114, 126, 157, 160–1, 175

Budget Buys 341

calculator memory 327, 337
calculators 19, 48, 65–6, 169
    graphical 137, 203, 207, 215, 216, 217
cancelling fractions 34, 47
cancelling ratios 6
capacity 235
car in a traffic jam 218
cause and effect relationship 404
centre of enlargement 234, 304, 305, 318
certainty 362
changing the subject 115, 166–7, 175, 203, 327
characters v
checking answers 4, 60
chords 231, 283, 284, 285
circles 231, 233, 247, 283, 292
    area 110, 235, 322, 332–3, 354
    inscribed 284
    intersecting 284, 285
    parts 231
circumcircle 284, 285
circumference 231, 235, 322, 332, 354
class intervals 359, 360, 377, 391
coffee, costings 76
coins, tossing 362, 423
collecting like terms 114, 157, 175
collection sheets 358
combinations, commutative 299, 318
common denominator 34, 47, 74, 83
common factors 34, 36, 47–8, 163
compensation 51
complements 51
compound interest 95, 109
compound measures 326, 327, 331, 354
computer games 417, 421, 430, 432
cones 112, 170, 281, 335
congruence 101, 174, 272, 284
congruent curves 212, 225
congruent shapes 231, 234, 299, 318
congruent triangles 231, 272–3, 284, 294
conjecture 371, 379
consistency as shown by data 361
constant difference 115, 183, 185, 198

constant multiplier 98, 99, 109
constant of proportionality k 103, 104, 110
construction, angle bisector 231
construction, perpendicular 231, 232
construction, quadrilaterals 232
construction, triangles 232, 270–1, 294
conventions 247, 264
converging sequences 180, 197
coordinate pairs 116, 136, 233, 315, 319
correlation between variables 361, 399, 413
cosine (cos) 341, 342, 343, 347, 355
counters, choosing 423
counting up 51
cross-section 159, 160, 233, 286, 295
cube numbers 65
cube roots 2, 38, 65
cubic functions 225
cubic graphs 208–12
cuboids 233, 235, 236, 281, 286
cumulative frequency graphs 391, 412, 415
cumulative frequency tables 390
currency exchange 53, 54, 101, 102, 202
cylinders 167, 168, 176, 311, 337, 341

data
    collecting 358, 377–8, 379, 380
    comparing 361, 391, 404–5
    continuous 23, 29, 359, 360, 385, 387
    discrete 23, 29, 359
        grouped 359, 385, 386
    displaying 359–61
    grouped 385, 386–7, 411
    interpreting 361, 404–5
    sources (primary and secondary) 358, 373, 379
decades 235
decimal form 15–16, 17, 29
decimals 4, 55
    dividing by 4, 59, 69
    to fractions 68
    to percentages 5, 68
    recurring 1, 85–6, 88
    rounding to 1, 25
definitions 247, 264
demonstrations 248, 264

Index

denominator 5, 34, 66, 73, 74, 78
    common 34, 47, 74, 83
density 326, 327
depression, angle of 350
derived properties 247, 264
diagonals 230, 247, 264
diameter 231, 283
dice 362, 367, 424, 430, 433
difference, constant 115, 183, 185, 198
difference table 115, 186, 187
differences, first and second 182, 183, 185, 186, 198, 248
direct proportion 98, 103, 105, 117, 218, 219, 326–7
displacement of water 327–8
distance charts 330
distributions 405
dividing 1, 57, 62, 69
    by 10, 100, 1000 etc 1, 16
    by decimals 4, 59, 69
    fractions 5, 81, 87
    using index laws 18, 41
    mentally 3, 51, 68
    in a given ratio 6, 315, 319
divisor 62
dog, tethered 294
dominoes 430
doubling and halving 3, 51

Earth 17, 18, 59, 66, 67, 330
Egyptian fractions 79
elevation (view) 233
elevation, angle of 350
elimination method 132–3, 134
employment changes 408–9
enlargement 234, 276, 304–7, 309, 311–12, 318
    centre of 234, 304, 305, 318
    inverse of 304, 305, 306, 318
    of shapes, 3-D 311
equally likely 362, 417
equations 18, 63–4, 113–14, 127, 203, 224
    linear 113, 127, 151
    simultaneous 132–4, 135, 136–7, 139–40, 152
equivalent calculations 4, 59, 69
equivalent fractions 5, 78, 83, 87
Escher, Maurits 269
estimating answers 3, 4, 57, 69
even chance 362
events 362
    compound, probability 421
    exhaustive 417
    independent 423, 424–5, 436
    mutually exclusive 362, 417, 421, 436
experimental results 362
experiments 373, 378

expressions 113
    factorising 114, 163, 175
    simplifying 114, 157, 175
    substituting into 114, 169, 176, 180

faces 172, 235
factor trees 2, 34
factorising expressions 114, 163, 175
factors 2, 3, 34, 51, 114
    see also scale factor
fans (sectors) 335
Fibonacci sequence 179, 189
filling containers 218
finding the length of a line 316–17, 318, 319, 343
finite length 229
first term 115, 180, 197
flight to Madrid 221
floor plan 298
flow chart 115, 175
formulae 151, 172–3
    acceleration 113, 168, 171, 125
    changing the subject 115, 166–7, 175, 203, 327
    density 327
    electrical 64, 115
    energy 118, 151
    Ohm's Law 65
    pressure 118, 326, 327
    speed 18, 115, 166, 326, 327
    substituting into 115, 169, 176
    temperature conversion 168
fossil fuels 67
fraction sequences 189, 198
fractional indices 43–4
fractions 6, 55, 65, 68
    adding 5, 34, 47, 78, 83, 87
    cancelling 34, 47
    comparing 5, 74, 87
    to decimals 68, 87
    dividing 5, 81, 87
    Egyptian 79
    equivalent 5, 78, 83, 87
    multiplying 5, 81, 87
    to percentages 5, 68, 73, 74
    simplifying 2, 163
    subtracting 5, 34, 47, 78, 83, 87
    unit 79
frequency diagrams 360, 395, 412
frequency polygons 395–6, 412
frequency tables 359, 360, 383, 384
function graph 193
function machines 113, 116, 193, 199
    inverse 194, 199
function table 193

functions 113, 116, 151, 193–4, 199
    cubic 225
    quadratic 194, 208, 210–12, 225
    reciprocal 210, 212, 225

games 433
    computer 417, 421, 430, 432
    fair and unfair 433, 437
    for 2 players 435
geometrical reasoning 230, 250, 265
global warming 67
goat, tethering 336
golden rati 90
golden rectangle 90
gradient of graphs 116, 121, 203, 204, 224
gradient of parallel lines 224
gradient of perpendicular lines 224
graphical calculators 137, 203, 207, 215, 216, 217
graphs 116–17
    axes of symmetry 212, 225
    cubic 208–12
    cumulative frequency 391, 412, 415
    functions 113, 193, 199
        quadratic 194, 199, 208, 210–12, 225
        reciprocal 210, 225
    gradient of 116, 121, 203, 204, 224
    real-life 117, 217
    scatter 361, 399, 400, 401, 402, 413
    sketching 214, 217
    straight-line 110, 113, 116, 117, 203, 224
    turning points 194, 199, 208, 211
greenhouse effect 67
grid method 160–1, 175
growth rate 66, 188, 192

half-life 17, 20
heat loss 61
hectares 235
hexagonal numbers 33
hexagons 33, 230, 238, 259, 265
highest common factor (HCF) 2, 34, 36, 47, 48, 114
hyperbolas 210, 212, 225
hypotenuse 232, 254, 279, 316, 343, 344

identities 125, 151
identity function 193, 199

identity transformation 301
image 299, 234, 304, 306, 318
imperial/metric equivalents 235
impossibility 362
impossible picture 269
independent events 423, 424–5, 436
index laws 18, 41, 44, 48, 114, 175
  in algebraic expressions 36, 42, 48
index notation 15–19, 29, 34, 65–6, 95, 109
indicators for selected countries 403, 409–10
indices 1, 2, 38, 41, 43–4, 65
inequalities 23, 113, 142–4, 145, 152, 323
  quadratic 150, 152
infinite length 229
input 113, 116
integers 1
interest 95, 109, 126, 167
interior angles 229, 230, 248, 265
interquartile range 390, 391, 412
inverse 81, 87, 113, 114, 163
inverse enlargement 304, 305, 306, 318
inverse function machines 194, 199
inverse operations 4, 5, 63, 91, 113, 193
inversely proportional 104, 105, 110
isosceles trapezium 275
isosceles triangles 230

kaleidoscope 303
kites 228, 230, 247

ladders 291
large numbers, rounding 26
last digits 4
lawnmowing business 322
laws of indices 18, 41, 44, 48, 114, 175
  in algebraic expressions 36, 42, 48
leap years 235
left-handedness 419
length 235
  finding 278, 282
like terms, collecting 114, 157, 175
line of best fit 400, 402, 413
line segments 229, 231, 232, 233
linear equations 113, 127, 151
linear expressions 113
linear graphs 110, 113, 116, 117, 203, 224

linear inequalities in two variables 146, 147–8
linear (arithmetic) sequences 115, 182, 183, 185
lines 229
  finding length 316–17, 318, 319, 343
  parallel to x and y axes 117, 147, 289
  of symmetry 199
loci 233, 287, 292, 295
London Eye 246
lowest common multiple (LCM) 2, 34, 36, 78, 87, 126

magic squares 157
mapping diagrams 116, 193, 199
mapping on to 234
maps 308, 312
mass 235
maximum points 199, 208, 211
mean 360, 361, 382, 383, 395, 412
  assumed 361
  for grouped data 385, 411
measurements, range of 323, 353
measures 235
  compound 326, 327, 331, 354
measuring 323, 353
median 360, 361, 382, 384, 391, 395
  from graphs 412
  for grouped data 385, 386, 411
memory of calculators 65, 66
mental calculation 2–3, 51, 55, 68
Mercury 31, 102, 328
metric conversions 235, 331, 354
metric/imperial equivalents 235
mid-point of line segments 233
minimum points 199, 208, 211
mirror line 234
mixed numbers 5, 65, 81
modal class 360, 382, 411
mode 360, 361, 382
Moon 18, 66
multiples 2
multiplying 1, 4, 57, 69
  by 0.1, 0.01, 0.001 etc 1, 16
  decimals 4
  fractions 5, 81, 87
  grid method 160–1, 175
  using index laws 18–19, 41
  mentally 2–3, 51, 68
mutual exclusivity 362, 417, 421, 436
mystery numbers 50

nearly numbers 3, 51
negative enlargement 304, 305, 306, 318
negative indices 44

negative numbers 1, 2, 38, 65, 144, 152
negative powers 15
nets 60, 233, 337
nomograms 73
nth term, rule for 115, 180, 185, 186, 197, 198
number lines 1, 143, 144, 145, 152
number systems 14
numbers
  mixed 5, 65, 81
  negative 1, 2, 38, 65, 144, 152
  very large and very small 15
numerator 5, 34, 66, 73, 78, 163

object (enlargement) 306
octahedron 286
Ohm's Law 65
opposite side 343, 344
order of operations 3, 51, 68
order of rotation symmetry 234, 242
outcome 362, 417
  favourable 417, 424
  possible 362, 417, 424, 433
output 113, 116

parabolas 194, 208, 210–12, 225
parallel lines 233
  angles made with 229, 248
  gradient of 204, 224
  to x- and y-axes 117, 147, 289
parallelograms 106, 230, 235, 247
partitioning 3, 51
pendulums 84, 171
pentagonal numbers 33
pentagons 33, 101, 242, 349
"per" 326, 354
percentages 6, 55, 68
  change 6, 91, 95, 108, 109
  to fractions and decimals 5, 68
  scale 73
perimeter 235, 332–3, 354
perpendicular bisector 231, 233, 287
perpendicular, constructing 232
perpendicular lines, gradient 204, 224
pi (π) 65
pie charts 359, 415
place holders 26
place value 1, 2, 51
plan view 233
planes 295
  of symmetry 234, 303
Planet X 14
planets 20, 31, 102, 328, 330
polygons 230, 231, 264, 277
population 16, 21, 23, 31, 92, 403
  pyramids 406–7

Index

powers 1, 2, 15, 38, 41, 43–4, 65
   see also index
pricing 67, 71
prime factors 2, 34
prisms 101, 159, 160, 245, 263
   volume 172, 236, 337
probability 362
   calculating 362, 417
   comparing theoretical and
      experimental 430
   estimating 362, 429–30, 436–7
   of compound events 421
   scale 362
   sum of 362, 417
proof 248, 264
properties 230, 247, 254, 264
proportion 6, 55, 74, 278, 417
   comparing 74, 75, 87
   direct 98, 105, 103, 117, 218,
      219, 326–7
   inverse 104, 105, 110
   problems 6, 98, 109–110
   to squares 104, 110
proportional relationships 103
proportionality, constant of k 103,
      104, 110
Proxima Centauri 20, 21
puzzles 131
pyramids 233, 286, 337, 339
Pythagoras' theorem 254–6, 265,
      279, 316, 319, 350
   converse 266
Pythagorean triples 259, 260, 262,
      266, 279
   converse 262

quadratic functions 225
   graphs (parabolas) 194, 208,
      210–12, 225
quadratic inequalities 150, 152
quadratic sequences 115, 182–3,
      185–7, 198
quadrilaterals, construction 230,
      232, 247, 248
quadrillion 21
quartiles 390, 391, 412
questionnaires 358, 369, 377, 380

radius 231
random selection 362, 375, 379,
      419, 423
random variation 362
range 360, 361, 382, 384
   for grouped data 386–7, 411–12
range of measurements 323, 353
rates 326, 354
ratio 6, 98–9, 103, 295, 315, 319
   golden 90
recipes 12, 102, 111
reciprocal functions 212, 225

reciprocal graphs 210
reciprocals 62, 63–4, 65, 69
rectangles 90, 230, 269
recurring decimals 1, 85–6, 88
reflection 247, 272
   symmetry 234
relative frequency 362, 429, 430,
      436–7
remainder 4, 59
rhombuses 230, 247
roots, other than square or cube
      38, 65
rotation 247, 272, 299
   symmetry 234, 242, 303
rounding 1, 3, 23, 24–6, 30, 57
rules
   for function machines 116
   of indices 18, 41, 44, 48, 114,
      175
      in algebraic expressions 36,
      42, 48
   for *n*th term 115, 180, 185, 186,
      197, 198
   for quadratic sequences 185–7
   term-to-term 115, 180, 197

sale prices 95
sample space 362, 433
samples 358, 374, 379
satellite 330
Saturn 330
scale 234
scale drawings 234, 241
scale factor 279
   for area and volume 311–12,
      319
   for enlargement 234, 278, 304,
      305, 306–7, 318
   for length 319
scatter graphs 361, 399, 400, 401,
      402, 413
sector of a circle 231, 332–3, 354
segments of a circle 231, 283
self-inverses 194
semicircles 231, 283
sequences 173
   constant difference 115, 183,
      185, 198
   first and second differences
      182, 183, 185, 186, 187, 198,
      248
   converging 180, 197
   Fibonacci 179, 189
   first term 115, 180, 197
   fraction 189, 198
   generating 180, 182–3, 197
   linear (arithmetic) 115, 182,
      183, 185
   quadratic 115, 182–3, 185–7,
      198

rule for *n*th term 115, 180, 185,
      186, 197, 198
   of square numbers 182
   term-to-term rule 115, 180, 197
set square 232
shapes
   2-D 230–1
   3-D (solids) 233, 234, 235, 236,
      286, 295
      enlargement 311
   congruent 318, 231, 234, 299
   similar 276, 278–9, 295, 304,
      306
significant figures (s.f.), rounding
      to 25–6, 30, 65–6
similar triangles 278–9, 283, 294,
      315, 319, 342
simplifying
   expressions 42, 48, 114, 157,
      175
   fractions 2
   ratios 6
   surds 46
simultaneous equations 132–4,
      135, 136–7, 139–40, 152
sine (sin) 341, 342, 343, 344, 355
skeleton 156
SOHCAHTOA 343
spatial patterns 191
speed of animals 329, 331
speed formula 326, 327
speed of light 17, 20
spheres 27, 28, 31, 169, 172
spinners 362, 421, 425
spirals 259, 416
square numbers 65, 182
square roots 2, 38, 65, 114, 259
squares (shapes) 230
standard form 15–19, 29, 34,
      65–6, 95, 109
stars 20, 21
stem-and-leaf diagrams 361
straight-line graphs 203, 224
"stretching" pictures 309
substituting into expressions 114,
      169, 176, 180
substituting into formulae 115,
      169, 176
substitution method 135
subtracting 1
   decimals 4
   fractions 5, 34, 47, 78, 87
   mentally 51, 68
Sun 17, 18, 30, 59, 330
surds 45–6, 48
surface area 28, 169, 170, 235, 355
surveys 358, 379, 371, 373, 410
symbols 4, 23, 103, 113, 125, 262,
      295
   on Planet X 14

symmetry 234, 242, 303

tangent to circles 231, 283, 295
tangent (trigonometry) 343, 344, 350, 355
tangram pieces 399
temperature conversion 168
tending to a limit 430, 433, 437
terms, like, collecting 114, 157, 175
term-to-term rule 115, 180, 197
tessellation 234
time 235
    use by Britons 407–8, 415
tonnes 235
transformations 233–4
    combining 299, 306
    commutative 299, 318
    identity 301
transforming both sides 113, 115, 151
translation 272
trapeziums 112, 230, 235, 275
trial and improvemen 39
trials 362, 429, 430, 436–7

triangles 230
    angles of 229, 248
    area 235
    congruent 231, 272–3, 284, 294
    construction 232, 270–1, 294
    equilateral 230
    isosceles 230
    properties 230
    right-angled 230, 232, 254, 343
    scalene 230
    similar 278–9, 283, 294, 315, 319, 342
    unique 271
triangular numbers 191
trigonometry 341–4, 350
trillion 21
turning points on graphs (vertices) 194, 199, 208, 211
two-way tables 358

unit fractions 79
unitary method 91, 99, 109

variables, relationship between 217–18

vectors 305
Venn diagrams 2, 34, 36, 48
vertices 194, 199, 208, 211, 229, 342
volume 112, 171, 236
    of cylinders 167, 168, 176, 311, 337, 341
    of prisms 172, 236, 337, 355
    of pyramids 337, 339
    of spheres 27, 31, 172

wavelength 106, 107
websites 179
weekly earnings 77
width of a river 282, 353
wordsearch 381

x-axis, lines parallel to 117, 147, 289
x-intercept 209, 210

y-axis, lines parallel to 117, 147
y-intercept 203, 116, 224
$y = mx + c$ 116, 203, 224